Model-based Clustering and Classification
for Data Science

Cluster analysis consists of methods for finding groups in data automatically. Most methods have been heuristic and leave open such central questions as: How many clusters are there? Which clustering method should I use? How should I handle outliers? Classification involves assigning new observations to groups given previously classified observations, and also has open questions about parameter tuning, robustness and uncertainty assessment. This book frames cluster analysis and classification in terms of statistical models, thus yielding principled estimation, testing and prediction methods, and soundly-based answers to the central questions. It develops the basic ideas of model-based clustering and classification in an accessible but rigorous way, using extensive real-world data examples and providing R code for many methods, and describes modern developments for high-dimensional data and for networks. It explains recent methodological advances, such as Bayesian regularization methods, non-Gaussian model-based clustering, cluster merging, variable selection, semi-supervised classification, robust classification, clustering of functional data, text and images, and co-clustering. Written for advanced undergraduates and beginning graduate students in data science, as well as researchers and practitioners, it assumes basic knowledge of multivariate calculus, linear algebra, probability and statistics.

CHARLES BOUVEYRON is Professor of Statistics at Université Côte d'Azur and the Chair of Excellence in Data Science at Inria Sophia-Antipolis. He has published extensively on model-based clustering, particularly for networks and high-dimensional data.

GILLES CELEUX is Director of Research Emeritus at Inria Saclay Île-de-France. He is one of the founding researchers in model-based clustering, having published extensively in the area for 35 years.

T. BRENDAN MURPHY is Professor of Statistics at University College Dublin. His research interests include model-based clustering, classification, network modeling and latent variable modeling.

ADRIAN E. RAFTERY is Professor of Statistics and Sociology at University of Washington, Seattle. He was one of the founding researchers in model-based clustering, having published in the area since 1984.

CAMBRIDGE SERIES IN STATISTICAL AND PROBABILISTIC MATHEMATICS

This series of high-quality upper-division textbooks and expository monographs covers all aspects of stochastic applicable mathematics. The topics range from pure and applied statistics to probability theory, operations research, optimization and mathematical programming. The books contain clear presentations of new developments in the field and also of the state of the art in classical methods. While emphasizing rigorous treatment of theoretical methods, the books also contain applications and discussions of new techniques made possible by advances in computational practice.

A complete list of books in the series can be found at www.cambridge.org/statistics.
Recent titles include the following:

Model-Based Clustering and Classification for Data Science

With Applications in R

Charles Bouveyron
Université Côte d'Azur

Gilles Celeux
Inria Saclay Île-de-France

T. Brendan Murphy
University College Dublin

Adrian E. Raftery
University of Washington

CAMBRIDGE
UNIVERSITY PRESS

CAMBRIDGE
UNIVERSITY PRESS

University Printing House, Cambridge CB2 8BS, United Kingdom

One Liberty Plaza, 20th Floor, New York, NY 10006, USA

477 Williamstown Road, Port Melbourne, VIC 3207, Australia

314–321, 3rd Floor, Plot 3, Splendor Forum, Jasola District Centre, New Delhi – 110025, India

79 Anson Road, #06–04/06, Singapore 079906

Cambridge University Press is part of the University of Cambridge.

It furthers the University's mission by disseminating knowledge in the pursuit of
education, learning, and research at the highest international levels of excellence.

www.cambridge.org
Information on this title: www.cambridge.org/9781108494205
DOI: 10.1017/9781108644181

First published 2019

Printed in Singapore by Markono Print Media Pte Ltd

A catalogue record for this publication is available from the British Library.

Library of Congress Cataloging-in-Publication Data
Names: Bouveyron, Charles, 1979– author. | Celeux, Gilles, author. |
Murphy, T. Brendan, 1972– author. | Raftery, Adrian E., author.
Title: Model-based clustering and classification for data science : with applications in R /
Charles Bouveyron, Université Côte d'Azur, Gilles Celeux, Inria Saclay Île-de-France,
T. Brendan Murphy, University College Dublin, Adrian E. Raftery, University of Washington.
Description: Cambridge ; New York, NY : Cambridge University Press, 2019. |
Series: Cambridge series in statistical and probabilistic mathematics |
Includes bibliographical references and index.
Identifiers: LCCN 2019014257 | ISBN 9781108494205 (hardback)
Subjects: LCSH: Cluster analysis. | Mathematical statistics. |
Statistics–Classification. | R (Computer program language)
Classification: LCC QA278.55 .M63 2019 | DDC 519.5/3–dc23
LC record available at https://lccn.loc.gov/2019014257

ISBN 978-1-108-49420-5 Hardback

Contents

5 Semi-supervised Clustering and Classification 134
5.1 Semi-supervised Classification 134
5.2 Semi-supervised Clustering 141
5.3 Supervised Classification with Uncertain Labels 144
5.4 Novelty Detection: Supervised Classification with Unobserved Classes 154
5.5 Bibliographic Notes 160

6 Discrete Data Clustering 163
6.1 Example Data 163
6.2 The Latent Class Model for Categorical Data 165
6.3 Model-based Clustering for Ordinal and Mixed Type Data 185
6.4 Model-based Clustering of Count Data 190
6.5 Bibliographic Notes 197

7 Variable Selection 199
7.1 Continuous Variable Selection for Model-based Clustering 199
7.2 Continuous Variable Regularization for Model-based Clustering 208
7.3 Continuous Variable Selection for Model-based Classification 210
7.4 Categorical Variable Selection Methods for Model-based Clustering 211
7.5 Bibliographic Notes 215

8 High-dimensional Data 217
8.1 From Multivariate to High-dimensional Data 217
8.2 The Curse of Dimensionality 221
8.3 Earlier Approaches for Dealing with High-dimensional Data 227
8.4 Subspace Methods for Clustering and Classification 238
8.5 Bibliographic Notes 257

9 Non-Gaussian Model-based Clustering 259
9.1 Multivariate t-Distribution 259
9.2 Skew-normal Distribution 267
9.3 Skew-t Distribution 270
9.4 Box–Cox Transformed Mixtures 278
9.5 Generalized Hyperbolic Distribution 282
9.6 Example: Old Faithful Data 285
9.7 Example: Flow Cytometry 287
9.8 Bibliographic Notes 288

10 Network Data 292
10.1 Introduction 292
10.2 Example Data 294
10.3 Stochastic Block Model 298
10.4 Mixed Membership Stochastic Block Model 304
10.5 Latent Space Models 312
10.6 Stochastic Topic Block Model 320
10.7 Bibliographic Notes 329

Expanded Contents

Preface

About this book

The century that is ours is shaping up to be the century of the data revolution. Our numerical world is creating masses of data every day and the volume of generated data is estimated to be doubling every two years. This wealth of available data offers hope for exploitation that may lead to great advances in areas such as health, science, transportation and defense. However, manipulating, analyzing and extracting information from those data is made difficult by the volume and nature (high-dimensional data, networks, time series, etc) of modern data.

Within the broad field of statistical and machine learning, model-based techniques for clustering and classification have a central position for anyone interested in exploiting those data. This textbook focuses on the recent developments in model-based clustering and classification while providing a comprehensive introduction to the field. It is aimed at advanced undergraduates, graduates or first-year Ph.D. students in data science, as well as researchers and practitioners. It assumes no previous knowledge of clustering and classification concepts. A basic knowledge of multivariate calculus, linear algebra and probability and statistics is needed.

The book is supported by extensive examples on data, with 72 listings of code mobilizing more than 30 software packages, that can be run by the reader. The chosen language for codes is the R software, which is one of the most popular languages for data science. It is an excellent tool for data science since the most recent statistical learning techniques are provided on the R platform (named CRAN). Using R is probably the best way to be directly connected to current research in statistics and data science through the packages provided by researchers.

The book is accompanied by a dedicated R package (the `MBCbook` package) that can be directly downloaded from CRAN within the R software or at the following address: `https://cran.r-project.org/package=MBCbook`. We also encourage the reader to visit the book website for the latest information: `http://math.unice.fr/~cbouveyr/MBCbook/`.

This book could be used as one of the texts for a graduate or advanced undergraduate course in multivariate analysis or machine learning. Chapters 1 and 2, and optionally a selection of later chapters, could be used for this purpose. The book as a whole could also be used as the main text for a one-quarter or one-semester course in cluster analysis or unsupervised learning, focusing on the model-based approach.

Acknowledgements

This book is a truly collaborative effort, and the four authors have contributed equally. Each of us has contributed to each of the chapters.

We would like to thank Chris Fraley for initially developing the `mclust` software and later R package, starting in 1991. This software was of extraordinary quality from the beginning, and without it this whole field would never have developed as it did. Luca Scrucca took over the package in 2007, and has enhanced it in many ways, so we also owe a lot to his work. We would also like to thank the developers and maintainers of `Rmixmod` software: Florent Langrognet, Rémi Lebret, Christian Poli, Serge Iovleff, Anwuli Echenim and Benjamin Auder.

The authors would also like to thank the participants in the Working Group on Model-based Clustering, which has been gathering every year in the third week of July since 1994, first in Seattle and then since 2007 in different venues around Europe and North America. This is an extraordinary group of people from many countries, whose energy, interactions and intellectual generosity have inspired us every year and driven the field forward. The book owes a great deal to their insights.

Charles Bouveyron would like to thank in particular Stéphane Girard, Julien Jacques and Pierre Latouche, for very fruitful and friendly collaborations. Charles Bouveyron also thanks his coauthors on this topic for all the enjoyable collaborations: Laurent Bergé, Camille Brunet-Saumard, Etienne Côme, Marco Corneli, Julie Delon, Mathieu Fauvel, Antoine Houdard, Pierre-Alexandre Mattei, Cordelia Schmid, Amandine Schmutz and Rawya Zreik. He would like also to warmly thank his family, Nathalie, Alexis, Romain and Nathan, for their love and everyday support in the writing of this book.

Gilles Celeux would like to thank his old and dear friends Jean Diebolt and Gérard Govaert for the long and intensive collaborations. He also thanks his coauthors in the area Jean-Patrick Baudry, Halima Bensmail, Christophe Biernacki, Guillaume Bouchard, Vincent Brault, Stéphane Chrétien, Florence Forbes, Raphaël Gottardo, Christine Keribin, Jean-Michel Marin, Marie-Laure Martin-Magniette, Cathy Maugis-Rabusseau, Abdallah Mkhadri, Nathalie Peyrard, Andrea Rau, Christian P. Robert, Gilda Soromenho and Vincent Vandewalle for nice and fruitful collaborations. Finally, he would like to thank Maïlys and Maya for their love.

Brendan Murphy would like to thank John Hartigan for introducing him to clustering. He would like to thank Claire Gormley, Paul McNicholas, Luca Scrucca and Michael Fop with whom he has collaborated extensively on model-based clustering projects over a number of years. He also would like to thank his students and coauthors for enjoyable collaborations on a wide range of model-based clustering and classification projects: Marco Alfò, Francesco Bartolucci, Nema Dean, Silvia D'Angelo, Gerard Downey, Bailey Fosdick, Nial Friel, Marie

Galligan, Isabella Gollini, Sen Hu, Neil Hurley, Dimitris Karlis, Donal Martin, Tyler McCormick, Aaron McDaid, Damien McParland, Keefe Murphy, Tin Lok James Ng, Adrian O'Hagan, Niamh Russell, Michael Salter-Townshend, Lucy Small, Deirdre Toher, Ted Westling, Arthur White and Jason Wyse. Finally, he would like to thank his family, Trish, Áine and Emer for their love and support.

Adrian Raftery thanks Fionn Murtagh, with whom he first encountered model-based clustering and wrote his first paper in the area in 1984, Chris Fraley for a long and very fruitful collaboration, and Luca Scrucca for another very successful collaboration. He warmly thanks his Ph.D. students who have worked with him on model-based clustering, namely Jeff Banfield, Russ Steele, Raphael Gottardo, Nema Dean, Derek Stanford and William Chad Young for their collaboration and all that he learned from them. He also thanks his other coauthors in the area, Jogesh Babu, Jean-Patrick Baudry, Halima Bensmail, Roger Bumgarner, Simon Byers, Jon Campbell, Abhijit Dasgupta, Mary Emond, Eric Feigelson, Florence Forbes, Diane Georgian-Smith, Ken Lo, Alejandro Murua, Nathalie Peyrard, Christian Robert, Larry Ruzzo, Jean-Luc Starck, Ka Yee Yeung, Naisyin Wang and Ron Wehrens for excellent collaborations.

Raftery would like to thank the Office of Naval Research and the Eunice Kennedy Shriver National Institute of Child Health and Human Development (NICHD grants R01 HD054511 and R01 HD070936) for sustained research support without which this work could not have been carried out. He wrote part of the book during a fellowship year at the Center for Advanced Study in the Behavioral Sciences (CASBS) at Stanford University in 2017–2018, which provided an ideal environment for the sustained thinking needed to complete a project of this kind. Finally he would like to thank his wife, Hana Ševčíková, for her love and support through this project.

1

Introduction

Cluster analysis and classification are two important tasks which occur daily in everyday life. As humans, our brain naturally clusters and classifies animals, objects or even ideas thousands of times a day, without fatigue. The emergence of science has led to many data sets with clustering structure that cannot be easily detected by the human brain, and so require automated algorithms. Also, with the advent of the "Data Age," clustering and classification tasks are often repeated large numbers of times, and so need to be automated even if the human brain could carry them out.

This has led to a range of clustering and classification algorithms over the past century. Initially these were mostly heuristic, and developed without much reference to the statistical theory that was emerging in parallel. In the 1960s, it was realized that cluster analysis could be put on a principled statistical basis by framing the clustering task as one of inference for a finite mixture model. This has allowed cluster analysis to benefit from the inferential framework of statistics, and provide principled and reproducible answers to questions such as: how many clusters are there? what is the best clustering algorithm? how should we deal with outliers?

In this book, we describe and review the model-based approach to cluster analysis which has emerged in the past half-century, and is now an active research field. We describe the basic ideas, and aim to show the advantages of thinking in this way, as well as to review recent developments, particularly for newer types of data such as high-dimensional data, network data, textual data and image data.

1.1 Cluster Analysis

The goal of cluster analysis is to find meaningful groups in data. Typically, in the data these groups will be internally cohesive and separated from one another. The purpose is to find groups whose members have something in common that they do not share with members of other groups.

1.1.1 From Grouping to Clustering

The grouping of objects according to things they have in common goes back at least to the beginning of language. A noun (such as "hammer") refers to any one of a set of different individual objects that have characteristics in common. As

Greene (1909) remarked, "naming is classifying." Plato was among the first to formalize this with his Theory of Forms, defining a Form as an abstract unchanging object or idea, of which there may be many instances in practice. For example, in Plato's *Cratylus* dialogue, he has Socrates giving the example of a blacksmith's tool, such as a hammer. There are many hammers in the world, but just one Platonic Form of "hammerness" which is the essence of all of them.

Aristotle, in his *History of Animals*, classified animals into groups based on their characteristics. Unlike Plato, he drew heavily on empirical observations. His student Theophrastus did something similar for plants in his *Enquiry Into Plants*.

An early and highly influential example of the systematic grouping of objects based on measured empirical characteristics is the system of biological classification or taxonomy of Linnaeus (1735), applied to plants by Linnaeus (1753) and to animals by Linnaeus (1758). For example, he divided plants into 24 classes, including flowers with one stamen (Monandria), flowers with two stamens (Diandria) and flowerless plants (Cryptogamia). Linnaeus' methods were based on data but were subjective. Adanson (1757, 1763) developed less subjective methods using multiple characteristics of organisms.

Cluster analysis is something more: the search for groups in quantitative data using systematic numerical methods. Perhaps the earliest methods that satisfy this description were developed in anthropology, and mainly consisted of defining quantitative measures of difference and similarity between objects (Czekanowski, 1909, 1911; Driver and Kroeber, 1932). Most of the early clustering methods were based on measures of similarity between objects, and Czekanowski (1909) seems to have been the first to define such a measure for clustering purposes.

Then development shifted to psychology, where Zubin (1938) proposed a method for rearranging a correlation matrix to yield clusters. Stephenson (1936) proposed the use of factor analysis to identify clusters of people, while, in what seems to have been the first book on cluster analysis, Tryon (1939) proposed a method for clustering variables similar to what is now called multiple group factor analysis. Cattell (1944) also introduced several algorithmic and graphical clustering methods.

In the 1950s, development shifted again to biological taxonomy, the original problem addressed by the ancient Greeks and the eighteenth-century scientists interested in classification. It was in this context that the single link hierarchical agglomerative clustering method (Sneath, 1957), the average link method and the complete link method (Sokal and Michener, 1958) were proposed. These are sometimes thought of as marking the beginning of cluster analysis, but in fact they came after a half-century of previous, though relatively slow development. They did mark the takeoff of the area, though. The development of computational power and the publication of the important book of Sokal and Sneath (1963) led to a rapid expansion of the use and methodology of cluster analysis, which has not stopped in the past 60 years.

Many of the ensuing developments in the 1970s and 1980s were driven by applications in market research as well as biological taxonomy. From the 1990s there was a further explosion of interest fueled by new types of data and questions, often involving much larger data sets than before. These include finding groups of genes or people using genetic microarray data, finding groups and patterns in retail barcode data, finding groups of users and websites from Internet use data, and automatic document clustering for technical documents and websites.

Another major area of application has been image analysis. This includes medical image segmentation, for example for finding tumors in digital medical images such as X-rays, CAT scans, MRI scans and PET scans. In these applications, a cluster is typically a set of pixels in the image. Another application is image compression, using methods such as color image quantization, where a cluster would correspond to a set of color levels. For a history of cluster analysis to 1988, see Blashfeld and Aldenderfer (1988).

1.1.2 Model-based Clustering

Most of the earlier clustering methods were algorithmic and heuristic. The majority were based on a matrix of measures of similarity between objects, which were in turn derived from the objects' measured characteristics. The purpose was to divide or partition the data into groups such that objects in the same group were similar, and were dissimilar from objects in other groups. A range of automatic algorithms for doing this was proposed, starting in the 1930s.

These developments took place largely in isolation from mainstream statistics, much of which was based on a probability distribution for the data. At the same time, they left several practical questions unresolved, such as which of the many available clustering methods to use? How many clusters should we use? How should we treat objects that do not fall into any group, or outliers? How sure are we of a clustering partition, and how should we assess uncertainty about it?

The mainstream statistical approach of specifying a probability model for the full data set has the potential to answer these questions. The main statistical model for clustering is a finite mixture model, in which each group is modeled by its own probability distribution.

The first successful method of this kind was developed in sociology in the early 1950s for multivariate discrete data, where multiple characteristics are measured for each object, and each characteristic can take one of several values. Data of this kind are common in the social sciences, and are typical, for example, of surveys. The model proposed was called the latent class model, and it specified that within each group the characteristics were statistically independent (Lazarsfeld, 1950a,c). We discuss this model and its development in Chapter 6.

The dominant model for clustering continuous-valued data is the mixture of multivariate normal distributions. This seems to have been first mentioned by Wolfe (1963) in his Master's thesis at Berkeley. John Wolfe subsequently developed the first real software for estimating this model, called NORMIX, and also developed related theory (Wolfe, 1965, 1967, 1970), so he has a real claim to be called the inventor of model-based clustering for continuous data. Wolfe proposed estimating the model by maximum likelihood using the EM algorithm, which is striking since he did so ten years before the article of Dempster et al. (1977) that popularized the EM algorithm. This remains the most used estimation approach in model-based clustering. We outline the early history of model-based clustering in Section 2.9, after we have introduced the main ideas.

Basing cluster analysis on a probability model has several advantages. In essence, this brings cluster analysis within the range of standard statistical methodology and makes it possible to carry out inference in a principled way. It turns out that many of the previous heuristic methods correspond approximately to particular clustering models, and so model-based clustering can provide a way of choosing between clustering methods, and encompasses many of them in its framework. In our experience, when a clustering method does not correspond to any probability model, it tends not to work very well. Conversely, understanding what probability model a clustering method corresponds to can give one an idea of when and why it will work well or badly.

It also provides a principled way to choose the number of clusters. In fact, the choice of clustering model and of number of clusters can be reduced to a single model selection problem. It turns out that there is a trade-off between these choices. Often, if a simpler clustering model is chosen, more clusters are needed to represent the data adequately.

Basing cluster analysis on a probability model also leads to a way of assessing uncertainty about the clustering. In addition, it provides a systematic way of dealing with outliers by expanding the model to account for them.

1.2 Classification

The problem of classification (also called discriminant analysis) involves classifying objects into classes when there is already information about the nature of the classes. This information often comes from a data set of objects that have already been classified by experts or by other means. Classification aims to determine which class new objects belong to, and develops automatic algorithms for doing so. Typically this involves assigning new observations to the class whose objects they most closely resemble in some sense.

Classification is said to be a "supervised" problem in the sense that it requires the supervision of experts to provide some examples of the classes. Clustering, in contrast, aims to divide a set of objects into groups without any examples of the "true" classes, and so is said to be an "unsupervised" problem.

1.2.1 *From Taxonomy to Machine Learning*

The history of classification is closely related to that of clustering. Indeed, the practical interest of taxonomies of animals or plants is to use them to recognize samples on the field. For centuries, the task of classification was carried out by humans, such as biologists, botanists or doctors, who learned to assign new observations to specific species or diseases. Until the twentieth century, this was done without automatic algorithms.

The first statistical method for classification is due to Ronald Fisher in his famous work on discriminant analysis (Fisher, 1936). Fisher asked what linear combination of several features best discriminates between two or more populations. He applied his methodology, known nowadays as Fisher's discriminant analysis or linear discriminant analysis, to a data set on irises that he had obtained from the botanist Edgar Anderson (Anderson, 1935).

In a following article (Fisher, 1938), he established the links between his discriminant analysis method and several existing methods, in particular analysis of variance (ANOVA), Hotelling's T-squared distribution (Hotelling, 1931) and the Mahalanobis generalized distance (Mahalanobis, 1930). In his 1936 paper, Fisher also acknowledged the use of a similar approach, without formalization, in craniometry for quantifying sex differences in measurements of the mandible.

Discriminant analysis rapidly expanded to other application fields, including medical diagnosis, fault detection, fraud detection, handwriting recognition, spam detection and computer vision. Fisher's linear discriminant analysis provided good solutions for many applications, but other applications required the development of specific methods.

Among the key methods for classification, logistic regression (Cox, 1958) extended the usual linear regression model to the case of a categorical dependent variable and thus made it possible to do binary classification. Logistic regression had a great success in medicine, marketing, political science and economics. It remains a routine method in many companies, for instance for mortgage default prediction within banks or for click-through rate prediction in marketing companies.

Another key early classification method was the perceptron (Rosenblatt, 1958). Originally designed as a machine for image recognition, the perceptron algorithm is supposed to mimic the behavior of neurons for making a decision. Although the first attempts were promising, the perceptron appeared not to be able to recognize many classes without adding several layers. The perceptron is recognized as one of the first artificial neural networks which recently revolutionized the classification field, partly because of the massive increase in computing capabilities. In particular, convolutional neural networks (LeCun et al., 1998) use a variation of multilayer perceptrons and display impressive results in specific cases.

Before the emergence of convolutional neural networks and deep learning, support vector machines also pushed forward the performances of classification at the end of the 1990s. The original support vector machine algorithm or SVM (Cortes and Vapnik, 1995), was invented in 1963 and it was not to see its first implementation until 1992, thanks to the "kernel trick" (Boser et al., 1992). SVM is a family of classifiers, defined by the choice of a kernel, which transform the original data in a high-dimensional space, through a nonlinear projection, where they are linearly separable with a hyperplane. One of the reasons for the popularity of SVMs was their ability to handle data of various types thanks to the notion of kernel.

As we will see in this book, statistical methods were able to follow the different revolutions in the performance of supervised classification. In addition, some of the older methods remain reference methods because they perform well with low complexity.

1.2.2 Model-based Discriminant Analysis

Fisher discriminant analysis (FDA, Fisher (1936)) was the first classification method. Although Fisher did not describe his methodology within a statistical modeling framework, it is possible to recast FDA as a model-based method. Assuming normal distributions for the classes with a common covariance matrix yields a classification rule which is based on Fisher's discriminant function. This classification method, named linear discriminant analysis (LDA), also provides a way to calculate the optimal threshold to discriminate between the classes within Fisher's discriminant subspace (Fukunaga, 1999).

An early work considering class-conditional distributions in the case of discriminant analysis is due to Welch (1939). He gave the first version of a classification rule in the case of two classes with normal distributions, using either Bayes' theorem (if the prior probabilities of the classes are known) or the Neyman–Pearson lemma (if these prior probabilities have to be estimated). Wald (1939, 1949) developed the theory of decision functions which offers a sound statistical framework for further work in classification. Wald (1944) considered the problem of assigning an individual into one of two groups under normal distributions with a common covariance matrix, the solution of which involves Fisher's discriminant function. Von Mises (1945) addressed the problem of minimizing the classification error in the case of several classes and proposed a general solution to it. Rao (1948, 1952, 1954) extended this to consider the estimation of a classification rule from samples. See Das Gupta (1973) and McLachlan (1992) for reviews of the earlier development of this area.

Once the theory of statistical classification had been well established, researchers had to face new characteristics of the data, such as high-dimensional data, low sample sizes, partially supervised data and non-normality. Regarding high-dimensional data, McLachlan (1976) realized the importance of variable selection to avoid the curse of dimensionality in discriminant analysis. Banfield and Raftery (1993) and Bensmail and Celeux (1996) proposed alternative approaches using constrained Gaussian models. About partially supervised classification, McLachlan and Ganesalingam (1982) considered the use of unlabeled data to update a classification rule in order to reduce the classification error. Regarding non-normality, Celeux and Mkhadri (1992) proposed a regularized discriminant analysis technique for high-dimensional discrete data, while Hastie and Tibshirani (1996) considered the classification of non-normal data using mixtures of Gaussians. In Chapter 4 specific methods for classification with categorical data are presented. These topics will be developed in this book, in Chapters 4, 5 and 8.

1.3 Examples

We now briefly describe some examples of cluster analysis and discriminant analysis.

Example 1: Old Faithful geyser data

The Old Faithful geyser in Yellowstone National Park, Wyoming erupts every 35–120 minutes for about one to five minutes. It is useful for rangers to be able to predict the time to the next eruption. The time to the next eruption and its duration are related, in that the longer an eruption lasts, the longer the time until the next one. The data we will consider in this book consist of observations on 272 eruptions Azzalini and Bowman (1990). Data on two variables were measured: the time from one eruption to the next one, and the duration of the eruption. These data are often used to illustrate clustering methods.

Example 2: Diagnosing type of diabetes

Figure 1.1 shows measurements made on 145 subjects with the goal of diagnosing diabetes and, for diabetic patients, the type of diabetes present. The data consist of the area under a plasma glucose curve (glucose area), the area under a plasma insulin curve (insulin area) and the steady-state plasma glucose response (SSPG) for 145 subjects. The subjects were subsequently clinically classified into three groups: chemical diabetes (Type 1), overt diabetes (Type 2), and normal (non-diabetic). The goal of our analysis is either to develop a method for grouping patients into clusters corresponding to diagnostic categories, or to learn a classification rule able to predict the status of a new patient. These data were described and analyzed by Reaven and Miller (1979).

Example 3: Breast cancer diagnosis

In order to diagnose breast cancer, a fine needle aspirate of a breast mass was collected, and a digitized image of it was produced. The cells present in the image were identified, and for each cell nucleus the following characteristics were measured from the digital image: (a) radius (mean of distances from center to points on the perimeter); (b) texture (standard deviation of gray-scale values); (c) perimeter; (d) area; (e) smoothness (local variation in radius lengths); (f) compactness ($perimeter^2$ / area $-$ 1); (g) concavity (severity of concave portions of the contour); (h) concave points (number of concave portions of the contour); (i) symmetry; (j) fractal dimension. The mean, standard deviation and extreme values of the 10 characteristics across cell nuclei were then calculated, yielding 30 features of the image (Street et al., 1993; Mangasarian et al., 1995).

A pairs plot of three of the 30 variables is shown in Figure 1.2. These were selected as representing a substantial amount of the variability in the data, and in fact are the variables with the highest loadings on each of the first three principal components of the data, based on the correlation matrix.

This looks like a more challenging clustering/classification problem than the first two examples (Old Faithful data and diabetes diagnosis data), where clustering

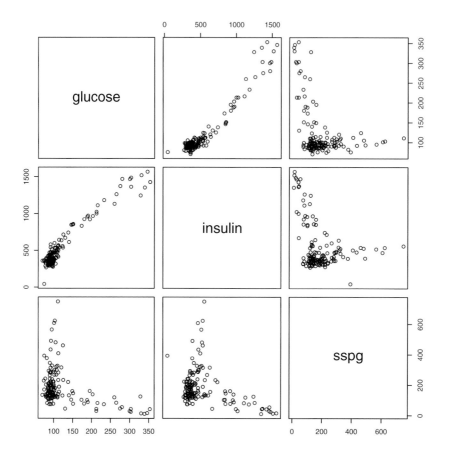

Figure 1.1 Diabetes data pairs plot: three measurements on 145 patients. *Source:* Reaven and Miller (1979).

was apparent visually from the plots of the data. Here it is hard to discern clustering in Figure 1.2. However, we will see in Chapter 2 that in this higher dimensional setting, model-based clustering can detect clusters that agree well with clinical criteria.

Example 4: Wine varieties

Classifying food and drink on the basis of characteristics is an important use of cluster analysis. We will illustrate this with a data set giving up to 27 physical and chemical measurements on 178 wine samples (Forina et al., 1986). The goal of an analysis like this is to partition the samples into types of wine, and potentially also by year of production. In this case we know the right answer: there are three types of wine, and the year in which each sample was produced is also known.

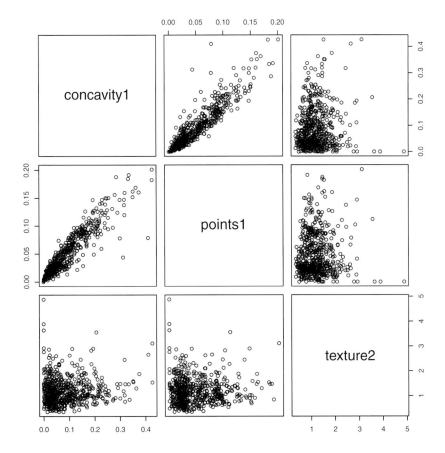

Figure 1.2 Pairs plot of three of the 30 measurements on breast cancer diagnosis images.

Thus we can assess how well various clustering methods perform. We will see in Chapter 2 that model-based clustering is successful at this task.

Example 5: Craniometric analysis

Here the task is to classify skulls according to the populations from which they came, using cranial measurements. We will analyze data with 57 cranial measurements on 2,524 skulls. As we will see in Chapter 2, a major issue is determining the number of clusters, or populations.

Example 6: Identifying minefields

We consider the problem of detecting surface-laid minefields on the basis of an image from a reconnaissance aircraft. After processing, such an image is reduced to a list of objects, some of which may be mines and some of which may be "clutter"

Minefield data   **True classification**

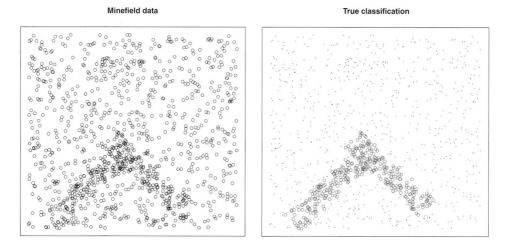

Figure 1.3 Minefield data. Left: observed data. Right: true classification into mines and clutter.

or noise, such as other metal objects or rocks. The objects are small and can be represented by points without losing much information. The analyst's task is to determine whether or not minefields are present, and where they are. A typical data set is shown in Figure 1.3.[1] The true classification of the data between mines and clutter is shown in the right panel of Figure 1.3. These data are available as the `chevron` data set in the `mclust` R package.

This problem is challenging because the clutter form over two-thirds of the data points and are not separated from the mines spatially, but rather by their density.

Example 7: Vélib data

This data set has been extracted from the Vélib large-scale bike sharing system of Paris, through the open-data API provided by the operator JCDecaux. The real time data are available at `https://developer.jcdecaux.com/` (with an api key) . The data set consists of information (occupancy, number of broken docks, ...) about bike stations collected on the Paris bike sharing system over five weeks, between February 24 and March 30, 2014. Figure 1.4 presents a map of the Vélib stations in Paris (left panel) and loading profiles of some Vélib stations (right panel). The red dots correspond to the stations for which the loading profiles are displayed on the right panel.

The information can be analyzed in different ways, depending on the objective or the data type. For instance, the data were first used in Bouveyron et al. (2015) in the context of functional time series clustering, in order to recover the temporal pattern of use of the bike stations. This data set will be analyzed in this book

---

[1] Actual minefield data were not available, but the data in Figure 1.3 were simulated according to specifications developed at the Naval Coastal Systems Station, Panama City, Florida, to represent minefield data encountered in practice (Muise and Smith, 1992).

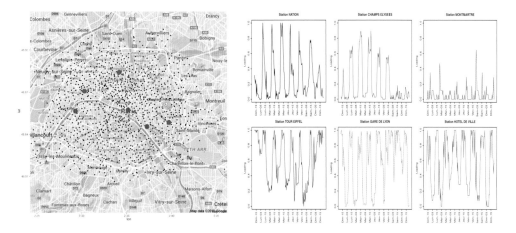

Figure 1.4 Map of the Vélib stations in Paris (left panel) and loading profiles of some Vélib stations (right panel). The red dots correspond to the stations for which the loading profiles are displayed on the right panel.

alternatively as count data (6), as functional data and as multivariate functional data (12). These data are available as the `velib` data set in the `funFEM` package.

Example 8: Chemometrics data

High-dimensional data are more and more frequent in scientific fields. The NIR data set (Devos et al., 2009) is one example, from a problem in chemometrics of discriminating between three types of textiles. The 202 textile samples are analyzed here with a near-infrared spectrometer that produces spectra with 2,800 wavelengths. Figure 1.5 presents the textile samples as 2,800-dimensional spectra, where the colors indicate the textile types. This is a typical situation in chemometrics, biology and medical imaging, where the number of samples is lower than the number of recorded variables. This situation is sometimes called "ultra-high dimensionality". This is a difficult situation for most clustering and classification techniques, as will be discussed in Chapter 8. Nevertheless, when using appropriate model-based techniques, it is possible to exploit the blessing of high-dimensional spaces to efficiently discriminate between the different textile types. The data set is available in the `MBCbook` package.

Example 9: Zachary's karate club network data

The Zachary's karate club data set (Zachary, 1977) consists of the friendship network of 34 members of a university-based karate club. It was collected following observations over a three-year period in which a factional division caused the members of the club to formally separate into two organizations. While friendships within the club arose organically, a financial dispute regarding the pay of part-time instructor Mr. Hi tested these ties, with two political factions developing. Key to the dispute were two members of the network, Mr. Hi and club president John A. The dispute eventually led to the dismissal of Mr. Hi by John A., and Mr. Hi's

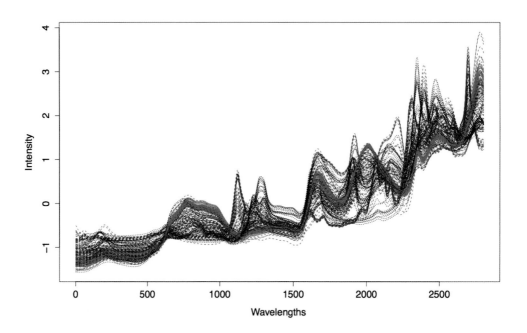

Figure 1.5 Some textile samples of the three-class NIR data set.

supporters resigned from the karate club and established a new club, headed by Mr. Hi. The data set exhibits many of the phenomena observed in social networks, in particular clustering, or community structure. The data are shown in Figure 1.6 where the friendship network is shown and the locations of Mr. Hi and John A. within the network are highlighted.

1.4 Software

We will give examples of software code to implement our analyses throughout the book. Fortunately, model-based clustering is well served by good software, mostly in the form of R packages. We will primarily use the R packages `mclust` (Scrucca et al., 2016) and `Rmixmod` (Langrognet et al., 2016), each of which carries out general model-based clustering and classification and has a rich array of capabilities. The capabilities of these two packages overlap to some extent, but not completely. An advantage of using R is that it allows one to easily use several different packages in the same analysis.

We will also give examples using several other R packages that provide additional model-based clustering and classification capabilities. These include `FlexMix` (Leisch, 2004), `fpc` (Hennig, 2015a), `prabclus` (Hennig and Hausdorf, 2015), `pgmm` (McNicholas et al., 2018), `tclust` (Iscar et al., 2017), `clustMD` (McParland and Gormley, 2017) and `HDclassif` (Bergé et al., 2016).

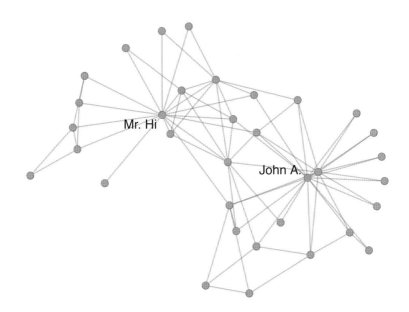

Figure 1.6 The friendship network of Zachary's karate club. The two key members in the dispute within the club, Mr. Hi and John A., are labeled and colored differently.

1.5 Organization of the Book

In Chapter 2 we introduce the basic ideas of model-based clustering. In Chapter 3 we discuss some common difficulties with the framework, namely outliers, degeneracies and non-Gaussian mixture components, and we describe some initial, relatively simple strategies for overcoming them.

In Chapter 4 we describe model-based approaches to classification. This differs from clustering in that a training set with known labels or cluster memberships is available. In Chapter 5 we extend this to discuss semi-supervised classification, in which unlabeled data are used as part of the training data.

Chapter 6 is devoted to model-based clustering involving discrete data, starting with the latent class model, which was the earliest instance of model-based clustering of any kind. We also consider ordinal data, and data of mixed type, i.e. that include both discrete and continuous variables.

In Chapter 7 we consider the selection of variables for clustering, for both continuous and discrete data. This is important because if variables are used that are not useful for clustering, the performance of algorithms can be degraded.

In Chapter 8 we describe methods for model-based clustering for high-dimensional data. Standard model-based clustering methods can be used in principle regardless of data dimension, but if the dimension is high their performance can decline. We describe a range of dimension reduction, regularization and subspace methods.

Chapter 9 describes ways of clustering data where the component distributions are non-Gaussian by modeling them explicitly, in contrast with the component merging methods of Chapter 3. In Chapter 10 we describe model-based approaches to clustering nodes in networks, with a focus on social network data. Chapter 11 treats methods for model-based clustering with covariates, with a focus on the mixture of experts model.

Finally, in Chapter 12, we describe model-based clustering methods for a range of less standard kinds of data. These include functional data, textual data and images. They also include data in which both the rows and the columns of the data matrix may have clusters and we want to detect them both in the same analysis. The methods we describe fall under the heading of model-based co-clustering.

1.6 Bibliographic Notes

A number of books have been written on the topic of cluster analysis and mixture modeling and these books are of particular importance in the area of model-based clustering.

Hartigan (1975) wrote an early monograph on cluster analysis, which included detailed chapters on k-means clustering and Gaussian model-based clustering. Everitt and Hand (1981) wrote a short monograph on finite mixture models including Gaussian, exponential, Poisson and binomial mixture models. Titterington et al. (1985) is a detailed monograph on mixture models and related topics. McLachlan and Basford (1988) is a detailed monograph on finite mixture modeling and clustering. Everitt (1993) is a textbook on cluster analysis which covers a broad range of approaches. Lindsay (1995) gives an in-depth overview of theoretical aspects of finite mixture modeling. Gordon (1999) gives a thorough overview of clustering methods, including model-based approaches. McLachlan and Peel (2000) is an extensive book on finite mixture modeling and clustering. Frühwirth-Schnatter (2006) is a detailed monograph on finite mixtures and Markov models with a particular focus on Bayesian modeling. McNicholas (2016a) gives a thorough overview of model-based clustering approaches developed from a range of different finite mixtures.

Hennig et al. (2015) is an edited volume on the topic of cluster analysis including chapters on model-based clustering. Mengersen et al. (2011) and Celeux et al. (2018a) are edited volumes on the topic of mixture modeling which include chapters on model-based clustering and applications.

2

Model-based Clustering: Basic Ideas

In this chapter we review the basic ideas of model-based clustering. Model-based clustering is a principled approach to cluster analysis, based on a probability model and using standard methods of statistical inference. The probability model on which it is based is a finite mixture of multivariate distributions, and we describe this in Section 2.1, with particular emphasis on the most used family of models, namely mixtures of multivariate normal distributions.

In Section 2.3 we describe maximum likelihood estimation for this model, with an emphasis on the Expectation-Maximization (EM) algorithm. The EM algorithm is guaranteed to converge to a local optimum of the likelihood function, but not to a global maximum, and so the choice of starting point can be important. We describe some approaches to this in Section 2.4. In Section 2.5, examples with known numbers of clusters are presented.

Two persistent questions in the practical use of cluster analysis are: how many clusters are there? and, which clustering method should be used? It turns out that determining the number of clusters can be viewed as a model choice problem, and the choice of clustering *method* is often at least approximately related to the choice of probability *model*. Thus both of these questions are answered simultaneously in the model-based clustering framework by the choice of an appropriate probability model, and we discuss model selection in Section 2.6. We will work through some examples in Section 2.7.

2.1 Finite Mixture Models

Suppose we have data that consist of n multivariate observations, $y_1, \ldots, y_n$, each of dimension d, so that $y_i = (y_{i,1}, \ldots, y_{i,d})$. A finite mixture model represents the probability distribution or density function of one multivariate observation, y_i, as a finite mixture or weighted average of G probability density functions, called *mixture components*:

$$p(y_i) = \sum_{g=1}^{G} \tau_g f_g(y_i \mid \theta_g). \tag{2.1}$$

In Equation (2.1), τ_g is the probability that an observation was generated by the gth component, with $\tau_g \geq 0$ for $g = 1, \ldots, G$, and $\sum_{g=1}^{G} \tau_g = 1$, while $f_g(\cdot \mid \theta_g)$ is the density of the gth component given the values of its parameters θ_g.

Density for 1–dim 2–component normal mixture model

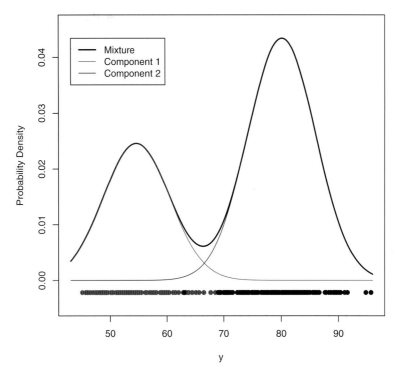

Figure 2.1 Probability density function for a one-dimensional univariate finite normal mixture with two mixture components. The individual component densities multiplied by their mixture probabilities are shown in red and blue, respectively, and the resulting overall mixture density (which is the sum of the red and blue curves) is the black curve. The dots show a sample of size 272 simulated from the density, with the colors indicating the mixture component from which they were generated.

Most commonly, f_g is a multivariate normal distribution. In the univariate case, when y_i is one-dimensional, $f_g(y_i|\theta_g)$ is a $N(\mu_g, \sigma_g^2)$ density function, and $\theta_g = (\mu_g, \sigma_g)$, consisting of the mean and standard deviation for the gth mixture component.

Figure 2.1 shows an example of the density function for a univariate finite normal mixture model with two mixture components, together with a sample simulated from it. The model parameters were selected by estimating such a model for the Waiting Time data from the Old Faithful data set in Example 1, yielding $\mu_1 = 54.7$, $\mu_2 = 80.1$, $\sigma_1 = 5.91$, $\sigma_2 = 5.87$, $\tau_1 = 0.362$ and $\tau_2 = 0.638$. The R code to produce Figure 2.1 is shown in Listing 2.1.

The density is clearly bimodal, and the density is lower in the middle between the two humps, so there is some separation between the two mixture components.

Listing 2.1: R code for Figure 2.1

```
# Estimate parameters of 1d finite normal mixture model
# with 2 components for Waiting time observations from
# the Old Faithful data set:
library(mclust)
waiting <- faithful$waiting
n <- length (waiting)
waiting.Mclust <- Mclust (waiting,model="V",G=2)

# Plot densities:
x <- seq (from=min(waiting),to=max(waiting),length
    =1000)
den1 <- dnorm (x,mean=waiting.Mclust$parameters$mean
    [1],
  sd=sqrt(waiting.Mclust$parameters$variance$sigmasq
    [1]))
den2 <- dnorm (x,mean=waiting.Mclust$parameters$mean
    [2],
  sd=sqrt(waiting.Mclust$parameters$variance$sigmasq
    [2]))
tau1 <- waiting.Mclust$parameters$pro[1]
tau2 <- waiting.Mclust$parameters$pro[2]
dens <- tau1*den1 + tau2*den2
plot (x,dens,type="l",xlab="y",ylab="Probability
    Density",
  ylim=c(-max(dens)/10,1.05*max(dens)),
  main="Density for 1-dim 2-component normal mixture
    model",lwd=2)
lines (x,tau1*den1,col="red")
lines (x,tau2*den2, col="blue")
legend (x=min(x),y=max(dens),legend=c("Mixture","
    Component 1","Component 2"),
  col=c("black","red","blue"),lty=c(1,1,1),lwd=c(2,1,1)
    )

# Simulate and plot a sample from the distribution:
sim.results <- simV (waiting.Mclust$parameters,n,seed
    =0)
ysim <- sim.results[,2]
groupsim <- sim.results[,"group"]
ysim1 <- ysim[groupsim==1]
ysim2 <- ysim[groupsim==2]
points (ysim1,rep(-max(dens)/20,length(ysim1)),col="red
    ",pch=19)
points (ysim2,rep(-max(dens)/20,length(ysim2)),col="
    blue",pch=19)
```

However, the separation is not total. Also, note that there is some overlap between the points from the two mixture components. In particular, one of the blue points is to the left of many of the red points. So even in this fairly clear situation we would be uncertain about which components the points in the middle belong to, if they were not conveniently colored. Assessing this kind of uncertainty is one of the things that model-based clustering allows us to do.

When the data are multivariate, f_g is often the multivariate normal or Gaussian density ϕ_g, parameterized by its mean vector, μ_g and by its covariance matrix Σ_g, and has the form

$$\phi_g(y_i|\mu_g, \Sigma_g) = |2\pi\Sigma_g|^{-\frac{1}{2}} \exp\left\{-\tfrac{1}{2}(y_i - \mu_g)^T \Sigma_g^{-1}(y_i - \mu_g)\right\}. \qquad (2.2)$$

Data generated by mixtures of multivariate normal densities are characterized by groups or clusters centered at the means μ_g, with increased density for points nearer the mean. The corresponding surfaces of constant density are ellipsoidal.

Figure 2.2 shows the density contours for a two-dimensional finite bivariate normal mixture model with two mixture components. The parameters were those estimated from the two-dimensional Old Faithful data (Example 1), namely $\mu_1 = (4.29, 79.97)$, $\mu_2 = (2.04, 54.48)$, $\tau_1 = 0.644$, $\tau_2 = 0.356$,

$$\Sigma_1 = \begin{bmatrix} 0.170 & 0.938 \\ 0.938 & 36.017 \end{bmatrix}, \text{ and } \quad \Sigma_2 = \begin{bmatrix} 0.069 & 0.437 \\ 0.437 & 33.708 \end{bmatrix}.$$

We can see the ellipsoidal nature of the contours. The values simulated from the two mixture components do not overlap in this bivariate setting. The six contours shown were chosen so as to contain 5%, 25%, 50%, 75%, 95% and 99% of the probability, respectively. The R code used to produce the plot is shown in Listing 2.2.

2.2 Geometrically Constrained Multivariate Normal Mixture Models

The full multivariate normal finite mixture model specified by Equations (2.1) and (2.2) has been found useful in many applications. However, it has $(G - 1) + Gd + G\{d(d + 1)/2\}$ parameters, and this can be quite a large number. In our simplest example, the Old Faithful data (Example 1), with $d = 2$ and $G = 2$, this is 11 parameters, which is manageable. However, for the more complicated Italian wine data, with $d = 27$ and $G = 3$, this is 1,217 parameters, which is very large. Such large numbers of parameters can lead to difficulties in estimation, including lack of precision or even degeneracies. They can also lead to difficulties in interpreting the results.

In order to alleviate this problem, it is common to specify more parsimonious versions of the model. One way to do this is via the eigenvalue decomposition of the group covariance matrices Σ_g, in the form

$$\Sigma_g = \lambda_g D_g A_g D_g^T. \qquad (2.3)$$

In Equation (2.3), D_g is the matrix of eigenvectors of Σ_g, $A_g = \text{diag}\{A_{1,g}, \ldots, A_{d,g}\}$

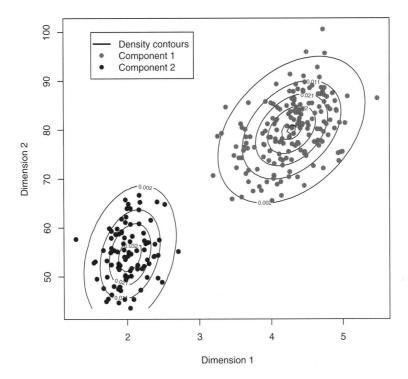

Figure 2.2 Contours of the density function for a two-dimensional finite bivariate normal mixture model with two mixture components (see text for parameter values). The dots show a sample of size 272 simulated from the density, with the colors indicating the mixture component from which they were generated.

is a diagonal matrix whose elements are proportional to the eigenvalues of Σ_g in descending order, and λ_g is the associated constant of proportionality.

Each factor in this decomposition corresponds to a particular geometric property of the gth mixture component. The matrix of eigenvectors D_g governs its orientation in R^d. The diagonal matrix of scaled eigenvalues A_g determines its shape. For example, if $A_{1,g} \gg A_{2,g}$, then the gth mixture component is tightly concentrated around a line in R^d. If $A_{1,g} \approx A_{2,g} \gg A_{3,g}$, then the gth mixture component is tightly concentrated around a two-dimensional plane in R^d. If all the values of $A_{j,g}$ are approximately equal, then the gth mixture component is roughly spherical.

The constant of proportionality λ_g determines the volume occupied by the gth mixture component in R^d, which is proportional to $\lambda_g^d |A_g|$. It is common to constrain the matrix A_g so that its determinant is equal to 1, in which case λ_g directly determines the volume of the gth component, which is then proportional

Listing 2.2: R code for Figure 2.2

```
# Estimate parameters of 2-component mixture model for the
# 2d Old Faithful data and plot the density
library(mclust)
n <- length (faithful[,1])
faithful.Mclust2den <- densityMclust (faithful, model="VVV",G=2)

# Simulate a sample from the density
sim.results <- simVVV (faithful.Mclust2den$parameters,n,seed=0)
ysim <- sim.results[,c(2,3)]
groupsim <- sim.results[,"group"]
ysim1 <- ysim[groupsim==1,]
ysim2 <- ysim[groupsim==2,]

# Plot the density and simulated points
plot (faithful.Mclust2den, col="black",xlab="Dimension 1",ylab="
    Dimension 2", xlim=c(min(ysim[,1]),max(ysim[,1])), ylim = c(min(
    ysim[,2]),max(ysim[,2])), levels = round (quantile(
    faithful.Mclust2den$density, probs = c(0.05, 0.25, 0.5, 0.75, 0
    .95, 0.99)),3))
points (ysim1,col="red",pch=19)
points (ysim2,col="blue",pch=19)
legend (x=1.5,y=100, legend=c("Density contours","Component 1","
    Component 2"),col=c("black","red","blue"),lty=c(1,0,0), lwd=c
    (2,0,0), pch=c(NA,19,19) )
```

to λ_g^{d}.[1] As a result of these geometric interpretations, the decomposition (2.3) is sometimes called the geometric decomposition or Volume-Shape-Orientation (VSO) decomposition.

Parsimony of the finite mixture model given by (2.1) and (2.2) can be enforced in various ways using the decomposition (2.3). Any or all of the geometric properties, volume, shape or orientation, can be constrained to be constant across mixture components. Also, the covariance matrix for any mixture component can be forced to be spherical (i.e. proportional to the identity matrix I), or diagonal.

This allows for two possible univariate models, and for 14 possible models in the multivariate case. These are shown in Table 2.1. Each of the multivariate models is denoted by a three-letter identifier, where the first letter is "E" if the volumes of the clusters are constrained to be equal, and "V" if not ("E" stands for "equal" and "V" stands for "varying"). Similarly, the second letter is "E" if the shape matrices A_g are constrained to be equal across clusters, so that $A_g \equiv A$ for $g = 1, \ldots, G$, "V" if they are not so constrained and "I" if the clusters are spherical, in which case $A_g = I$ for $g = 1, \ldots, G$. Finally, the third letter is equal to "E" if the matrices D_g of eigenvectors specifying the cluster orientations are constrained to be equal, so that $D_g \equiv D$ for $g = 1, \ldots, G$, to "V" if they are not so constrained, and to "I" if the clusters are spherical, in which case $D_g = I$ for $g = 1, \ldots, G$.

---

[1] In earlier work (Banfield and Raftery, 1993), λ_g was taken to be equal to the first eigenvalue of Σ_g.

Table 2.1 *Parameterizations of the covariance matrix Σ_g in model-based clustering. A denotes a diagonal matrix.*

Identifier	Model	Distribution	Volume	Shape	Orientation
E		Univariate	Equal		
V		Univariate	Variable		
EII	λI	Spherical	Equal	Equal	NA
VII	$\lambda_g I$	Spherical	Variable	Equal	NA
EEI	λA	Diagonal	Equal	Equal	Axis-aligned
VEI	$\lambda_g A$	Diagonal	Variable	Equal	Axis-aligned
EVI	λA_g	Diagonal	Equal	Variable	Axis-aligned
VVI	$\lambda_g A_g$	Diagonal	Variable	Variable	Axis-aligned
EEE	Σ	Ellipsoidal	Equal	Equal	Equal
VEE	$\lambda_g DAD^T$	Ellipsoidal	Variable	Equal	Equal
EVE	$\lambda DA_g D^T$	Ellipsoidal	Equal	Variable	Equal
EEV	$\lambda D_g AD_g^T$	Ellipsoidal	Equal	Equal	Variable
VVE	$\lambda_g DA_g D^T$	Ellipsoidal	Variable	Variable	Equal
EVV	$\lambda D_g A_g D_g^T$	Ellipsoidal	Equal	Variable	Variable
VEV	$\lambda_g D_g AD_g^T$	Ellipsoidal	Variable	Equal	Variable
VVV	Σ_g	Ellipsoidal	Variable	Variable	Variable

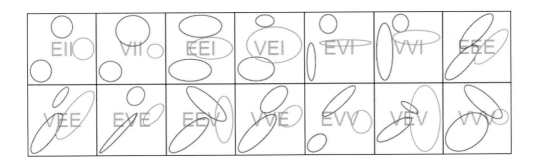

Figure 2.3 Models used in model-based clustering: examples of contours of the bivariate normal component densities for the 14 parameterizations of the covariance matrix used in model-based clustering.

Figure 2.3 shows examples of contours of the component densities for the various models in the two-dimensional case with two mixture components.

Sometimes these constrained models can have far fewer parameters that need to be independently estimated than the unconstrained model, while fitting the

Table 2.2   *Numbers of parameters needed to specify the covariance matrix for models used in model-based clustering.*

Model	# Covariance Parameters		
	General	$d = 2, G = 2$	$d = 27, G = 3$
EII	1	1	1
VII	G	2	3
EEI	d	2	27
VEI	$G + (d-1)$	3	29
EVI	$1 + G(d-1)$	3	79
VVI	Gd	4	81
EEE	$d(d+1)/2$	3	378
VEE	$G + (d+2)(d-1)/2$	4	380
EVE	$1 + (d+2G)(d-1)/2$	4	430
EEV	$1 + (d-1) + G[d(d-1)/2]$	4	1080
VVE	$G + (d+2G)(d-1)/2$	5	432
EVV	$1 + G(d+2)(d-1)/2$	5	1132
VEV	$G + (d-1) + G[d(d-1)/2]$	5	1082
VVV	$G[d(d+1)/2]$	6	1134

sample data almost as well. When this is the case, they can yield more precise estimates of model parameters, better out-of-sample predictions, and more easily interpretable parameter estimates. All the models have Gd parameters for the component means $\boldsymbol{\mu}_g$, and $(G-1)$ parameters for the mixture proportions τ_g.

Table 2.2 shows the numbers of parameters needed to specify the covariance matrix for each model in the two-dimensional two-component case, $d = 2, G = 2$, and the 27-dimensional three-component case, $d = 27, G = 3$. These results are obtained by noting that for one mixture component, the volume is specified by 1 parameter, the shape by $(d-1)$ parameters, and the orientation by $d(d-1)/2$ parameters.

It is clear that the potential gain in parsimony as measured by number of parameters is small for the two-dimensional case. But for higher-dimensional cases, the gain can be large. In the most extreme case in Table 2.2, in the 27-dimensional case with 3 mixture components, the VVV model requires 1,134 parameters to represent the covariance matrices, whereas the EII model requires only one.

Also, more parameters are required to specify the shape than the volume, and far more again to specify the orientation. Thus big gains in parsimony are achieved by the models that require the orientations to be equal across mixture components, and the largest gains come from requiring the component densities to be diagonal. However, as we will see, these most parsimonious models do not always fit the data adequately.

2.3 Estimation by Maximum Likelihood

We will now describe how the models in Section 2.1 can be estimated by maximum likelihood. The most common way to do this is via the Expectation-Maximization, or EM, algorithm (Dempster et al., 1977; McLachlan and Krishnan, 1997). This is a general approach to maximum likelihood for problems in which the data can be viewed as consisting of n multivariate observations (y_i, z_i), in which y_i is observed and z_i is unobserved. If the (y_i, z_i) are independent and identically distributed (iid) according to a probability distribution f with parameters θ, then the *complete-data likelihood* is

$$\mathcal{L}_C(y, z \mid \theta) = \prod_{i=1}^{n} f(y_i, z_i \mid \theta), \qquad (2.4)$$

where $y = (y_1, \ldots, y_n)$ and $z = (z_1, \ldots, z_n)$. The *observed data likelihood*, $\mathcal{L}_O(y \mid \theta)$, can be obtained by integrating the unobserved data z out of the complete-data likelihood,

$$\mathcal{L}_O(y \mid \theta) = \int \mathcal{L}_C(y, z \mid \theta) \, dz. \qquad (2.5)$$

The maximum likelihood estimator (MLE) of θ based on the observed data maximizes $\mathcal{L}_O(y \mid \theta)$ with respect to θ.

The observed data likelihood, also called the mixture likelihood, or just the likelihood, can also be written as

$$\begin{aligned}
\mathcal{L}_O(y \mid \theta) &= \prod_{i=1}^{n} \sum_{g=1}^{G} \tau_g f_g(y_i \mid \theta_g) \\
&= \prod_{i=1}^{n} \sum_{g=1}^{G} \tau_g \phi_g(y_i \mid \mu_g, \Sigma_g).
\end{aligned} \qquad (2.6)$$

The EM algorithm alternates between two steps. The first is an "E-step," or expectation step, in which the conditional expectation of the complete data log-likelihood given the observed data and the current parameter estimates is computed. The second is an "M-step," or maximization step, in which parameters that maximize the expected log-likelihood from the E-step are determined. In its application to MLE for mixture models, the unobserved portion of the data, z, involves quantities that are introduced in order to reformulate the problem for EM. Under fairly mild regularity conditions, EM can be shown to converge to a local maximum of the observed-data likelihood (Dempster et al., 1977; Boyles, 1983; Wu, 1983; McLachlan and Krishnan, 1997). The EM algorithm has been widely used for maximum likelihood estimation of mixture models, with good results.

In EM for mixture models, the "complete data" are considered to be (y_i, z_i), where $z_i = (z_{i,1}, \ldots, z_{i,G})$ is the unobserved portion of the data, with

$$z_{i,g} = \begin{cases} 1 & \text{if } y_i \text{ belongs to group } g \\ 0 & \text{otherwise.} \end{cases} \qquad (2.7)$$

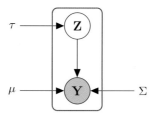

Figure 2.4 Graphical model representation of a Gaussian mixture model (GMM). Random variables and vectors are nodes represented by circles. Observed random variables are shown in green, whereas latent (unobserved) variables are shown in white. The plate, or red rectangle, indicates that the nodes within it consist of n conditionally independent observations or instances. Thus $Y = (y_1, \ldots, y_n)$ and $Z = (z_1, \ldots, z_n)$. Unknown model parameters are nodes shown only by their names.

We assume that the $\mathbf{z}_i$ are independent and identically distributed, each according to a multinomial distribution of one draw from G categories with probabilities $\tau_1, \ldots, \tau_G$. We also assume that the density of an observation y_i given $\mathbf{z}_i$ is given by $\prod_{g=1}^{G} f_g(y_i \mid \theta_g)^{z_{i,g}}$. Then the resulting complete-data log-likelihood is

$$\ell_C(\theta_g, \tau_g, z_{i,g} \mid y) = \sum_{i=1}^{n} \sum_{g=1}^{G} z_{i,g} \log \left[\tau_g f_g(y_i \mid \theta_g) \right]. \tag{2.8}$$

It can be helpful in understanding this to represent the Gaussian mixture model as a graphical model, using a directed acyclic graph (DAG), as shown in Figure 2.4. In a DAG, the absence of an edge between two nodes indicates that the corresponding variables are conditionally independent given the intervening variables.

The E-step of the sth iteration of the EM algorithm for mixture models is given by

$$\hat{z}_{i,g}^{(s)} = \frac{\hat{\tau}_g^{(s-1)} f_g(y_i \mid \hat{\theta}_g^{(s-1)})}{\sum_{h=1}^{G} \hat{\tau}_h^{(s-1)} f_h(y_i \mid \hat{\theta}_h^{(s-1)})}, \tag{2.9}$$

where $\hat{\tau}_g^{(s)}$ is the value of τ_g after the sth EM iteration. The M-step involves maximizing (2.8) in terms of τ_g and θ_g with $z_{i,g}$ fixed at the values computed in the E-step, namely $\hat{z}_{i,g}$.

The quantity $\hat{z}_{i,g}^{(s)} = E[z_{i,g} | y_i, \theta_1, \ldots, \theta_G]$ for the model (2.1) is the conditional expectation of $z_{i,g}$ given the parameter values at the $(s-1)$th iteration and the observed data y. The value $\hat{z}_{i,g}$ of $z_{i,g}$ at a maximum of (2.6) is the estimated conditional probability that observation i belongs to group g. The maximum likelihood classification of observation i is $\{h \mid \hat{z}_{ih} = \max_g \hat{z}_{i,g}\}$. As a result, $(1 - \max_g \hat{z}_{i,g})$ is a measure of the uncertainty in the classification of observation i (Bensmail et al., 1997).

For multivariate normal mixtures, the E-step is given by (2.9) with f_g replaced by ϕ_g as defined in (2.2), regardless of the parameterization. For the M-step, estimates of the means and probabilities have simple closed-form expressions involving the data and $\hat{z}_{i,g}$ from the E-step, namely:

$$\hat{\tau}_g^{(s)} = \frac{\hat{n}_g^{(s-1)}}{n}; \quad \hat{\mu}_g^{(s)} = \frac{\sum_{i=1}^n \hat{z}_{i,g}^{(s-1)} y_i}{\hat{n}_g^{(s-1)}}; \quad \hat{n}_g^{(s-1)} = \sum_{i=1}^n \hat{z}_{i,g}^{(s-1)}. \qquad (2.10)$$

Computation of the covariance estimate $\hat{\Sigma}_g^{(s)}$ depends on its parameterization. For details of the M-step for Σ_k parameterized by the eigenvalue decomposition (2.3) see Celeux and Govaert (1995). For instance, for the more general model "VVV," where no constraints are placed on the volumes, shapes and orientations of the mixture covariance matrices, it is

$$\hat{\Sigma}_g^{(s)} = \frac{\sum_{i=1}^n \hat{z}_{i,g}^{(s-1)} (y_i - \hat{\mu}_g^{(s)})(y_i - \hat{\mu}_g^{(s)})^T}{\hat{n}_g^{(s-1)}}. \qquad (2.11)$$

Convergence of the EM algorithm to a local maximum of the log-likelihood function can be assessed in several ways. In essence, these consist of seeing whether the algorithm has been moving slowly in the latest iterations. One possible criterion is that the log-likelihood has changed very little between the last two iterations; a typical threshold is a change of less than 10^{-5}. Another possible criterion is that the model parameters have changed little between the last two iterations. The R package `mclust` uses the log-likelihood criterion by default.

As we will see in Section 3.2, the EM algorithm can also be used in a Bayesian context to find the posterior mode when there is a prior distribution on the parameters.

Example 1 (ctd.)

We illustrate the EM algorithm for multivariate normal mixture models using the Old Faithful data. We used the EM algorithm to carry out maximum likelihood estimation for the model with two mixture components and unconstrained covariance matrices (model "VVV"). We initialized the EM algorithm using a random partition (other initialization strategies will be discussed in Section 2.4).

A trace of the values of the component means, μ_1 and μ_2, at each iteration of the EM algorithm is shown in Figure 2.5. The algorithm converged within 50 iterations, and the converged estimates are shown by solid circles. Because the algorithm was initialized with a random partition, the starting estimates of μ_1 and μ_2 were very similar. R code to produce Figure 2.5 is shown in Listings 2.3 and 2.4;; the other figures in this example can be produced in a similar way.

Figure 2.6 (left) shows the log-likelihood values for each iteration of the EM algorithm. As the theory predicts, the log-likelihood increases at each iteration. For the first 30 iterations, the log-likelihood remains relatively constant as the algorithm struggles to escape from the initial random partition, which gives little information about the two groups in the data. Then it detects the separation between the groups, and the log-likelihood increases rapidly between iteration 30 and iteration 40 as the algorithm refines the partition. From iteration 40 onwards

Mean Parameters EM Trace

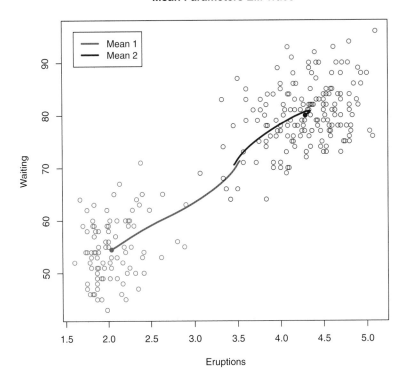

Figure 2.5 Estimates of the component mean parameter vectors μ_1 and μ_2 for the two-component bivariate normal mixture model with unconstrained variance matrix (**VVV**) for the Old Faithful data at each iteration of the EM algorithm. The red and blue solid curves show the estimates of the means of the first and second mixture components, respectively. The solid circles show the final estimates. The open circles show the data points, with colors indicating the final classification from the EM algorithm.

it stabilizes and remains essentially constant, increasing by tiny amounts at each iteration. By iteration 50 it has fairly clearly converged.

Figure 2.6 (right) shows the estimates of the mixing proportion for the first component, $\hat{\tau}_1^{(s)}$, at each iteration. A rapid change between iterations 30 and 40 is again apparent. Similar plots for the elements of the component covariance matrices, Σ_g, are shown in Figure 2.7. The parameters for component 1 show the same rapid change between iterations 30 and 40, while the parameters for component 2 show a more gradual evolution.

Figure 2.8 shows the evolution of the classification and the fitted model over the course of the EM iterations. The top left panel shows the random initial partition, which gives no indication of grouping in the data. The fit after one iteration is shown in the top right panel. This shows the bivariate normal distributions for the

Listing 2.3: R code for Figure 2.5 (part 1)

```
# Loading the mclust library
library(mclust)

# Random initial partition
n <- length (faithful[,1])
set.seed (0)
z.init <- rep (1,n)
n2 <- rbinom (n=1,size=n,prob=0.5)
s2 <- sample (1:n, size=n2)
z.init[s2] <- 2

# EM algorithm step-by-step
# Initialize EM algorithm
itmax <- 50
G <- 2
zmat <- matrix (rep (0,n*G), ncol=G)
for (g in (1:G)) zmat[,g] <- z.init==g
mstep.out <- vector ("list",itmax)
estep.out <- vector ("list",itmax)

# EM iterations
for (iter in 1:itmax) {
  # M-step
  mstep.tmp <- mstep (modelName="VVV",data=faithful,z=zmat)
  mstep.out[[iter]] <- mstep.tmp
  # E-step
  estep.tmp <- estep (modelName="VVV",data=faithful,parameters=
      mstep.tmp$parameters)
  estep.out[[iter]] <- estep.tmp
  zmat <- estep.tmp$z
}

# Extract classification
EMclass <- rep (NA,n)
for (i in 1:n) {
 zmati <- zmat[i,]
 zmat.max <- zmati==max(zmati)
 zmat.argmax <- (1:G)[zmat.max]
 EMclass[i] <- min(zmat.argmax)
}
```

two components (shown by ellipses that represent constant density contours) to overlap almost completely. Almost all the observations are classified into cluster 2, reflecting the fact that $\hat{z}_{i,2}^1 > \hat{z}_{i,1}^1$ for most i. This means that after one iteration, the estimated posterior probability of having been generated by component 2 is higher than that of coming from component 1 for almost all observations. In fact, however, the difference is small and the classification is far from decisive. The slight preference for component 2 is due largely to the fact that $\hat{\tau}_2^1 > \hat{\tau}_1^1$, i.e. that after one iteration the estimated mixture probability is higher for component 2 than for component 1.

The bottom left panel of Figure 2.8 shows the situation after the 33rd iteration,

Listing 2.4: R code for Figure 2.5 (part 2)

```
# Two-dimensional trace plot of component means
# First, extract the EM iteration estimates of mu_{g,d}
d <- length (faithful[1,]) # data dimension
mu1 <- matrix (rep(NA,itmax*d),ncol=2)
mu2 <- matrix (rep(NA,itmax*d),ncol=2)
for (iter in (1:itmax)) {
  mu1[iter,] <- estep.out[[iter]]$parameters$mean[,1]
  mu2[iter,] <- estep.out[[iter]]$parameters$mean[,2]
}

# Plot the traces of mu_{g,d}
plot (faithful, type="n", xlab="Eruptions", ylab="Waiting", main="
    Mean Parameters EM Trace")
y1 <- faithful[EMclass==2,]
y2 <- faithful[EMclass==1,]
lines (mu1[,1],mu1[,2],type="l",lwd=3,col="red")
lines (mu2[,1],mu2[,2],type="l",lwd=3,col="blue")
points (y1, col="red", pch=1)
points (y2, col="blue", pch=1)
points (mu1[itmax,1],mu1[itmax,2],col="red",pch=19)
points (mu2[itmax,1],mu2[itmax,2],col="blue",pch=19)
legend (x=min(faithful[,1]),y=max(faithful[,2]), legend=c("Mean 1",
    "Mean 2"), col=c("red","blue"),lty=c(1,1), lwd=c(3,3))
```

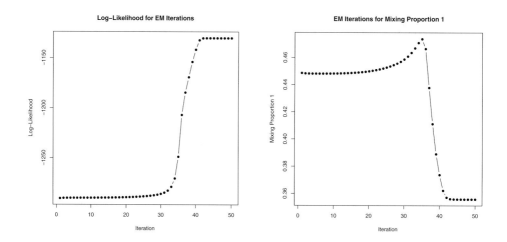

Figure 2.6 Left: log-likelihood, and Right: mixing proportion, τ_1, for component 1, at each iteration of the EM algorithm for the two-component bivariate normal mixture model with unconstrained variance matrix (VVV) for the Old Faithful data.

when the algorithm has started to detect the separation between the two clusters. It is apparent that the classification is still inadequate. The bottom right panel shows the result of the 50th iteration, at which stage the algorithm has converged. The two clusters have been satisfactorily separated.

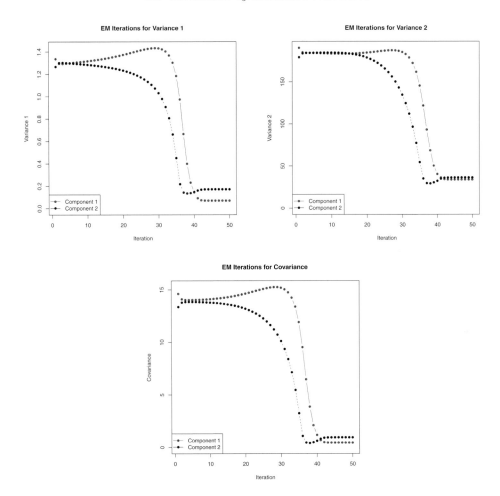

Figure 2.7 Estimates of the variance matrices Σ_1 and Σ_2 for the two-component bivariate normal mixture model with unconstrained variance matrix (VVV) for the Old Faithful data at each iteration of the EM algorithm. Top left: variance of variable 1. Top right: variance of variable 2. Bottom: covariance of variables 1 and 2.

The probability density function of the data evaluated at the parameter estimates is a finite mixture distribution, and this is depicted in several ways in Figure 2.9, using the R code in Listing 2.5.

The uncertainty associated with the allocation of the ith observation's cluster membership is measured by

$$\text{Uncer}_i = 1 - \max_{g=1,\ldots,G} \hat{z}_{i,g}. \tag{2.12}$$

The uncertainty for data point i, Uncer_i, will be largest for data points i for which the $\hat{z}_{i,1}, \ldots, \hat{z}_{i,G}$ are equal to one another, and thus $\text{Uncer}_i = 1/G$ for those data points. It will be smallest when one of the $\hat{z}_{i,1}, \ldots, \hat{z}_{i,G}$ is close to 1, in which case

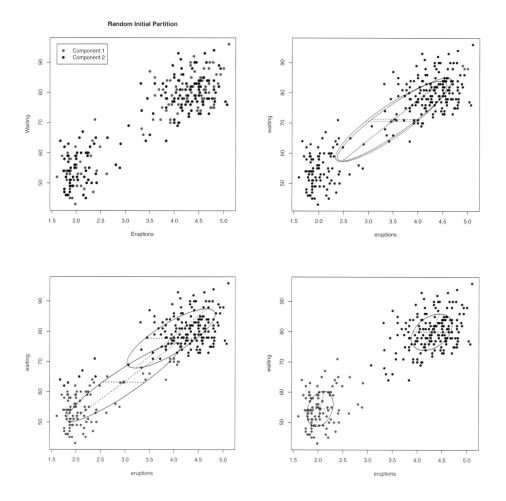

Figure 2.8 Classifications and fitted models for the two-component bivariate normal mixture model with unconstrained variance matrix (VVV) for the Old Faithful data during the EM algorithm. Top left: random initial partition. Top right: iteration 1. Bottom left: iteration 33. Bottom right: iteration 50 (converged).

it will be close to zero. The uncertainty values for the Old Faithful data are shown in Figure 2.10, produced using the R code in Listing 2.6.

In the left panel of Figure 2.10, the 5% of the observations with the greatest uncertainty are shown by large black circles; these are the ones that are roughly equidistant between the two clusters. The next 20% of the observations in terms of uncertainty are shown by smaller gray circles, and these are closer to the cluster centers than the most uncertain points. The least uncertain 75% of the observations are shown by small gray dots, and these are the ones that are closest to the cluster centers. In this data set the clusters are well separated, and so there is little uncertainty about the allocation of most of the data points.

Listing 2.5: Code for density plots

```
# Loading the mclust library
library(mclust)
data(faithful)

# Clustering with Mclust
faithful.Mclust2 <- Mclust (faithful , model="VVV",G=2)

# Plotting the densities as contour
surfacePlot (faithful,faithful.Mclust2$parameters ,type="contour",
    what="density")
points (faithful)

# Alternative ways of plotting the densities
surfacePlot (faithful,faithful.Mclust2$parameters ,type="image",
    what="density")
surfacePlot (faithful,faithful.Mclust2$parameters ,type="persp",
    what="density")
```

Listing 2.6: Code for Old Faithful uncertainty plot

```
# Loading the mclust library
library(mclust)
data(faithful)

# Clustering with Mclust
faithful.Mclust2 <- Mclust (faithful , model="VVV",G=2)

# Display uncertainty on cluster assignments
plot(faithful.Mclust2,what="uncertainty")
surfacePlot(faithful,faithful.Mclust2$parameters ,type="contour",
    what="uncertainty")
points (faithful)
```

The right panel of Figure 2.10 shows contours of uncertainty level. These are slightly nonlinear, reflecting the fact that the two clusters have different covariance matrices. They also fall away quickly from the contour of highest uncertainty, reflecting the fact that most of the data points are assigned to clusters with little uncertainty.

2.4 Initializing the EM Algorithm

The likelihood for a mixture model is in general not convex. As a result there can be local maxima; an example is given in Biernacki et al. (2003, Section 1). One consequence is that the converged estimate can depend on the initial value chosen.

For normal mixture models things are even trickier, as there are typically multiple paths in parameter space along which the likelihood can tend to infinity at parameter values on the edge of the parameter space (Titterington et al.,

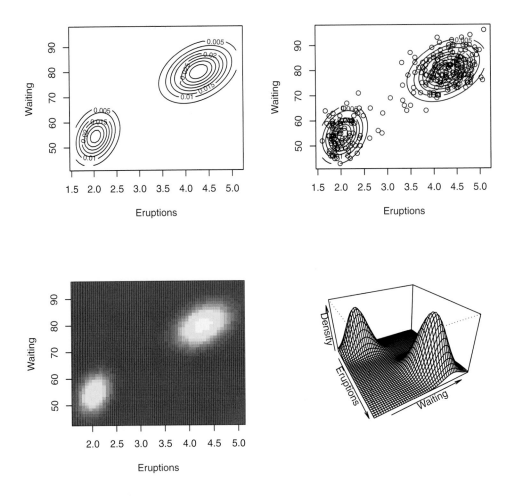

Figure 2.9 Two-dimensional probability density function of the data at the parameter estimates for the Old Faithful data: contour plot, contour plot with observations, image plot, and perspective plot.

1985). For example, for a one-dimensional mixture of two normal distributions, the likelihood tends to infinity as $\sigma_1 \longrightarrow 0$ when $\mu_1 = y_i$ for any i. These infinite solutions are in general degenerate and do not yield estimates with good properties. The solutions with good properties are local maxima of the likelihood function that are internal to the parameter space (Redner and Walker, 1984). Thus in general we are searching for the largest internal local maximum of the likelihood function.

Because of these properties of the likelihood function, the result can depend critically on the initial values chosen. In Example 1 with the Old Faithful data,

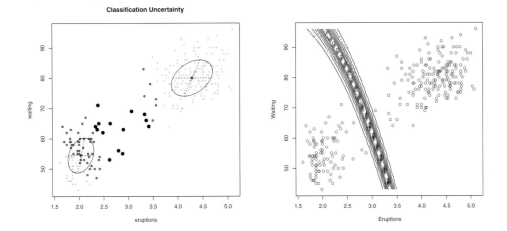

Figure 2.10 Uncertainty values for the Old Faithful data with two clusters. Left: uncertainty values for individual observations. The large black circles indicate the top 5% of the observations in terms of uncertainty, the smaller gray circles indicate the next 20%, and the light gray dots indicate the lowest 75%. Right: contours of uncertainty level.

we saw in Section 2.3 that the EM algorithm gave satisfactory performance when initialized with a random partition. This often does not work well, however.

Here we will describe two methods for selecting initial values that have performed well in a variety of settings: hierarchical model-based clustering, and the so-called smallEM method.

2.4.1 Initialization by Hierarchical Model-based Clustering

In its first decades, cluster analysis was done mainly using hierarchical clustering methods, also called agglomerative hierarchical clustering methods. The starting point for these methods is a set of pairwise dissimilarities between all objects. If these are not given or measured directly, they must be computed before the cluster analysis is carried out. The methods are generally initialized with the trivial partition where each object is its own cluster. Then, at each stage two clusters are merged. The pair of clusters to be merged is chosen so as to minimize a criterion over all possible pairs. Different merging criteria define different hierarchical clustering methods. If there are n observations, then there are $n-1$ stages in the algorithm, yielding n different clusterings, with $n, n-1, \dots, 1$ clusters respectively.

Popular criteria include the single link criterion, defined as the smallest dissimilarity between a member of one cluster and a member of the other cluster (Sneath, 1957), the average link criterion, defined as the average between-cluster dissimilarity (Sokal and Michener, 1958), the complete link criterion, defined as the largest between-cluster dissimilarity (Defays, 1978), and the sum of squares criterion, defined as the total within-group sum of squares about the group means (Ward,

1963). These criteria are all special cases of a more general, computationally efficient algorithm proposed by Lance and Williams (1967).

In hierarchical model-based clustering (Banfield and Raftery, 1993), the criterion used for deciding which clusters to merge at each stage is no longer based on dissimilarities. Instead the criterion used is the *classification likelihood*,

$$\mathcal{L}_{CL}(\theta, z|y) = \prod_{i=1}^{n} f_{z_i}(y_i|\theta_{z_i}). \tag{2.13}$$

Note that this is different from the observed data (or mixture) likelihood given by Equation (2.5) in that the set of cluster memberships z is an argument of the classification likelihood, whereas these are integrated out in the observed data likelihood. It is also different from the complete-data likelihood given by Equation (2.4), in that it does not include the mixing probabilities τ_g corresponding to the cluster memberships z_i.

Methods that attempt to maximize the classification likelihood (2.13) attempt to estimate the cluster memberships and the model parameters simultaneously and are not in general asymptotically consistent in the sense of being guaranteed to give estimates that tend to the true values as sample size increases (Marriott, 1975). However, they are often much more computationally efficient and so can work well for initialization. Clustering methods using the classification and mixture likelihoods were compared by Celeux and Govaert (1993).

Another method aimed at maximizing the classification likelihood is the *Classification EM* (CEM) algorithm (Celeux and Govaert, 1992).This CEM algorithm incorporates a classification step (C-step) between the E-step and the M-step of the EM algorithm. The C-step consists of replacing the unobserved labels $h_i, i = 1, \ldots, n$, defining a partition into G clusters of the observations by $\{h_i^{(s)} \mid \hat{z}_{ih}^{(s)} = \max_g \hat{z}_{i,g}^{(s)}\}$. Thus, the M-step reduces to an estimation of the proportions, means and covariance matrices of the resulting clusters. An advantage of this CEM algorithm, which is a k-means-like algorithm, is that it converges in a finite number of iterations. In most situations, this number of iterations is small. Thus, it is recommended and not expensive to run the CEM algorithm many times from random initial positions to get a good maximizer of the classification likelihood.

Hierarchical model-based clustering proceeds by successively merging pairs of clusters corresponding to the greatest increase in the classification likelihood (2.13) among all possible pairs. In the absence of any information about groupings, the procedure starts by treating each observation as a singleton cluster. When the probability model in (2.13) is multivariate normal with the equal-volume spherical covariance λI (the EI model), model-based hierarchical clustering is equivalent to hierarchical clustering with the sum of squares criterion.

An advantage of initialization by hierarchical model-based clustering is that it is often computationally efficient, in that a single pass over the data gives initial clusterings for all models and numbers of clusters. As we will see, model-based

Listing 2.7: Initialization with hierarchical model-based clustering

```
# Calling hc for initializing mclust
faithful.hmc <- hc(modelName="VVV",faithful)
faithful.hmclass <- c(hclass (faithful.hmc,2))
```

Listing 2.8: Initialization using a subset in hierarchical model-based clustering

```
# Initialization by sub-sampling
S <- sample (1:nrow(faithful), size=100)
Mclust (faithful, initialization=list(subset=S))
```

clustering can involve fitting large numbers of different models, so obtaining initial partitions for all of them from a single run is appealing.

By default, in the `mclust` R package, the EM algorithm is initialized by running the hierarchical model-based clustering with the VVV model. This gives a partition for each number of clusters. These partitions are used as the initial points for the algorithm.

Example 1 (ctd.)

In the Old Faithful data, a hierarchical model-based clustering of the data is carried out using the R commands shown in Listing 2.7.

The results are shown in Figure 2.11. This is close, but not identical, to the converged maximum likelihood solution shown in the bottom left panel of Figure 2.8. It is close enough to provide a good starting value for the EM algorithm. This is the default initialization in the `mclust` package.

A potential disadvantage of hierarchical model-based clustering is that, in its most basic form, it can take a long time to converge when the data set is large. This is because it involves computing and maintaining a matrix of the classification likelihood increase for all pairs of clusters that could potentially be merged. Since initially there is one cluster per observation, this matrix is of size $O(n^2)$. If n is large enough, this matrix does not fit in memory, and manipulating it involves swapping in and out of memory. When that threshold is passed, computing time can increase hugely.

A simple fix is to base the initialization by hierarchical model-based clustering on a random subset of the data. Experience suggests that there are few gains to be made by increasing the size of the subset beyond about 2,000 (Wehrens et al., 2004). Code to do this for the Old Faithful data with a subset size of 100 is shown in Listing 2.8.

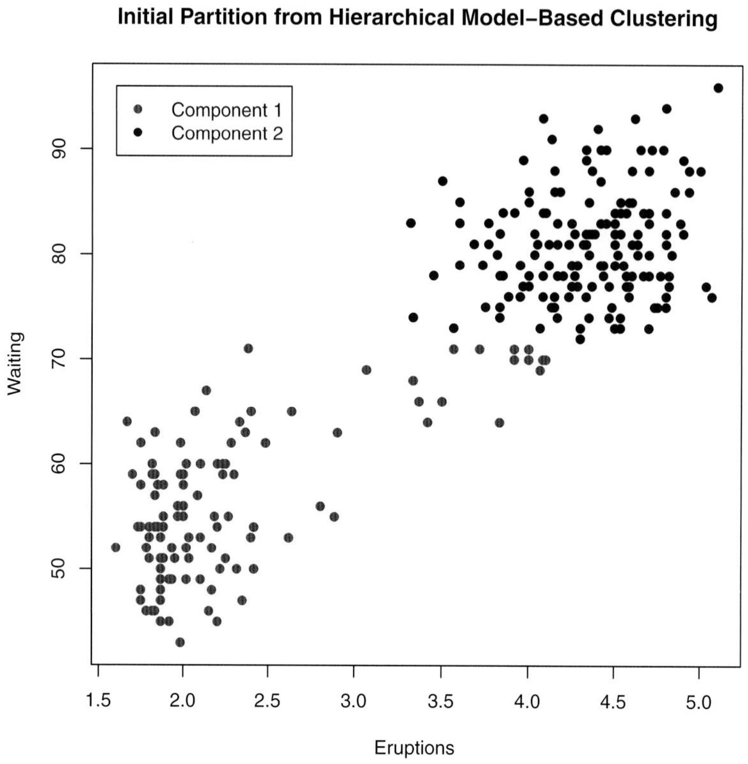

Initial Partition from Hierarchical Model–Based Clustering

Figure 2.11 Partition of the Old Faithful data into two clusters using hierarchical model-based clustering with the unconstrained variance (VVV) model.

2.4.2 Initialization Using the smallEM Strategy

The "smallEM" strategy (Biernacki et al., 2003) is implemented separately for each clustering model and number of clusters considered. It consists of the following steps:

1 Repeat the following M times:

  1 Simulate a random set of parameter values θ.

  2 Carry out a short run of the EM algorithm starting at the simulated set of parameter values. This is not run to convergence but instead is terminated at the first iteration s where

$$\frac{\ell_O(y|\hat{\theta}^{(s)}) - \ell_O(y|\hat{\theta}^{(s-1)})}{\ell_O(y|\hat{\theta}^{(s)}) - \ell_O(y|\hat{\theta}^{(1)})} < \varepsilon,$$

where $\ell_O(y|\hat{\theta}^{(s)}) = \log \mathcal{L}_O(y|\hat{\theta}^{(s)})$ is the mixture (or observed data) log-likelihood after the sth EM iteration.

Listing 2.9: Initialization with smallEM

```
# Load the Rmixmod library
library(Rmixmod)

# Run six smallEMs (here limited to two iterations)
nb = 6
par(mfrow=c(2,3))
res = list(); likelihood = c()
myModel = mixmodGaussianModel(listModels='Gaussian_pk_Lk_Ck')
for (i in 1:nb){
  res[[i]] = mixmodCluster(faithful,nbCluster=2,model=myModel,
     strategy=mixmodStrategy(initMethod='random' ,nbTryInInit=1,
     nbIterationInInit=1,nbIterationInAlgo=1))
  likelihood[i] = res[[i]]@bestResult@likelihood
  plot(faithful,col=res[[i]]@bestResult@partition+1,pch=c(17,19)[
     res[[i]]@bestResult@partition])
  title(main=paste('Likelihood =',round(likelihood[i]))) }

# Run EM initialized with the best smallEM solution
par(mfrow=c(1,1))
final.res = mixmodCluster(faithful,nbCluster=2,model=myModel)
plot(faithful,col=final.res@bestResult@partition+1,pch=c(17,19)[
   final.res@bestResult@partition])
title(main=paste('Likelihood =',round(
   final.res@bestResult@likelihood)))
```

3 Record the mixture log-likelihood at the stopping point of the short EM run.

2 Among the stopping points of the short EM runs, choose the one with the highest mixture likelihood. Using it as a starting point, run the EM algorithm to convergence.

Biernacki et al. (2003) recommended using $\varepsilon = 0.01$. They compared the smallEM strategy with several other possible initialization strategies and found it to work at least as well as the other ones they considered. The smallEM strategy with $M = 5$ is the default initialization method in the Rmixmod package.

Note that smallEM starts with random sets of parameter values. By contrast, for our illustration of the EM algorithm in Section 2.3 we used a random partition of the data points. This was a bad initial value, although the EM algorithm was able to recover from it and find a good estimate eventually. Random parameter values tend to provide better starting values than random partitions of the data points.

Example 1 (ctd.)

Results from the smallEM initialization method for the Old Faithful data are shown in Figures 2.12 and 2.13. They were produced using the R commands in Listing 2.9.

The smallEM solutions shown in Figure 2.12 are quite similar, even though

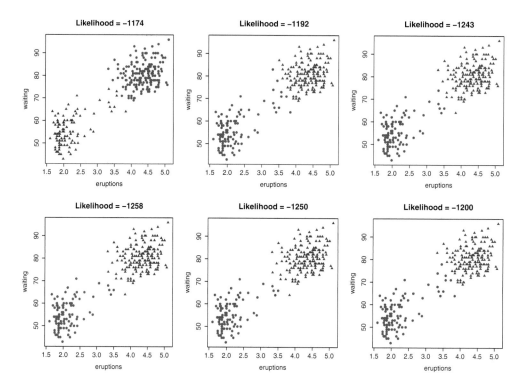

Figure 2.12 Initialization with smallEM on the Old Faithful Geyser data: six runs of two-iteration EM to select the best initialization. Here, the best initialization is the top-left panel with log-likelihood equal to $-1,174$. "Likelihood" refers to the maximized mixture log-likelihood.

each one is the result of only two EM iterations. However, the differences in log-likelihood between the solutions are quite substantial.

The final estimate from smallEM shown in Figure 2.13 is the same classification as that from the random partition initial value shown in Figure 2.5. This is in spite of the fact that the initial values are dramatically different.

Finally, especially in the clustering context, the CEM algorithm could be a good alternative to the smallEM strategy since the CEM algorithm is expected to converge rapidly and can be repeated many times at small cost (Biernacki et al., 2003).

We also illustrate the possible differences between the partitions derived from the EM and the CEM algorithms with the Old Faithful geyser data. If the number of clusters is fixed at $G = 2$, the CEM algorithm with the VVV model from a random initial solution gives exactly the same partition, displayed in Figure 2.13, that we get from EM initialized with the smallEM strategy. This is not surprising since with $G = 2$ the two mixture components are well separated. But with $G = 3$ and the VVV model, the EM and the CEM algorithms yield different partitions; see Figure 2.14.

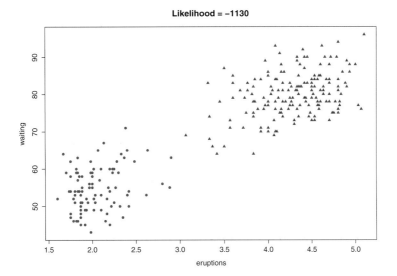

Figure 2.13 Initialization with smallEM on Old Faithful geyser data: final result of the EM algorithm from the best two-iteration run. "Likelihood" refers to the maximized mixture log-likelihood.

2.5 Examples with Known Number of Clusters

In this section, we will analyze several of the data sets introduced in Chapter 1. In each case, we will take the number of clusters to be known in advance. This is often not the case in practice, and in Section 2.6 we will describe methods for inferring the number of clusters from the data. We will analyze the data using model-based clustering, and also using some other popular methods for comparison.

Example 2 (ctd.): Diabetes diagnosis

We will now analyze the diabetes measurements shown in Figure 1.1. To carry out model-based clustering, we will take the number of clusters to be equal to 3, the number in the clinical classification. We will also use the unconstrained covariance model (**VVV**), in the absence of any information about likely constraints. In Section 2.6, we will see how to relax the assumption that the number of clusters is known, and instead select the clustering model using the data.

Figure 2.15 shows the result of model-based clustering for these data. The clusters clearly have quite different volumes, shapes and orientations. The compact blue cluster captures the non-diabetic (normal) subjects, and corresponds to relatively low values of all three measurements. The red cluster largely captures the chemical diabetics, who can have a wide range of levels of SSPG, up to very high ones. The green cluster corresponds to the subjects suffering from overt diabetes, who have a wide range of both glucose and insulin measurements, again up to very high levels. Insulin and glucose levels are highly correlated with one another for the overt diabetics, but less so for the other groups.

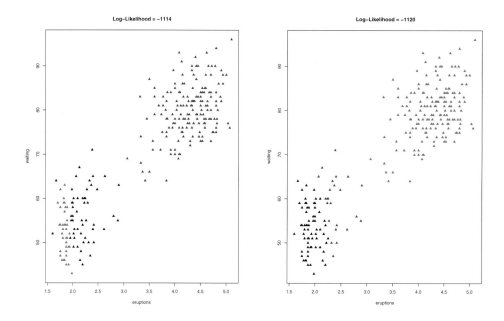

Figure 2.14 Classification plots for the Old Faithful data with the VVV model and $G = 3$. Left: EM results. Right: CEM results.

The model-based clustering classification corresponds quite well to the clinical classification, as can be seen from Figure 2.16. Seventeen of the 145 subjects were misclassified, an error rate of 11.7%. Perhaps more importantly, the general configuration of all three groups was well captured; this is important for classifying future subjects.

We compared model-based clustering to several other popular clustering methods. The results are shown in Figure 2.17. Single link clustering gave a very poor result, clustering all the points except two into a single cluster. Average link clustering, complete link clustering, and k-means all combined the normal subjects with those suffering from chemical diabetes. They also split the overt diabetes subjects into two clusters. Thus these other four methods failed to capture the general nature of the clustering. In the case of the last three methods, this is due to the fact that they are designed to find tight clusters, and as a result tend to find clusters that are roughly spherical. Model-based clustering seeks instead to find clusters with distributions that fit the data well.

Table 2.3 shows a quantitative comparison between the different clustering methods. We compared model-based clustering with single link clustering, average link clustering, complete link clustering and k-means clustering. We show two measures of performance. The first is the classification error rate (CER), which is the proportion of points that are misclassified. The minimum over all permutations of cluster labels is taken. The smaller the CER, the better.

The second measure is the adjusted Rand index (ARI). The Rand index (Rand,

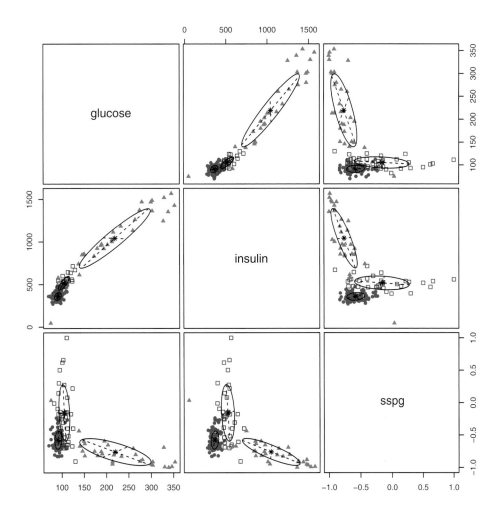

Figure 2.15 Model-based clustering of the diabetes data with $G = 3$ groups using the unconstrained covariance (VVV) model. The colors indicate the classifications of the points into the three clusters. The ellipses show contours of the fitted multivariate component densities corresponding to one standard deviation along each principal component. The dashed black lines show the principal component directions of each mixture component.

1971) of the similarity of two partitions of a data set is the proportion of pairs of data points that are in the same group in both partitions. The adjusted Rand index (ARI) (Hubert and Arabie, 1985) is the chance-adjusted value of the Rand index. When the two partitions are statistically independent, its expected value is zero, and when the two partitions are identical its value is 1. The *higher* the ARI, the better.

The classification error rate for model-based clustering for this data set was

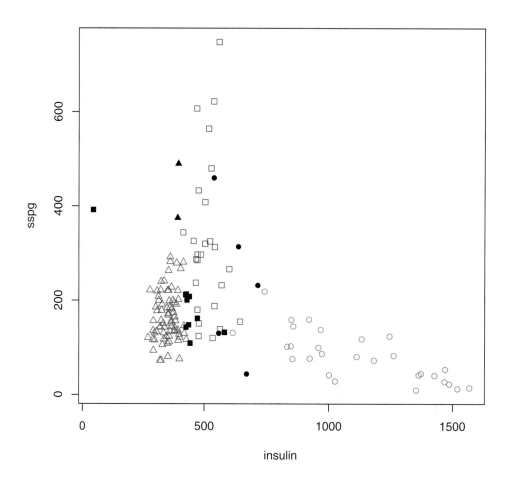

Figure 2.16 Model-based classification of diabetes data, with errors marked. The errors are shown by black symbols, and the clinically determined class is shown by the symbol (triangle, square or circle).

11.7%, and the ARI was 0.719. Single link clustering performed poorly for this data set, putting all the data points except two into the same cluster. Average link clustering, complete link clustering and k-means performed similarly, as one might expect from Figure 2.17, with classification error rates between 27.6% and 29.7%, and ARI values between 0.318 and 0.375. These are all substantially worse than model-based clustering, as they fail to capture the strongly non-spherical nature of the clusters.

Model-based clustering also provides a quantification of the uncertainty about the cluster allocation of each data point, Uncer$_i$, defined by Equation (2.12). These

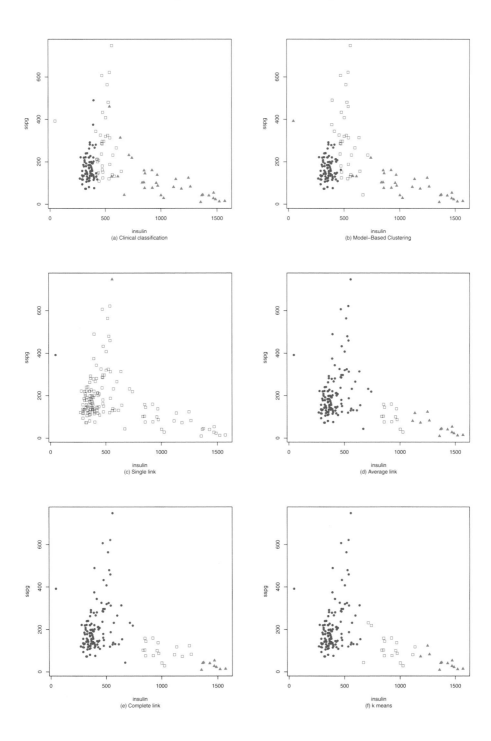

Figure 2.17 Clustering results for diabetes data, shown for two of the three variables, insulin and SSPG. (a) Clinical classification; (b) model-based clustering; (c) single link; (d) average link; (e) complete link; (f) k-means.

Table 2.3　*Performance of different clustering methods on the diabetes data: classification error rate (CER: smaller is better) and Adjusted Rand Index (ARI: larger is better).*

Clustering Method	CER %	ARI
Model-based $G = 3$	11.7	.719
Single link	46.2	.013
Average link	29.7	.318
Complete link	29.7	.321
k-means	27.6	.375

are summarized in the uncertainty plot in Figure 2.18, which plots the uncertainty values in increasing order, and shows the classification errors by vertical lines. Most of the classification errors correspond to higher levels of uncertainty.

The uncertainty values are shown in a different way in Figure 2.19, where the top 5% of points by uncertainty are shown by solid black circles, and the next 20% by solid gray circles. The points with the largest uncertainty are between clusters; the points close to cluster centers have low uncertainty.

R commands to analyze the diabetes diagnosis data using model-based clustering are shown in Listing 2.10. R commands to analyze the same data using the non-model-based clustering methods are shown in Listing 2.11.

Example 3 (ctd.): Breast cancer diagnosis

We carried out model-based clustering of the breast cancer diagnosis data, which has 569 subjects and 30 variables. We know the number of known classes to be $G = 2$, and so we will restrict attention to models with two clusters. We will also refrain, for now, from imposing any constraints on the component covariance matrices, and so we will use the VVV model. This is done using the command:

```
Mclust (diagnostic.data, G=2, modelNames="VVV")
```

We compared the results with the clinical classification, which we will take here as being correct. Of the 569 subjects, 541 were correctly classified, a 95.1% correct classification rate, or, equivalently, a 4.9% classification error rate. To depict this, we computed the first two principal components of the data, using the correlation matrix (i.e. after first scaling the variables by their standard deviations). The model-based clustering classification and the classification errors are shown in Figure 2.20. The errors are all on the boundary between the two components in the plot of the first two principal components.

Figure 2.21 shows a comparison of the model-based clustering with the clinical classification and with other widely used clustering methods, again in terms of the first two principal components. As for the diabetes diagnosis data set, single link clustering performed very poorly, putting almost all the subjects into one cluster. Average link clustering and complete link clustering gave results that were very similar to each other, classifying too few subjects into the Malignant category.

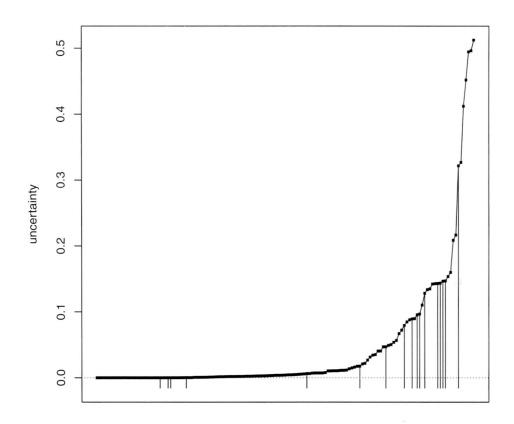

observations in order of increasing uncertainty

Figure 2.18 Uncertainty plot for model-based clustering of the diabetes data. The vertical lines indicate errors in the model-based classification.

The k-means method also classified too few subjects into the Malignant category, but more than average link and complete link.

Table 2.4 shows a quantitative comparison between the different clustering methods in terms of classification error rate (CER) and adjusted Rand index (ARI). Single link clustering had an ARI that was essentially zero, indicating that its classification was not better than random chance, while average link and complete link clustering had slightly better performances. The k-means method performed better, but it had a classification error rate that was three times greater than model-based clustering, which performed best for this data set.

The uncertainty plot is shown in Figure 2.22. It can be seen that the majority

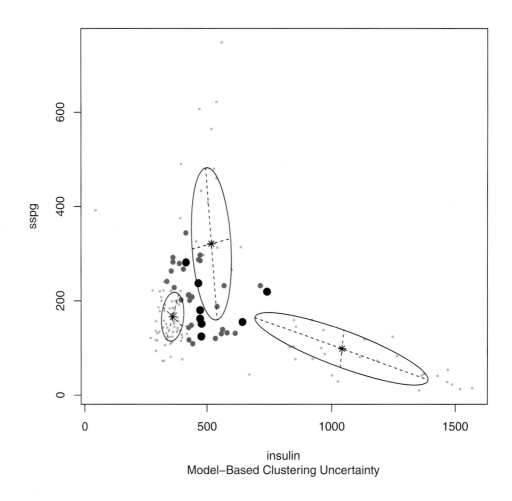

Figure 2.19 Uncertainty for model-based clustering of the diabetes data. The solid black circles show the top 5% of data points by uncertainty, the solid gray circles show the next 20%, and the least uncertain 75% of data points are shown by gray dots.

of the errors corresponded to the subjects with the highest uncertainty, so in this case the assessment of uncertainty captured the likelihood of error fairly well, at least in a qualitative sense.

2.6 Choosing the Number of Clusters and the Clustering Model

Two basic issues arising in applied cluster analysis are selection of the clustering method and determination of the number of clusters. In the mixture modeling approach, these questions can be reduced to a single problem, that of model

Listing 2.10: R code for model-based cluster analysis of the diabetes data

```
# Initial setup
library (mclust)
data (diabetes)
diabetes.data <- diabetes[2:4]
diabetes.class <- unlist (diabetes[1])

# Model-based clustering fitting and plotting
diabetes.Mclust <- Mclust (diabetes.data,modelNames="VVV",G=3)
plot(diabetes.Mclust, what = "classification")

# Classification error measures
classError (diabetes.Mclust$classification, truth=diabetes.class)
adjustedRandIndex (diabetes.Mclust$classification, diabetes.class)

# Plot clinical classification
coordProj (diabetes.data, dimens=c(2,3), what="classification",
 classification=diabetes.class,
 col=c("red2","dodgerblue2","green3"), symbols=c(0,16,17),
  sub="(a) Clinical classification")

# Plot mclust classification
coordProj (data=diabetes.data, dimens=c(2,3), what="classification"
 ,
 classification=diabetes.Mclust$classification,
  sub="(b) Model-Based Clustering")


# Uncertainty plot
uncerPlot (z=diabetes.Mclust$z,truth=diabetes.class)

# Plot mclust uncertainty
coordProj (data=diabetes.data, dimens=c(2,3), what="uncertainty",
  parameters=diabetes.Mclust$parameters, z=diabetes.Mclust$z,
  truth=diabetes.class)
```

selection. Recognizing that each combination of a number of groups and a clustering model corresponds to a different statistical model for the data reduces the problem to comparison among the members of a set of possible models.

There are trade-offs between the choice of the number of clusters and that of the clustering model. If a simpler model is used, more clusters may be needed to provide a good representation of the data. If a more complex model is used, fewer clusters may suffice. As a simple example, consider the situation where there is a single Gaussian cluster whose covariance matrix corresponds to a long, thin ellipsoid. If a model with equal-volume spherical components (the model underlying Ward's method and k-means) were used to fit these data, more than one hyperspherical cluster would be needed to approximate the single elongated ellipsoid.

One approach to the problem of model section in clustering is based on Bayesian model selection, via Bayes factors and posterior model probabilities (Kass and

Listing 2.11: R code for cluster analysis of diabetes data using non-model-based methods

```
# Loading libraries and data
library(MBCbook)
data(diabetes)
diabetes.data = diabetes[,-1]
diabetes.class = diabetes$class

# Single link clustering
tmp <- hclust (dist(diabetes.data),method="single")
diabetes.single <- rect.hclust (tmp, k=3)
n.diabetes <- length (diabetes.data[,1])
diabetes.single.class <- rep (NA, n.diabetes)
for (g in 1:3) diabetes.single.class[diabetes.single[[g]]] <- g
coordProj (diabetes.data, dimens=c(2,3), what="classification",
 col=c("dodgerblue2","green3","red2"), symbols=c(16,17,0),
 classification=diabetes.single.class, sub="(c) Single link")
classError (diabetes.single.class, truth=diabetes.class)
adjustedRandIndex (diabetes.single.class, diabetes.class)

# Average link clustering
tmp <- hclust (dist(diabetes.data),method="average")
diabetes.average <- rect.hclust (tmp, k=3)
n.diabetes <- length (diabetes.data[,1])
diabetes.average.class <- rep (NA, n.diabetes)
for (g in 1:3) diabetes.average.class[diabetes.average[[g]]] <- g
coordProj (diabetes.data, dimens=c(2,3), what="classification",
 classification=diabetes.average.class, sub = "(d) Average link")
coordProj (diabetes.data, dimens=c(2,3), what="errors",
 classification=diabetes.average.class, truth=diabetes.class)
classError (diabetes.average.class, truth=diabetes.class)
adjustedRandIndex (diabetes.average.class, diabetes.class)

# Complete link clustering
tmp <- hclust (dist(diabetes.data),method="complete")
diabetes.complete <- rect.hclust (tmp, k=3)
n.diabetes <- length (diabetes.data[,1])
diabetes.complete.class <- rep (NA, n.diabetes)
for (g in 1:3) diabetes.complete.class[diabetes.complete[[g]]] <- g
coordProj (diabetes.data, dimens=c(2,3), what="classification",
 classification=diabetes.complete.class,
 col=c("dodgerblue2","green3","red2"), symbols=c(16,17,0),
  sub="(e) Complete link")
classError (diabetes.complete.class, truth=diabetes.class)
adjustedRandIndex (diabetes.complete.class, diabetes.class)

# k-means clustering
diabetes.kmeans <- kmeans (diabetes.data, centers=3, nstart=20)
coordProj (diabetes.data, dimens=c(2,3), what="classification",
 classification=diabetes.kmeans$cluster,
 col=c("green3","red2","dodgerblue2"), symbols=c(17,0,16),
 sub="(f) k means")
classError (diabetes.kmeans$cluster, truth=diabetes.class)
adjustedRandIndex (diabetes.kmeans$cluster, diabetes.class)
```

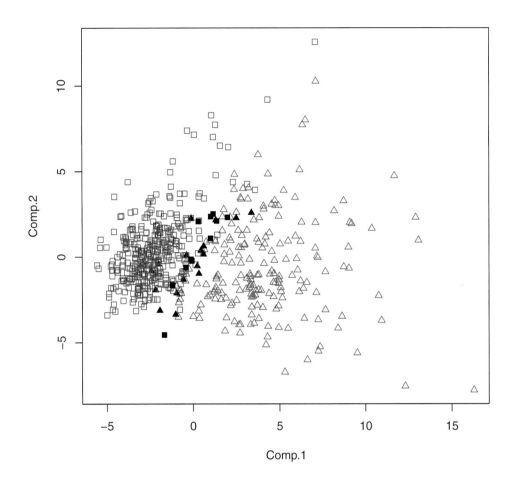

Figure 2.20 Model-based classification of breast cancer diagnosis data, with errors marked. The first two principal components of the data (with variables scaled) are shown. Benign = red squares, Malignant = blue triangles. The errors are shown by black symbols, and for them the clinically determined class is shown by the symbol (square or triangle).

Raftery, 1995). The basic idea is that if several statistical models, $M_1, \ldots, M_K$, are considered, with prior probabilities $p(M_k)$, $k = 1, \ldots, K$ (often taken to be equal), then by Bayes' theorem the posterior probability of model M_k given data D is proportional to the probability of the data given model M_k, times the model's prior probability, namely

$$p(M_k|D) \propto p(D|M_k)p(M_k). \tag{2.14}$$

When there are unknown parameters, then, by the law of total probability, $p(D|M_k)$

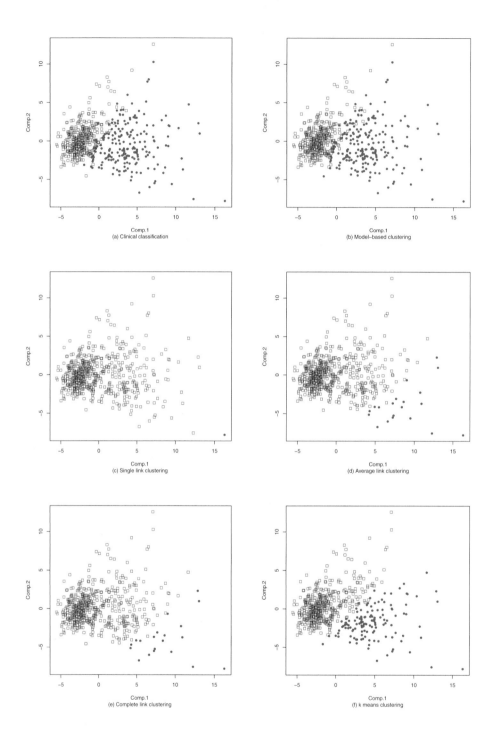

Figure 2.21 Clustering results for breast cancer diagnosis data, shown for the first two principal components of the data. (a) Clinical classification (Benign = red squares, Malignant = blue circles); (b) model-based clustering; (c) single link; (d) average link; (e) complete link; (f) k-means.

Table 2.4 *Performance of different clustering methods on the breast cancer diagnosis data: classification error rate (CER: smaller is better) and Adjusted Rand Index (ARI: larger is better). All methods take the number of clusters ($G = 2$) as known.*

Clustering Method	CER %	ARI
Model-based	4.9	.811
Single link	37.1	.002
Average link	33.7	.052
Complete link	33.7	.052
k-means	14.6	.491

is obtained by integrating (not maximizing) over the parameters, i.e.

$$p(D|M_k) = \int p(D|\theta_{M_k}, M_k) p(\theta_{M_k}|M_k) d\theta_{M_k}, \tag{2.15}$$

where $p(\theta_{M_k}|M_k)$ is the prior distribution of θ_{M_k}, the parameter vector for model M_k. The quantity $p(D|M_k)$ is known as the *integrated likelihood*, or *marginal likelihood*, of model M_k.

A natural Bayesian approach to model selection is then to choose the model that is most likely a posteriori. If the prior model probabilities, $p(M_k)$, are the same, this amounts to choosing the model with the highest integrated likelihood. For comparing two models, M_1 and M_2, the Bayes factor is defined as the ratio of the two integrated likelihoods, $B_{12} = p(D|M_1)/p(D|M_2)$, with the comparison favoring M_1 if $B_{12} > 1$, and conventionally being viewed as providing very strong evidence for M_1 if $B_{12} > 100$ (Jeffreys, 1961). Often, values of $2\log(B_{12})$ rather than B_{12} are reported, and on this scale, rounding, very strong evidence corresponds to a threshold of 10 (Kass and Raftery, 1995).

This approach is appropriate in the present context because it applies when there are more than two models, and can be used for comparing non-nested models. In addition to being a Bayesian solution to the problem, it has some desirable frequentist properties. For example, if one has just two models and they are nested, then basing model choice on the Bayes factor minimizes the total error rate, which is the sum of the Type I and Type II error rates (Jeffreys, 1961).

The main difficulty in the use of Bayes factors is the evaluation of the integral that defines the integrated likelihood. For regular models, the integrated likelihood can be approximated simply by the Bayesian Information Criterion or BIC:

$$2\log p(D|M_k) \approx 2\log p(D|\hat{\theta}_{M_k}, M_k) - \nu_{M_k}\log(n) = \mathrm{BIC}_{M_k}, \tag{2.16}$$

where ν_{M_k} is the number of independent parameters to be estimated in model M_k (Schwarz, 1978; Haughton, 1988). This approximation is particularly good when a unit information prior is used for the parameters, that is, a multivariate normal prior that contains the amount of information provided on average by one observation (Kass and Wasserman, 1995; Raftery, 1995).

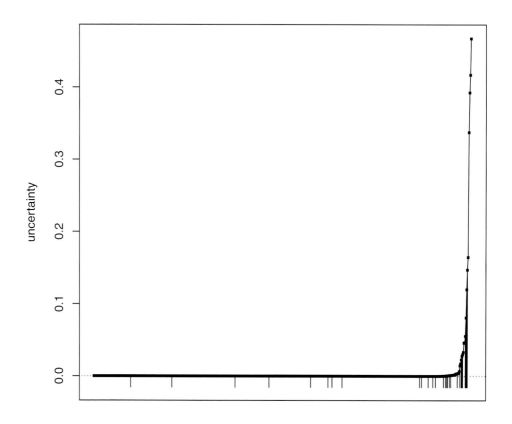

observations in order of increasing uncertainty

Figure 2.22 Uncertainty plot for model-based clustering of the breast cancer diagnosis data. The vertical lines indicate errors in the model-based classification.

The reasonableness of the unit information prior is discussed by Raftery (1999). For example, for a mean parameter it is a prior that spreads most of its weight over the range of the data but not beyond, and thus can be viewed as corresponding to the prior information of someone who knows something, but relatively little, about the problem at hand. Such a person would be likely to know the range of the data to be expected, at least roughly, and would likely be comfortable saying that the mean is likely to lie within this range, but would probably not have enough information to refine this range greatly. In this sense, it is as spread out as is reasonable: an analyst is unlikely to want to use a more spread out prior, but may well have information that would allow them to refine the prior. It thus seems a reasonable candidate to be used as a default prior.

Note that an informed analyst may well have a more precise prior or external information, leading to a more concentrated prior distribution. More concentrated prior distributions often tend to lead to Bayes factors that favor larger models (and the alternative hypothesis in two-hypothesis testing situations). Thus the unit information prior can be viewed as conservative, in the sense that informative priors are likely to favor the alternative hypothesis (or larger models) more. If external information tends to provide evidence against the null hypothesis, then its use is legitimate, but the fact that the conclusion is based on external information as well as the data at hand should be revealed to readers and users.

Finite mixture models do not satisfy the regularity conditions that underlie the published proofs of (2.16), but several results suggest its appropriateness and good performance in the model-based clustering context. Leroux (1992) showed that basing model selection on a comparison of BIC values will not underestimate the number of groups asymptotically. Keribin (1998) showed that BIC is consistent for the number of groups, subject to the assumption that the likelihood is bounded. The likelihood is not in fact bounded in general for Gaussian mixture models because of the degenerate paths of the likelihood to infinity as the variance matrix becomes singular. However, a very mild restriction, such as a very small lower bound on the smallest eigenvalue of the covariance matrices, is enough to ensure validity of the assumption. The main software used in model-based clustering incorporates such restrictions, so the assumption is valid in practice for all the models used.

Roeder and Wasserman (1997) showed that if a mixture of (univariate) normal distributions is used for one-dimensional nonparametric density estimation, then using BIC to choose the number of components yields a consistent estimator of the density. Finally, in a range of applications of model-based clustering, model choice based on BIC has given good results (Campbell et al., 1997, 1999; Dasgupta and Raftery, 1998; Fraley and Raftery, 1998; Stanford and Raftery, 2000).

For multivariate normal mixture models, the models to be compared correspond to different numbers of clusters and different sets of constraints on the covariance matrices, or clustering models. Each combination of a number of clusters and a clustering model defines a distinct model to be compared. Typically, models with $G = 1, \ldots, G_{\max}$ are considered, for some reasonable choice of the maximum number of clusters, $G_{\max}$. In the `mclust` R package, the default value of $G_{\max}$ is 9, but in specific applications the appropriate number might be much greater. For example, in the craniometric data of Example 5, there are known to be at least 28 populations represented, so in examples of this kind it would be wise to take $G_{\max}$ to be 30 or greater. By default, the `mclust` software considers 14 covariance models, leading to a default number of $G_{\max} \times 14 = 126$ models considered. The `Rmixmod` software considers 28 covariance models, leading to a larger total number of models considered.

For a finite mixture model, the usual likelihood function is also called the mixture likelihood, or, particularly in the context of the EM algorithm, the observed data likelihood. From (2.1), it has the form

$$p(y|\theta_{M_k}, M_k) = \prod_{i=1}^{n} \left(\sum_{g=1}^{G_{M_k}} \tau_g f_g(y_i|\theta_g) \right),\qquad (2.17)$$

where $y = (y_1, \ldots, y_n)$, G_{M_k} is the number of mixture components in model M_k, and θ_{M_k} consists of the parameters for model M_k. For a multivariate normal mixture model, the likelihood takes the form (2.17), with $f_g(y_i!\theta_g) = \phi_g(y_i|\mu_g, \Sigma_g)$, where $\phi_g(\cdot|\cdot)$ is the multivariate normal density defined by (2.2). The integrated likelihood is then equal to

$$p(y|M_k) = \int p(y|\theta_{M_k}, M_k) p(\theta_{M_k}|M_k) d\theta_{M_k}.\qquad (2.18)$$

If there are no cross-component constraints on the covariance matrices (as for example in models VII or VVV), then $\theta_{M_k} = \{(\tau_g, \theta_g) : g = 1, \ldots, G_{M_k}\}$, where θ_g is the parameter vector for the gth mixture component. In model VVV, $\theta_g = (\mu_g, \Sigma_g)$. For a model with cross-cluster constraints on the covariance matrix, the parameter set θ_{M_k} will reflect the constraints. For example, if M_k is a VEV model, then $\theta_{M_k} = \{(\tau_g, \lambda_g, D_g) : g = 1, \ldots, G_{M_k}; A\}$, where λ_g, D_g and A are defined as in (2.3).

The BIC for model M_k is then defined by (2.16), where the data D consist of the observations to be clustered y, $\hat{\theta}_{M_k}$ is the MLE of the model parameters, typically found by the EM algorithm, and ν_{M_k} is the number of free parameters in M_k, found for example from Table 2.2.

The BIC is designed to choose the number of components in a mixture model rather than the number of clusters in the data set per se. The difference is subtle but important. Model-based clustering was initially grounded in the hope that the number of mixture components is the same as the number of clusters, but this may not always be true. For example, one cluster may be strongly non-Gaussian, and may itself be best represented by a mixture of normal distributions. In this case, the number of mixture components would be greater than the number of clusters. In Section 3.3 we will describe how this problem can be addressed by merging mixture components, but for now we focus on a different solution that uses a different model selection criterion, designed for clustering rather than selecting the model that best fits the data.

When interest is primarily focused on clustering rather than finding the best mixture model to fit the data, one solution to this problem is to use, instead of BIC, the *Integrated Completed Likelihood* (ICL) (Biernacki et al., 2000). This is based on the joint likelihood of *both* the data y and the clustering z, defined by

$$p(y, z|\theta_{M_k}, M_k) = \prod_{i=1}^{n} \prod_{g=1}^{G_{M_k}} \{\tau_g f_g(y_i|\theta_g)\}^{z_{i,g}}.\qquad (2.19)$$

Integrating over the model parameter θ_{M_k} then gives

$$p(y, z|M_k) = \int p(y, z|\theta_{M_k}, M_k)p(\theta_{M_k}|M_k)d\theta_{M_k}. \tag{2.20}$$

We use a BIC-like approximation to the integral and take twice the logarithm of it, yielding

$$2\log p(y, z|M_k) \approx 2\log p(y, z|\hat{\theta}_{M_k}, M_k) - \nu_{M_k}\log(n). \tag{2.21}$$

The Integrated Completed Likelihood of model M_k is then defined as the right-hand side of (2.21) with z replaced by its maximum a posteriori (MAP) estimate. The MAP estimate of z, denoted by z^*, satisfies $z^*_{i,g} = 1$ if $\hat{z}_{i,g} = \arg\max_h \hat{z}_{i,h}$ and 0 otherwise, where

$$\hat{z}_{i,g} = \frac{\hat{\tau}_g f_g(y_i|\hat{\theta}_g)}{\sum_{h=1}^{G_{M_k}} \hat{\tau}_h f_h(y_i|\hat{\theta}_h)}. \tag{2.22}$$

Thus

$$\text{ICL} = 2\log p(y, z^*|\hat{\theta}_{M_k}, M_k) - \nu_{M_k}\log(n). \tag{2.23}$$

It turns out that

$$\text{ICL} = \text{BIC} - E(M_k), \tag{2.24}$$

where

$$E(M_k) = -\sum_{i=1}^{n}\sum_{g=1}^{G_{M_k}} \hat{z}_{i,g}\log(\hat{z}_{i,g}) \tag{2.25}$$

is the expected entropy of the classification from model M_k (Biernacki et al., 2000). Thus ICL is equal to the BIC penalized by the expected entropy of the classification. Entropy is high when there is large uncertainty about the classification, and it is highest when all the $\hat{z}_{i,g}$ are equal (i.e. all equal to $1/G_{M_k}$), at which point it attains the value $n\log(G_{M_k})$. It is lowest when all the $\hat{z}_{i,g}$ are equal to either 0 or 1, at which point it is equal to zero. As a result, ICL tends to favor models that produce more clearly separated clusters more than BIC does, and so in practice ICL usually chooses the same or smaller numbers of clusters than BIC.

Example 1 (ctd.): Old Faithful data

Figure 2.23 shows the BIC and ICL plots for the Old Faithful data, using the R commands in Listing 2.12. Values are shown for up to $G_{\max} = 9$ mixture components and for the 14 covariance models estimated in `mclust`, i.e. for $9 \times 14 = 126$ different competing models in all. A zoomed version of Figure 2.23 is shown in Figure 2.24.

BIC selects the model with three mixture components and the **EEE** covariance specification, in which all the covariance matrices are the same, so that $\Sigma_g \equiv \Sigma$ in (2.2). ICL selects the model with two mixture components and the equal-orientation

Listing 2.12: BIC and ICL plots

```
# Loading the mclust library
library(mclust)

# Clustering with Mclust and displaying BIC selection
faithful.Mclust <- Mclust (faithful)
plot (faithful.Mclust, what="BIC")

# Clustering with Mclust and displaying ICL selection
faithful.mclustICL <- mclustICL (faithful)
plot (faithful.mclustICL)
```

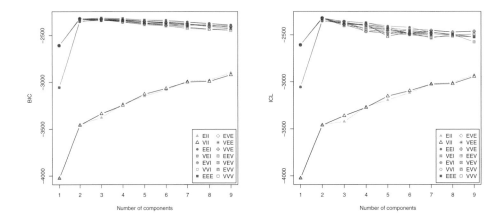

Figure 2.23 Model selection for the Old Faithful data: Left: BIC plot.
Right: ICL plot.

VVE covariance model. The model selected by BIC uses 11 free parameters to model the covariances, while the model selected by ICL has one fewer, at 10.

Figure 2.25 shows the classifications and the component density ellipses for the two selected models. Figure 2.26 shows the density contours for the overall estimated mixture density from the two selected models. It seems that the model selected by BIC may give a better estimate of the overall density of the data, but ICL yields a clustering that agrees better with the visual impression.

One possible conclusion from this analysis is that there are likely two clusters in the data, but that the cluster at the top right in the right panel of Figure 2.25 may well be non-Gaussian and could be better represented as a mixture of two Gaussian components than as by a single multivariate normal distribution. This suggests using BIC to select the number of mixture components, but then merging the two top right components into a single non-Gaussian cluster. We will discuss how to do this in Chapter 3.3. For the Old Faithful data, it can be seen from

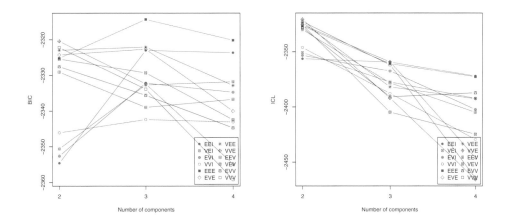

Figure 2.24 Zoomed version of Figure 2.23: BIC and ICL plots for Old Faithful data.

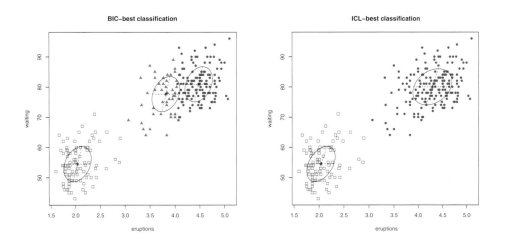

Figure 2.25 Classification plots for the Old Faithful data using selected models. Left: model selected by BIC with $G = 3$, model EEE. Right: model selected by ICL with $G = 2$, model VVV.

Figure 2.25 that this approach gives the same classification as the two-cluster Gaussian model selected by ICL.

Example 2 (ctd): Diabetes diagnosis

Figure 2.27 shows the BIC and ICL plots for the diabetes diagnosis data. These two criteria both select the model with three components and the unconstrained (VVV) covariance model, in both cases fairly decisively. The resulting classification is shown in the left panel of Figure 2.28.

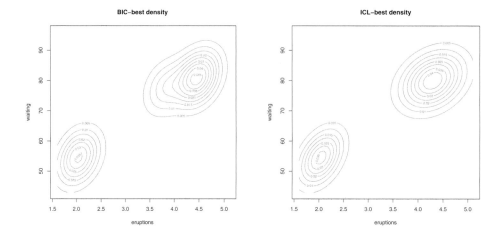

Figure 2.26 Density contour plots for the Old Faithful data using selected models. Left: Model selected by BIC with $G = 3$, model EEE. Right: Model selected by ICL with $G = 2$, model VVV.

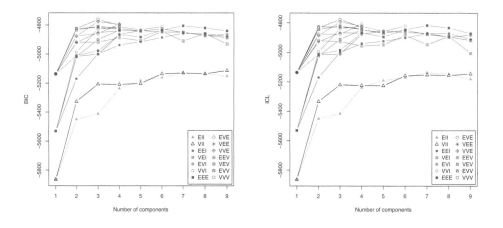

Figure 2.27 Model selection for the diabetes diagnosis data. Left: BIC. Right: ICL.

This is a satisfactory choice of number of clusters and of clustering model. The number of clusters selected corresponds to the clinical classification. Also, in the clinical classification, the volumes of the clusters differ: the cluster of non-diabetic or "normal" subjects is much more compact and has a much smaller volume than the other two clusters, and so the volume should differ between clusters. The shapes are also very different, with the normal cluster being compact and close to spherical, while the other two clusters are elongated. Finally, the orientations are different, in that the overt diabetes and chemical diabetes clusters point in

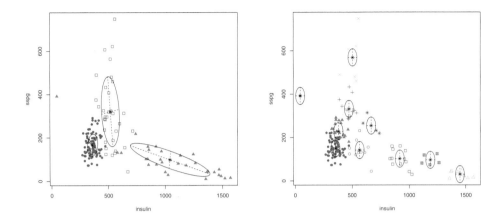

Figure 2.28 Classification plots for selected models for the diabetes diagnosis data. Left: three-component model with unconstrained covariance selected by BIC and ICL. Right: ten-component equal volume spherical model (EII).

very different directions. Thus, the cluster volume, shape and orientation all vary between clusters, so the VVV model is the most appropriate one, and it is the one selected by our two criteria.

One interesting aspect of model choice in model-based clustering is that there is a trade-off between complexity of clustering model and number of clusters. For example, a single elongated cluster could also be represented approximately by a larger number of spherical clusters. If we think of a pea plant, the single elongated cluster could be like a pod, and the approximating spherical clusters could be like the peas, so the problem could be viewed as one of distinguishing between the peas and the pod.

The best constant variance spherical model for these data has ten mixture components, with the EII model. The classification from this is shown in the right panel of Figure 2.28. The "peas versus pod" dynamic is apparent. For example, the long thin cluster in the left panel of Figure 2.28 is replaced in the ten-component spherical clustering by three small components in the right panel. In this case, BIC and ICL both choose the "pod" model over the "peas" model, selecting the three-component model.

For some purposes, such as grouping data into compact groups for data compression purposes, a model such as EII with a large number of clusters might be better suited than a selected more complex covariance model. In such a case, the spherical model may well be preferred, even if its fit is not as good according to model selection criteria. It is always important to bear the purpose of the data analysis in mind when making modeling decisions.

2.7 Illustrative Analyses

We now carry out more complete analyses of two substantial data sets.

2.7.1 Wine Varieties

We will analyze a data set giving 27 physical and chemical measurements on 178 wine samples (Forina et al., 1986). These data are available in the `pgmm` R package. There are three types of wine, Barolo, Grignolino and Barbera, and the type of wine is known for each of the samples. The year in which each sample was produced is also known. For now, we will use only the 27 measurements to cluster the samples, and assess to what extent they reproduced the known types.

Figure 2.29 shows a pairs plot for five of the 27 measurements. These were chosen because they were the variables with the highest loadings on each of the first five principal components, suggesting the possibility that they might be variables carrying information about clustering. In fact, only the plot of the first two variables (Flavanoids and Color Intensity) gives a strong visual suggestion of clustering.

Figure 2.30 shows the same pairs plot, but with the wine types shown in different colors. Note that the clustering method does not have access to this information. It shows that, while there is no clear separation between the wine types, the samples do tend to cluster, particularly again in the first two variables shown. This suggests that clustering could be successful.

Figure 2.31 shows the BIC and ICL plots for these data, with a zoomed version showing the most likely models and numbers of clusters in Figure 2.32. It can be seen from Figure 2.32 that there are two competing and quite different descriptions of the data between which the evidence is inconclusive. One description consists of the spherical models with scales in each dimension that vary between clusters, EVI and VVI, and three clusters. The other description consists of the spherical model with volumes that vary between clusters but scales that do not, VEI, and sevan clusters.

The top three covariance-model/number-of-cluster combinations are the same for the two criteria, but in different orders, as shown in Table 2.5. We will focus initially on the ICL solution, which gives the same number of clusters as wine types, and so gives what at first sight appears to be the "correct" number of clusters. However, we will see later that the BIC seven-cluster solution recovers some information about year as well as type that the three-cluster solution cannot do, and so which solution is better can be debated.

Figure 2.33 (left) shows the correct classification and the confusion matrix is shown in Table 2.6. It can be seen that the red group (Grignolino) is for the most part fairly compact, but also has about nine samples that are quite spread out and are separated from the core Grignolino samples. The right panel of Figure 2.33 shows the same information, but with classification errors from model-based clustering shown by filled black symbols. It is clear that all but one

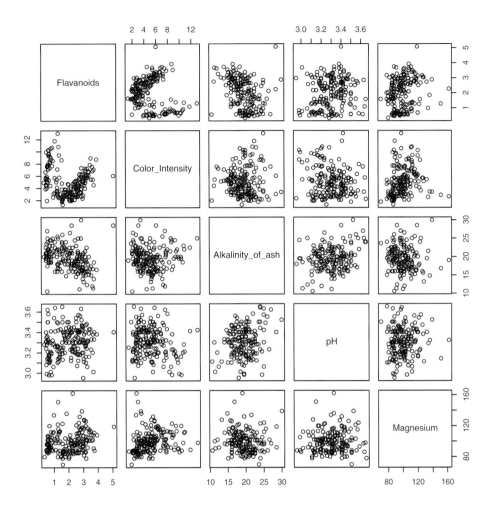

Figure 2.29 Pairs plot for five of the 27 physical and chemical measurements of wine samples

of the ten misclassified samples are among these Grignolino samples whose values are separated from the main Grignolino group, and so are classified as being of one of the other types. It would be hard for any clustering method to classify these samples correctly.

Table 2.7 shows the performance of different clustering methods for the wine data, when all are given the "correct" number of clusters, $G = 3$. The model-based method performs best by a wide margin, while the single link method again performs very poorly.

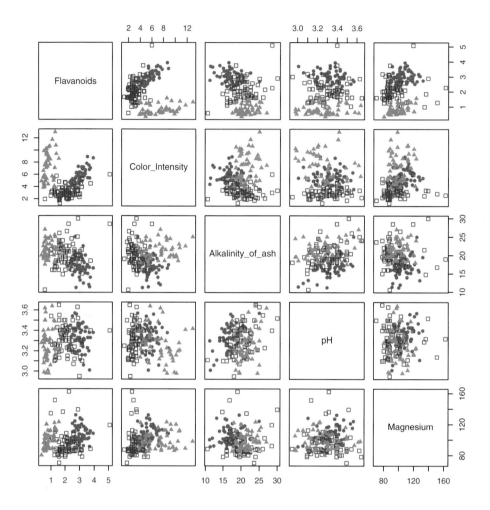

Figure 2.30 The same pairs plot for five of the 27 physical and chemical measurements of wine samples, with true wine types shown by color.

It remains the fact that with BIC, although not with ICL, model-based clustering has chosen seven clusters, even though there are in fact only three wine varieties. The confusion matrix for comparing the seven-cluster solution with the partition into three wine types is shown in Table 2.8. It can be seen that each of the three wine types has itself been divided into either two or three clusters by the clustering algorithm. If the clusters were merged correctly, the number of errors would be very similar to the three-cluster solution (11 versus 10). Not surprisingly, though, the classification error rate (52%) and the Adjusted Rand Index (0.418) are much worse than for the three-cluster solution. Note, however, that the ARI is still

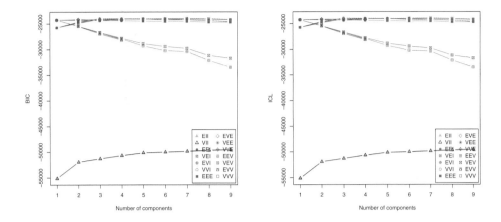

Figure 2.31 BIC (left) and ICL (right) for the wine data.

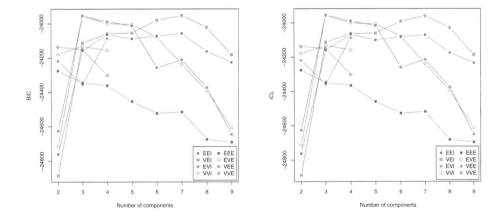

Figure 2.32 Zoomed version of the BIC (left) and ICL (right) plots for the wine data.

better than for the three heuristic methods, even though the latter were "told" the correct number of clusters.

There are 14 wine type/year combinations in the data set. The confusion matrix for the seven-cluster solution against those 14 type/year categories is shown in Table 2.9. Within each type, several of the clusters tend to separate out particular years. For example, cluster 2 corresponds to Barolo wines that are not from 1973, cluster 5 corresponds to later Grignolino wines, particularly from 1976, while cluster 7 is mostly made up of Barbera wines from 1978. Thus it seems that the

Table 2.5 *Wine data: top models by BIC and ICL. The number of clusters is denoted by G. The best model by each criterion is shown in bold font. 23,950 has been added to all BIC and ICL values for readability.*

G	Model	BIC	ICL
7	VEI	**−1.91**	−8.99
3	EVI	−3.87	**−5.05**
3	VVI	−4.37	−6.46

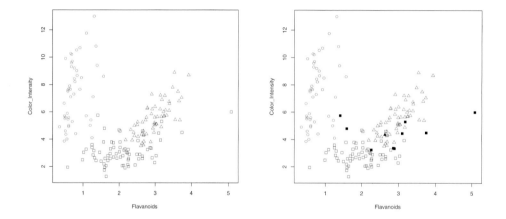

Figure 2.33 Wine data: true classification (left) and classification errors from model-based clustering (right), in clustering samples by wine type. The errors are shown as solid black symbols, with the shape corresponding to the true classification. Two of the 27 measurements, Flavanoids and Color Intensity, are shown.

more detailed seven-cluster clustering selected by BIC is at least partly capturing differences between years.

Table 2.6 *Wine data: confusion matrix for three-cluster model and wine types. For each cluster, the most represented wine type is marked by a box.*

Cluster	Wine type		
	Barolo	Grignolino	Barbera
1	58	7	0
2	1	62	0
3	0	2	48

Table 2.7   *Wine data: performance of different clustering methods for classifying wine types. classification error rate (CER – smaller is better) and Adjusted Rand Index (ARI – larger is better). All methods are based on G = 3 clusters.*

Method	CER %	ARI
Model-based (EVI,3)	5.6	0.830
Single link	59.0	0.001
Average link	32.6	0.400
Complete link	32.3	0.323
k-means	25.8	0.353

Table 2.8   *Wine data: confusion matrix for seven-cluster model chosen by BIC and wine types. For each cluster, the most represented wine type is marked by a box.*

Cluster	Wine type		
	Barolo	Grignolino	Barbera
1	31	0	0
2	25	0	0
3	3	17	0
4	0	28	0
5	0	18	0
6	0	8	21
7	0	0	27

2.7.2 Craniometric Analysis

We now consider a data set consisting of 57 different cranial measurements from 2,524 skulls. The goal is to classify the skulls according to the populations from which they came. The idea of using quantitative data and cluster analysis to classify skulls comes from Howells (1973, 1989, 1995). Howells (1995) used a sum of squares hierarchical clustering method, and was unable to find strong evidence for clustering of populations. Here we reanalyze the data using model-based clustering, drawing on the earlier analysis of Pontikos (2004, 2010). The skulls come from 30 different populations.

The first version of the data set we analyze here was documented by Howells (1996). The version we analyze here is the "test" data set assembled by Pontikos (2010), slightly modifying an earlier version considered by Pontikos (2004).

The data are roughly balanced between male and female skulls. Exploratory analysis indicated that the marginal distribution of the male skull measurements is statistically indistinguishable from that of the female skull measurements increased by 5%. We therefore multiplied the female measurements by 1.05 prior to analysis, and analyzed the male and female skulls together.

The BIC values for the model-based clustering of the resulting data are shown in Figure 2.34. The best model for the covariance structure is the **VEE** model, which says that the covariance matrices for the different clusters are equal up to

Table 2.9 *Wine data: confusion matrix for the seven-cluster model chosen by BIC against wine type and year. The horizontal lines group clusters that correspond largely to the same type.*

Cluster	Barolo			Grignolino							Barbera			
	1971	1973	1974	1970	1971	1972	1973	1974	1975	1976	1974	1976	1978	1979
1	6	17	8	0	0	0	0	0	0	0	0	0	0	0
2	13	1	11	0	0	0	0	0	0	0	0	0	0	0
3	0	2	1	2	4	1	0	5	4	1	0	0	0	0
4	0	0	0	4	3	3	9	8	1	0	0	0	0	0
5	0	0	0	0	0	1	0	3	4	10	0	0	0	0
6	0	0	0	3	2	2	0	0	0	1	9	5	5	2
7	0	0	0	0	0	0	0	0	0	0	0	0	24	3

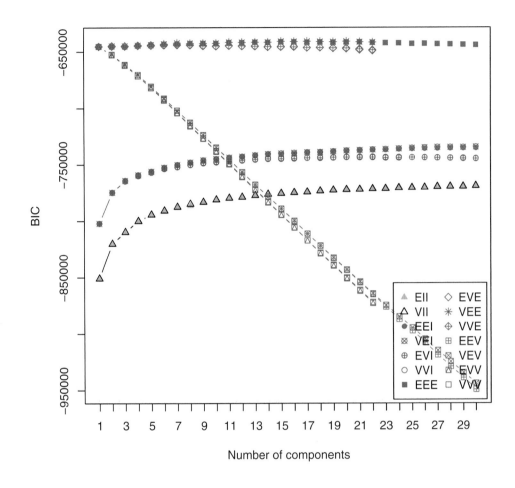

Figure 2.34 BIC plot for the Howells craniometric data.

a cluster-specific scale factor. The BIC values for different numbers of clusters for the VEE model are shown in Figure 2.35. The maximum is attained at $G = 15$ clusters.

The correspondence between the 15 clusters inferred by model-based clustering and the 30 populations is shown in Table 2.10. The number of clusters is smaller than the number of true populations, so there are several clusters that include the majority of skulls from more than one population. Apart from this, however, the correspondence between the clusters and the populations is close.

Seven of the 15 clusters correspond closely to a single population: Andaman, Berg, Buriat, Bushman, Eskimo, Santa Cruz and Tolai. Cluster 8 groups the Ainu (aboriginal Japanese) and Guam skulls. Cluster 9 groups most of the Arikara and

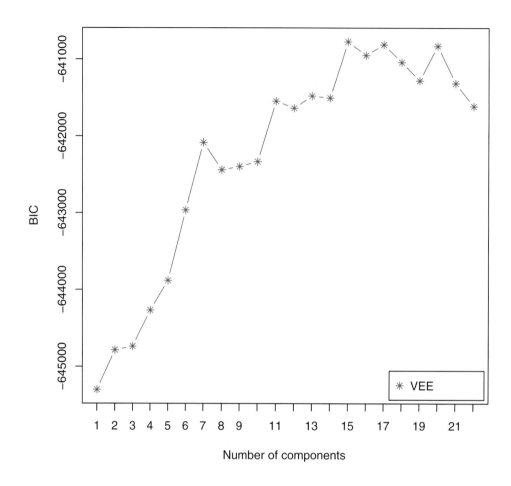

Figure 2.35 BIC plot for the VEE model only for the Howells craniometric data.

Peruvian skulls; these are both Native American populations. Cluster 10 groups (aboriginal) Australian and Tasmanian skulls; these are geographically proximate. Cluster 11 groups Easter Island and Mokapu skulls; again these are relatively geographically proximate. Cluster 12 groups Dogon from Mali, Teita from Kenya and Zulu from South Africa; these are all sub-Saharan African populations.

Cluster 13 groups (medieval) Norse, (Hungarian) Zalavar and ancient Egyptian skulls; these are all viewed as Caucasoid populations. Cluster 14 groups most of the North and South Maori and Moriori skulls; these are all from New Zealand. Finally, cluster 15 groups the Anyang and Hainan (both from China), Atayal

Table 2.10 *Craniometric data: confusion matrix for the 15 clusters inferred by model-based clustering and the 30 populations from which the skulls come.*

	1	2	3	4	5	6	7	8	9	10	11	12	13	14	15
ANDAMAN	59	0	0	1	0	0	0	1	0	0	0	6	2	0	1
BERG	0	98	0	1	0	1	0	0	1	0	0	0	8	0	0
BURIAT	0	2	105	0	0	0	0	1	0	0	0	0	0	0	1
BUSHMAN	0	0	0	80	0	0	0	0	0	1	0	8	0	0	1
ESKIMO	0	0	0	0	108	0	0	0	0	0	0	0	0	0	0
SANTA CRUZ	0	1	0	0	0	95	0	0	3	0	0	0	3	0	0
TOLAI	0	0	0	0	0	0	100	1	0	3	2	2	0	1	1
AINU	0	1	0	0	0	0	0	71	0	0	2	3	3	0	6
GUAM	0	0	0	0	0	0	0	45	1	0	0	0	0	0	11
ARIKARA	0	5	0	0	0	4	0	3	43	0	1	1	1	4	7
PERU	1	0	0	0	0	3	0	3	102	0	0	0	0	0	1
AUSTRALIA	0	0	0	0	0	0	2	0	0	98	0	0	1	0	0
TASMANIA	0	1	0	0	0	0	10	0	0	70	3	2	1	0	0
EASTER I	0	0	0	0	0	0	0	1	0	0	84	1	0	0	0
MOKAPU	0	0	0	0	0	0	0	4	0	0	92	0	0	3	1
DOGON	0	0	0	0	0	0	0	0	0	0	0	99	0	0	0
TEITA	0	0	0	0	0	0	0	0	0	1	0	80	2	0	0
ZULU	0	0	0	3	0	0	1	0	0	1	0	91	2	0	3
EGYPT	0	0	0	0	0	0	0	0	0	0	0	7	104	0	0
NORSE	0	15	0	0	0	2	0	5	0	0	0	1	85	1	1
ZALAVAR	0	32	0	0	1	0	0	4	1	1	0	1	57	0	1
MORIORI	0	0	0	0	0	0	0	2	0	0	5	0	0	100	1
N MAORI	0	0	0	0	0	0	0	0	0	0	4	0	0	6	0
S MAORI	0	0	0	0	0	0	0	2	0	0	3	0	0	5	0
ANYANG	0	0	0	0	0	0	0	5	1	0	0	0	0	0	36
ATAYAL	0	0	0	0	0	0	1	2	0	0	0	2	0	0	42
HAINAN	0	1	1	0	1	0	0	9	1	0	0	1	0	0	69
N JAPAN	1	0	0	0	0	0	0	12	1	0	0	3	0	0	70
S JAPAN	0	0	0	0	0	0	0	5	1	0	0	0	0	0	85
PHILIPPINES	1	1	0	0	0	0	0	6	1	0	1	4	2	0	34

(Taiwan aboriginal), North and South Japanese, and Philippine populations. These are all East Asian populations.

If we match each cluster with the population most represented within it, we calculate a misclassification rate of 44.7%, so that about 56% of the skulls were correctly classified. However, it seems more reasonable to match each cluster with the set of geographically proximate populations most represented within it, as discussed above. If we do this, the misclassification rate is only 12.3%. Overall, it seems that the clustering has largely kept skulls from the same populations together, but in some cases has been unable to distinguish between populations that are relatively close geographically or historically.

Since it is known that there are in fact 30 populations, it is of interest to run the analysis with the true number of groups, $G = 30$. This time the preferred model for the covariance structure is the equal-covariance model, EEE, which is closely related to the VEE model found best for $G = 15$ clusters. The confusion matrix is shown in Table 2.11.

When the algorithm is told to use the correct number of clusters, 30, the

Table 2.11 Craniometric data: Confusion matrix for model-based clustering when the true number of groups, 30, is assumed.

	1	2	3	4	5	6	7	8	9	10	11	12	13	14	15	16	17	18	19	20	21	22	23	24	25	26	27	28	29	30
AINU	69	0	0	0	0	0	0	0	0	0	0	0	0	0	0	1	1	0	0	1	1	0	1	1	2	3	2	1	3	0
ANDAMAN	0	63	0	0	0	0	0	0	0	0	0	0	0	0	0	0	0	0	2	1	0	1	0	0	0	0	1	1	0	2
BERG	0	0	73	0	1	0	2	0	0	0	1	0	1	0	0	0	0	0	2	0	28	0	0	0	0	0	0	0	1	0
BURIAT	0	0	0	106	0	0	0	0	0	0	1	0	1	0	0	0	0	0	2	0	0	0	0	0	2	0	0	0	1	0
BUSHMAN	0	0	0	0	81	0	0	0	0	0	0	0	0	0	0	0	0	0	1	0	1	1	1	5	0	0	0	1	0	0
ESKIMO	0	0	0	0	0	108	0	0	0	0	0	0	0	0	0	0	0	0	0	0	0	0	0	0	0	0	0	0	0	0
SANTA CRUZ	0	0	0	0	0	0	93	0	0	0	3	1	0	3	0	1	0	1	1	0	1	0	0	1	0	0	0	1	0	2
TOLAI	0	0	0	0	0	0	0	82	19	0	0	0	4	0	0	0	2	1	0	0	0	0	0	0	0	3	1	2	0	0
ARIKARA	0	0	1	0	0	0	5	0	0	45	4	0	4	3	0	0	0	0	0	0	2	0	0	1	2	3	0	0	0	0
PERU	0	1	0	0	0	0	3	0	0	1	37	27	37	0	0	0	2	0	0	0	0	0	0	0	1	0	0	1	0	2
AUSTRALIA	0	0	0	0	0	0	0	1	0	0	0	0	0	100	0	0	0	0	0	0	0	0	0	0	0	0	0	0	0	0
TASMANIA	0	0	0	0	0	0	0	8	4	0	0	0	0	3	65	2	1	0	0	0	2	1	0	0	0	0	0	0	0	0
EASTER I	0	0	0	0	1	0	0	0	0	0	0	0	0	0	0	84	2	0	0	0	0	0	0	0	0	0	0	0	0	0
MOKAPU	0	0	0	0	0	0	0	0	0	0	0	0	0	0	0	3	94	2	0	0	0	0	0	0	0	0	1	0	0	0
MORIORI	0	0	0	0	0	0	1	0	0	2	0	0	0	0	0	1	5	98	0	0	0	0	0	0	1	0	1	0	0	0
N MAORI	0	0	0	0	0	0	0	0	0	0	0	0	0	0	0	4	1	5	0	0	0	0	0	0	0	0	0	0	0	0
S MAORI	0	0	0	0	0	0	0	0	0	0	0	0	0	0	0	1	2	7	0	0	0	0	0	0	0	0	0	0	0	0
EGYPT	1	0	0	0	0	0	0	0	0	0	0	0	0	0	0	0	0	1	68	38	2	0	0	1	0	0	0	1	0	0
NORSE	0	0	1	0	0	0	2	0	0	1	0	0	0	0	0	0	0	0	67	4	34	0	1	1	0	1	0	0	0	0
ZALAVAR	1	0	4	0	0	0	0	0	0	0	0	0	0	0	1	0	0	1	12	5	71	0	0	1	0	1	0	0	0	0
DOGON	0	1	0	0	0	0	0	0	0	1	1	0	0	0	0	0	0	0	0	1	0	87	1	6	0	1	0	1	0	1
TEITA	1	0	0	0	1	0	0	0	0	0	0	0	0	0	0	0	0	0	2	0	0	3	66	9	0	0	0	0	0	1
ZULU	0	0	0	0	3	0	0	2	0	0	0	0	0	0	0	0	0	0	0	0	0	5	1	90	0	0	0	0	0	1
ANYANG	0	0	0	0	0	0	0	0	0	0	1	0	0	0	0	0	0	0	0	0	0	0	0	0	27	11	2	0	0	0
ATAYAL	0	0	0	0	0	0	0	0	0	0	0	0	0	0	0	0	0	0	0	0	0	0	0	0	1	28	51	16	2	0
GUAM	0	0	0	0	0	0	0	0	0	1	0	0	2	0	0	0	1	0	0	0	0	0	0	0	4	1	2	1	0	0
HAINAN	1	0	0	1	0	0	0	0	0	0	0	0	0	0	0	0	0	0	1	0	0	0	0	1	30	29	2	15	1	0
N JAPAN	1	0	0	0	0	0	0	0	0	1	0	0	0	0	0	0	0	0	0	1	0	0	0	0	26	15	3	1	40	0
PHILIPPINES	0	0	0	0	0	0	0	0	0	0	0	0	0	0	0	0	1	0	1	0	0	1	0	1	5	7	3	28	0	1
S JAPAN	0	0	0	0	0	0	0	0	0	0	0	0	1	0	0	0	0	0	0	0	0	0	0	0	55	25	1	3	6	0

Table 2.12 *Performance of different clustering methods on the craniometric data: classification error rate (CER – smaller is better) and Adjusted Rand Index (ARI – larger is better). Except for model-based clustering, the other methods are run with the true number of clusters, $G = 30$.*

Clustering Method	CER %	ARI
Model-based ($\hat{G} = 15$)	44.7	.463
Model-based ($G = 30$)	24.7	.612
Single link	94.5	0
Average link	80.2	.048
Complete link	73.1	.093
k-means	65.5	.160

misclassification rate is 24.7%, as compared to the 44.3% misclassification rate of the 15-cluster solution preferred by BIC. However, the misclassification rate of the 15-cluster solution is only 12.3% when compared with the aggregated populations it detects.

In the 30-cluster solution, several populations are split between multiple clusters, which does not happen in the 15-cluster solution. If one wishes to avoid splitting populations, one might therefore prefer the 15-cluster solution. On the other hand, in the 30-cluster solution, 17 of the 30 populations are largely concentrated in one cluster each, which might lead to a preference for the 30-cluster solution.

In any event, this analysis suggests that, even when the "true" number of groups is known, it is worth paying attention to the solution preferred by BIC or another statistical criterion if the estimated number of groups is lower than the "true" number. The BIC solution may be indicating the maximum number of clusters that can be reliably detected; increasing the number of clusters increases the risk of splitting up subpopulations in the data.

Different clustering methods are compared for these data in Table 2.12. The model-based approach gives the best results, even when it is not told the correct number of clusters.

2.8 Who Invented Model-based Clustering?

Who invented model-based clustering? And when did it happen? For the purposes of answering this question, we will define model-based clustering as involving the following three elements: (1) being based on a finite mixture probability model; (2) a systematic statistical method for estimating the parameters of the model; and (3) a systematic method for classifying the observations conditionally on the model. As it unfolded, there were two quite separate inventions, one for discrete data and one for continuous data.

The American and Austrian sociologist Paul F. Lazarsfeld (1901–1976) (Figure 2.36 (a)) invented model-based clustering for discrete data in a 110-page opus published in 1950. It appeared as two chapters in Volume IV (*Measurement and*

Figure 2.36 Left: Paul Lazarsfeld; Right: John H. Wolfe.

Prediction) of the massive series *Studies in Social Psychology in World War II* which described a range of social psychological studies carried out by interdisciplinary teams of researchers employed by the US War Department to study military personnel in World War II. (Lazarsfeld, 1950a,c). He called it latent class analysis, a special case of the broader latent structure analysis that he proposed in the same work. It was motivated by the analysis of a large survey aimed at understanding American soldiers who had fought in World War II.

It has been suggested that Lazarsfeld's fondness for the word "latent," which he was among the first to use in a statistical context, was due to his having spent his early adulthood in Vienna at the time when Sigmund Freud was active there, developing his theory of the latent content of dreams (Henry, 1999). Lazarsfeld also referred to observed data as "manifest data," again echoing Freud's theory of dreams, in which the manifest content of the dream is the dream as it is remembered, while the latent content of the dream is its hidden meaning.

This work is remarkable in several ways. It created latent class analysis from scratch, and included fully worked out versions of the three essential elements of model-based clustering we have identified, namely model, estimation and classification. There seem to be no real precursors, and the chapters cited almost no previous publications. It acknowledges conversations with sociologist Samuel Stouffer and statistician Frederick Mosteller, but their contributions seem to have been minor.

The probability model he used was a finite mixture of distributions, each one of which was a product of independent multinomial distributions, one for each variable (Lazarsfeld, 1950a, e.g. p.384). He proposed a method of moments estimation method for the model, where observed marginal counts of the multiway table are equated with their expected values under the model, and the resulting system of nonlinear equations is solved. Finally, in one of the earliest uses of the term, he provided what he called the "a posteriori probability" that someone with a particular response pattern came from a given latent class (Lazarsfeld, 1950c, p.430).

It took a long time after that for these ideas to be translated to model-based clustering for continuous data. Tiedeman (1955) described verbally the mixture of multivariate normal distributions as a model for the analysis of "types," classes or groups in psychology, and came very close to writing it down explicitly. He was considering the situation where the covariance matrix for each group is different, or what we would call the VVV model. However, he did not provide either a systematic method for estimating the parameters when the type labels were not available, or a systematic classification method. He correctly identified these as key requirements for an applicable method and encouraged research to develop them. So, Tiedeman can be viewed as a precursor of model-based clustering for continuous data, but not its inventor.

John H. Wolfe (Figure 2.36(b)), an American born in Buenos Aires in 1933, graduated with a B.A. in mathematics from Caltech in 1955, and then went to graduate school in Psychology at University of California, Berkeley. He worked with Robert Tryon, who as we have seen wrote the first book on cluster analysis in 1939. Tryon encouraged Wolfe to develop hierarchical clustering, but Wolfe recounts that he couldn't see a rationale for it without arbitrary assumptions at several stages. Around 1959, Paul Lazarsfeld visited Berkeley and gave a lecture on his latent class analysis, which fascinated Wolfe, and led him to start thinking about how one could do the same thing for continuous data.

Wolfe's 1963 M.A. thesis is an attempt to do this. He started by defining a "type" as one of the component distributions of a mixture of distributions. Then he said:

"The above definition makes it clear that the problem of identifying and describing types is merely an application of the statistical theory of estimation. The population from which a sample of objects is drawn is assumed to have a particular form of distribution (e.g. a mixture of r normal distributions with unknown parameters, such as means and variances for each type and percentage of objects of each type). The task is to devise a procedure for *estimating* the population parameters." (Wolfe, 1963, p.3)

This is effectively the model that underlies continuous model-based clustering to this day. However, the estimation method he used was an effort to use Mahalanobis distances, which was not very effective (Wolfe, 1963, chap. 2). Wolfe acknowledged this, saying that the methods in the thesis were bad. So his thesis can be viewed as a heroic failure. However, he did say at the end of the thesis that he had developed a much better method called "maximum likelihood analysis of types." He later said that his professor didn't believe him, but grudgingly approved the thesis anyway (Wolfe, 2018).

After graduating from Berkeley, Wolfe took a job with the US Navy in San

Figure 2.37 Cover page of Wolfe (1965), the paper in which model-based clustering for continuous data was invented.

Diego first as a computer programmer and then as an operations research analyst. He continued his research on clustering and in 1965 he published the paper that really did solve the problem (Wolfe, 1965) (see Figure 2.37). He used the same mixture of multivariate normals model as in his thesis, which we would call the VVV model. He estimated it by maximum likelihood using a Newton–Raphson method. A method of successive substitution is proposed as an alternative. Finally, he gives the correct explicit expression for the (posterior) probabilities of membership in each "type" (or cluster). Thus this paper for the first time proposes a method for model-based clustering of continuous data that includes all our key elements: model, estimation method and a method for classification.

Wolfe (1965) is an unusual paper, because it took the form of a Technical Bulletin meant to document a FORTRAN computer program, the listing of which is given in the paper, covering 22 pages (Figure 2.37). It is, however, a regular publication, submitted in 1963, revised in 1964, and approved by a sequence of scientific reviewers and supervisors. The fact that it was accompanied by the computer program to implement it was rare then, but of course is commonplace now, giving it in some ways a modern feel. It also may have led Wolfe to consider some issues which have subsequently been the subject of substantial threads of work in the model-based clustering literature.

One of these is initialization. Wolfe recognized the importance of this for his optimization methods, and proposed specific, and quite sensible, initialization methods. He also recommended using "any good prior information from theoretical hypotheses or any other classification-clustering procedures to provide initial estimates" (Wolfe, 1965, p.12). Initialization methods have been a significant consideration in the ensuing literature.

Secondly, he recognized the importance of variable selection, saying:

"By intuition, cluster analysis of variables, factor analysis, or any other means, reduce the number of variables to those that are really important for discriminating types." (Wolfe, 1965, p.12)

This is highly prescient, prefiguring the significant literature on variable selection in model-based clustering that got started almost 40 years later. His conjecture that reducing the number of variables to those that are important for clustering was borne out by subsequent research.

Thirdly, he recognized that computer time may become excessive if the sample is large. He gives the advice:

"If you have a large sample, say 10,000, don't run it all at once. Take a small sample, say 100 to 200 cases at random and run an experiment using a weak convergence criterion. Get an idea of how many types there are and some initial estimates from the small sample. Run the entire set of data using these initial estimates on *specified* numbers of types." (Wolfe, 1965, p.12)

What to do with large samples also emerged as a concern, and Wolfe's advice prefigured later findings remarkably accurately.

Fourthly, he recognized that with model-based clustering the choice of the number of clusters (or "types") is reduced to one of statistical hypothesis testing. He proposed a likelihood-ratio test of the hypothesis that there are r types against the alternative that there are $r - 1$ types. He wrote that the test statistic has a χ^2 distribution, which we now know not to be the case, but this was a key insight for subsequent work.

Finally, he evaluated his methods using a simulated example, which became a preferred approach in this literature.

On the basis of this one paper, we can conclude that John Wolfe was the inventor of model-based clustering for continuous data. He developed the ideas further in Wolfe (1967), and in Wolfe (1970) produced the first journal article describing his results, but the key insights were already present in his 1965 Technical Bulletin.

2.9 Bibliographic Notes

Ward (1963) proposed hierarchical agglomerative clustering using the sum of squares method. This involves choosing the cluster memberships $z_{i,g}$ so as to minimize the criterion $\mathrm{tr}(W)$ with respect to the $z_{i,g}$, where $W = \sum_{g=1}^{G} W_g$, and $W_g = \sum_{i=1}^{n} z_{i,g}(y_i - \hat{\mu}_g)(y_i - \hat{\mu}_g)^T$. This amounts to minimizing the sum of within-cluster sums of squared distances from the cluster means. MacQueen (1967) later proposed an iterative relocation algorithm for the same minimization problem, which became known as k-means.

Although Ward was not working in a model-based context, his clustering criterion later turned out to correspond to a specific probability model. Minimizing

Ward's sum of squares is equivalent to likelihood estimation for the multivariate normal mixture model with $\Sigma_g = \lambda I$, i.e. the EII model. The likelihood being maximized, however, was not the mixture likelihood (2.6), but instead the likelihood of the parameters and the cluster memberships jointly, namely $\prod_{i=1}^{n} \prod_{g=1}^{G} \phi(y_i|\mu_{z_{i,g}}, \Sigma_{z_{i,g}})^{z_{i,g}}$, called the *classification likelihood*. Note that the mixture proportions τ_g do not appear in this likelihood. Edwards and Cavalli-Sforza (1965) described a similar method. The authors of these papers did not put their methods in a model-based context, however, and so are not candidates to be considered as inventors of model-based clustering for continuous data.

Friedman and Rubin (1967) proposed a clustering method that corresponds to maximizing the classification likelihood for the $\Sigma_g = \Sigma$, or EEE model. Scott and Symons (1971) proposed a CEM algorithm for model-based clustering based on the model where the component-specific covariances, Σ_g, can vary freely between components, namely the VVV model.

Although John Wolfe deserves credit as the inventor of model-based clustering for continuous data, Day (1969), working independently, was the first to publish a journal article describing the approach. He proposed estimating a multivariate normal mixture model by maximum likelihood and clustering by assigning each object to the most likely cluster.

Binder (1978) developed a Bayesian estimation method for the VVV model. In this framework, the difference between the classification likelihood and mixture likelihood approaches disappears if the prior distribution for $z_{i,g}$ is multinomial with probabilities τ_g. In that case, the classification maximum likelihood is an approximation to the posterior mode of z under both approaches, as shown by Banfield and Raftery (1993).

Murtagh and Raftery (1984) appear to have been the first to develop a model-based clustering method based on the eigenvalue decomposition of the component covariance matrices. They developed a hierarchical agglomerative model-based clustering method based on the variable orientation, or EEV model, $\Sigma_g = \lambda D_g A D_g^T$, and applied it to problems in character recognition.

McLachlan and Basford (1988) was the first book on model-based clustering. This was a major advance, showing the potential of the approach, and spurred a great deal of interest.

Banfield and Raftery (1989, 1993) coined the term "model-based clustering" and, building on Murtagh and Raftery (1984), proposed the general eigenvalue decomposition-based geometric formulation of Σ_g and the corresponding family of model-based clustering models. They also proposed the normal-uniform mixture model for outliers in clustering. They introduced the idea of using approximate Bayes factors to choose the number of clusters using the approximate weight of evidence (AWE) criterion based on the classification likelihood. They also described the first version of the mclust software, written by Chris Fraley in FORTRAN and also available as an S-PLUS function, and published on the StatLib repository in 1991 (Fraley and Raftery, 1999, 2007b). Celeux and Govaert (1995) introduced maximum mixture likelihood for the Banfield–Raftery family of models using the EM algorithm.

Lavine and West (1992) and Bensmail et al. (1997) proposed methods for Bayesian estimation of the multivariate normal mixture model for clustering using Markov chain Monte Carlo (MCMC). They each did this for a different subset of the Banfield–Raftery family of models. Cheeseman and Stutz (1995) introduced the AUTOCLASS software, which carried out model-based clustering in a partly Bayesian way.

Alternative parsimonious covariance matrix structures have been proposed for model-based clustering. McNicholas and Murphy (2010a) developed a model-based clustering approach for longitudinal data based on the Cholesky decomposition of the covariance and Fop et al. (2018) proposed a sparse covariance structure for model-based clustering.

Celeux and Soromenho (1996) proposed a normalized entropy criterion (NEC) to select the number of clusters and the clustering model. The NEC criterion has been improved by Biernacki et al. (1999) to decide between one and more than one clusters. NEC is available in the MIXMOD software. There have been many comparisons of the various available model selection criteria in the context of model-based clustering, including, more recently, Steele and Raftery (2010).

Alternatives to model selection have also been explored. These include Bayesian model averaging, in which, instead of choosing a single model, predictive distributions are averaged over all the models considered, with weights that reflect their predictive performance over the data (Russell et al., 2015). They also include regularization methods (Witten and Tibshirani, 2010), which have been compared with model selection methods by Celeux et al. (2014).

Finding the best way to initialize the EM algorithm remains an open question. We have described the hierarchical model-based clustering and smallEM approaches, but improvements are possible. Improved versions of the initialization strategy have been described by Scrucca and Raftery (2015) and are implemented in the `mclust` R package. O'Hagan et al. (2012) developed a variant of the smallEM approach that discarded unpromising starting values incrementally. Michael and Melnykov (2016) and O'Hagan and White (2018) proposed an alternative approach to initialization based on model averaging, and reported good results.

Most of the heuristic clustering methods that predated model-based clustering were based on a matrix of similarities between objects, often derived from a matrix of measured characteristics of the objects, as discussed in Section 1.1. Model-based clustering, on the other hand, has for the most part modeled the full data on the measured characteristics directly. Sometimes, however, the full set of measured characteristics isn't available, or the similarities are what are measured directly, or it is desirable to model the similarities rather than the characteristics for computational reasons. For these situations, Oh and Raftery (2007) developed a model-based clustering method based on the similarity matrix rather than the measured characteristics, building on the Bayesian multidimensional scaling method of Oh and Raftery (2001).

Model-based clustering has been used in a wide range of application areas, several of which we illustrate in this book. Others include astronomy (Mukherjee et al., 1998), chemistry (Fraley and Raftery, 2006), political science (Ahlquist and

Breunig, 2012), education (Howard et al., 2018) and actuarial science (O'Hagan and Ferrari, 2017). It has been used in a variety of ways in the analysis of gene microarray data. For example, Yeung et al. (2001) used it to cluster genes according to their expression levels in a range of experiments, while Young et al. (2017) used it to remove artifacts caused by incorrect swapping of genes in paired experiments.

McLachlan and Peel (2000) and Fraley and Raftery (1998, 2002) gave overviews of the area to that point. More recent overviews have been provided by Ahlquist and Breunig (2012) and McNicholas (2016b,a).

3

Dealing with Difficulties

In this chapter, we will discuss some difficulties that the basic modeling strategy can have in specific contexts, and describe some relatively simple strategies for overcoming them, which are readily available and implemented in current software. More elaborate strategies will be described in later chapters.

One difficulty is that data sets often contain outliers, that is data points that do not belong to any cluster. We will discuss some ways of dealing with this issue in Section 3.1.

Another issue is that maximum likelihood estimates of the model parameters can be degenerate. Bayesian estimation can be useful for regularizing the inference in such situations, and avoiding these degeneracies. We discuss one Bayesian approach to this problem in Section 3.2. This is a simple approach designed to solve the degeneracy problem; when degeneracy is not an issue, it produces solutions similar to those produced by maximum likelihood.

It often happens that some clusters are not Gaussian, so the most common probability model in model-based clustering, the multivariate Gaussian, does not apply. Using it tends to lead to the representation of one cluster by a mixture of Gaussian components rather than just one. This still gives good estimates of the underlying overall probability density function, but can lead to overestimation of the number of clusters. In Section 3.3 we describe one way to overcome this difficulty, by merging components after an initial fit.

Finally, we briefly mention some other approaches to these issues in Section 3.4. Other approaches will be described in more detail later in the book.

3.1 Outliers

3.1.1 Outliers in Model-based Clustering

In general, in statistical analysis, an outlier is defined as one of a relatively small number of points that do not follow the same pattern as the main bulk of the data. In model-based clustering, outliers are points that do not belong to any of the clusters. In many statistical analyses, outliers are often data points that lie outside the main bulk of the data, and this can be true in cluster analysis too. However, they may also be "inliers" in the sense that they may be between clusters, and thus in the interior of the data, particularly if clusters are well-separated. Outliers can increase the estimated number of clusters beyond what is present in the data,

Dealing with Difficulties

Simulated data with outliers

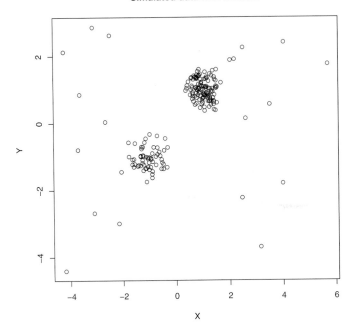

Figure 3.1 Simulated data set with two well-separated clusters and outliers.

because if not identified they can themselves be identified as separate clusters. Also, if not separated out or down-weighted, they can create error in the estimates of the cluster means, and inflate the estimated cluster variances.

A simple example with some simulated data is shown in Figure 3.1. In this data set, 180 two-dimensional data points were simulated from two well-separated clusters using the mixture model $\frac{2}{3}N((1,1)^T, 0.3^2 I) + \frac{1}{3}N((-1,-1)^T, 0.3^2 I)$. Then 20 noise points were added, simulated from a $N(0, 4I)$ distribution, but with points constrained to not overlap with the clusters.

Nineteen of the 20 outliers are fairly clear visually in Figure 3.1. However, there are several points in the vicinity of the clusters that are more visually ambiguous. Figure 3.2 shows which points were simulated as outliers; the outlier closest to the left cluster is hard to detect visually.

Sometimes model-based clustering can handle outliers without any special precautions, and indeed this turns out to be the case in our simulated data set. We will soon see that this does not always work, though, and it can be necessary to take precautions against outliers. Figure 3.3 shows the results from a routine application of model-based clustering. BIC chooses the unequal-variance spherical model, VII, quite decisively. It recovers the two actual clusters, and represents the outliers by a third, highly dispersed, mixture component. It is thus able to

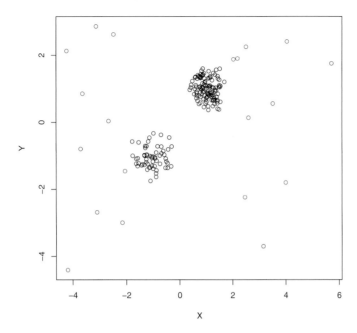

Figure 3.2 Simulated data set with outliers shown.

identify all the simulated outliers correctly, except for the one close to the left cluster, which is hard to identify visually anyway.

We will now describe two relatively straightforward ways to deal with outliers. One is to add an additional component to the mixture model to represent the outliers, typically a uniform component. A second approach is to remove observations identified as outliers at the estimation stage.

3.1.2 Mixture Modeling with a Uniform Component for Outliers

A straightforward but more general way to handle outliers, or "noise," in model-based clustering is simply to add an additional component to the mixture model to represent the outliers. Ideally, this should have outlier-like properties, being dispersed over the data region. One form that has been used successfully is a uniform distribution on the data region. This corresponds to the assumption that outliers are generated approximately by a homogeneous Poisson process on the data region. The Gaussian mixture model defined by (2.1) and (2.2) then becomes

$$p(y_i) = \frac{\tau_0}{V} + \sum_{g=1}^{G} \tau_g \phi_g(y_i | \mu_g, \Sigma_g), \qquad (3.1)$$

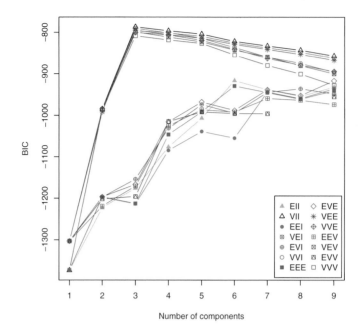

Figure 3.3 Simulated data with outliers: BIC plot for standard
model-based clustering.

where τ_0 is the expected proportion of outliers in the data and V is the volume of
the data region. Estimation and model selection proceed as before.

Operationally, V can be defined as the volume of the smallest axis-aligned
hypercube containing the data. Another definition is the volume of the smallest
hypercube aligned with the principal components of the data set that contains the
data. The `mclust` software uses the minimum of these two values as its default
value.

It might seem at first sight that the result would be sensitive to the value of
the volume of the data region. However, in practice the sensitivity typically does
not seem to be great. This is probably because with the model (3.1), a point
is classified as an outlier approximately if the uniform density is greater than
the density under the nearest mixture component, and so the cutoff point will
usually be in the tail of one of the mixture densities. Because the Gaussian density
declines fast as one moves away from its mode, the precise location of this cutoff
point is not too sensitive to the height of the uniform density.

The estimation results can be quite sensitive to the initialization of the EM
algorithm, however, particularly the initial assignment of noise points. One automatic way to initialize the outlier assignment is by the nearest neighbor cleaning
method of Byers and Raftery (1998). This essentially designates points as outliers

if they are far from other points. It does this by approximating the non-outlier distribution of the distance of a point to its Kth nearest neighbor by a gamma distribution with shape parameter K (this is exact if the data are generated by a homogeneous Poisson distribution). It then represents the distribution of Kth nearest neighbor distances by a mixture of two components, one corresponding to non-outliers and one to outliers, estimating the resulting mixture model, and designating as outliers the points in the cluster with larger values of the Kth nearest neighbor distance. The method is implemented in the `NNclean` function in the `prabclus` R library (Hennig and Hausdorf, 2015).

If the variables are measured on the same scale, such as spatial coordinates or scores on similar tests, it is not necessary to scale the data before applying the nearest neighbor cleaning method. Otherwise it is usually a good idea to scale the data first. This can be done, for example, by dividing each variable by its standard deviation, or by a robust measure of scale. Otherwise, the result is not invariant to rescaling of particular variables. A variable measured on a scale that gives it a large variance may dominate the calculations, which would be undesirable.

For our simulated data, this analysis proceeds as shown in Listing 3.1, with the results shown in Figure 3.4. The upper left panel shows the initialization using nearest neighbor cleaning. This slightly overestimates the number of outliers, identifying 25 outliers instead of the 20 that were simulated. However, this does not matter much, as this is only the initial assignment, which is then refined by the model-based clustering.

Figure 3.4(b) shows the BIC plot. This allows for the possibility of there being no clusters, i.e. that the data set consists only of Poisson noise, shown in the plot as "0 components." This possibility is strongly rejected by BIC for the simulated data. BIC chooses the equal-variance spherical model, `EII`, with two clusters and noise. This is indeed the model from which the data were simulated. Figure 3.4(c) shows the resulting classification. This correctly identifies the 20 outliers that were simulated, including the one that was visually ambiguous, with no false positives.

Finally Figure 3.4(d) shows the uncertainty plot. Uncertainty was relatively high on the periphery of the clusters, but 17 of the 20 simulated outliers were identified with little uncertainty.

The nearest neighbor cleaning method tends to identify more points as outliers than are actually present. This happened in the simulated data, where nearest neighbor cleaning identified 25 outliers compared with 20 that were actually simulated; the overestimation is often more severe. This usually does not matter too much because it is being used only for initialization; the mixture model estimation typically corrects things.

A modification of the nearest neighbor cleaning method that tends to identify fewer false positives is the nearest neighbor variance estimation method (NNVE). This adds simulated noise to the data before applying the nearest neighbor cleaning method; perhaps surprisingly this tends to reduce the number of false positives identified in the original data. It is implemented in the `covRobust` R package, and its application to our simulated data is shown in Listing 3.2, with the results shown in Figure 3.5.

Dealing with Difficulties

Listing 3.1: Outlier modeling for simulated data

```
# Loading libraries and data
library(prabclus)

# Simulating data with noise
simdata1 <- sim(modelName="EII",n=180,
     parameters=list(pro=c(.67,.33),mean=matrix(c(1,1,-1,-1),
     ncol=2),variance=list(sigmasq=0.09)))
x1N <- rnorm (60,sd=2); x2N <- rnorm (60,sd=2)

# Remove noise points that are close to the cluster centers
dist1 <- rep (NA,60); dist2 <- rep (NA,60)
for (k in 1:60) dist1[k] <- sqrt ((x1N[k]-1)^2 + (x1N[k]-1)^2)
for (k in 1:60) dist2[k] <- sqrt ((x1N[k]+1)^2 + (x1N[k]+1)^2)
close <- dist1<1.4 | dist2<1.4
noise <- cbind (x1N,x2N)
noise <- noise[!close,]
Nnoise <- dim (noise)[[1]]
simdata <- rbind (simdata1[,c(2,3)],noise)
n <- dim(simdata)[[1]]

# Plotting noisy data
plot (simdata, type="n")
points (simdata[1:180,])
points (simdata[181:n,], col="red")

# Noise detection and clustering with Mclust
NNclean.out <- NNclean (scale(simdata), k=12)
MclustN <- Mclust(simdata, initialization = list(noise = (
    NNclean.out$z==0)))

# Plotting results
plot (simdata, col=1+NNclean.out$z)
mclust2Dplot(simdata, parameters=MclustN$parameters, z=MclustN$z,
    what = "classification", PCH=20)
plot (MclustN, what="uncertainty")
```

NNVE identifies fewer outliers than nearest neighbor cleaning, in this case 23 compared with 25. While this does not seem like a big difference, the number of false positives is reduced from 5 to 3, i.e. by 40%. The final classification from the mixture model is the same as when it is initialized with nearest neighbor cleaning. The initial sets of outliers identified by nearest neighbor cleaning and NNVE both include all the simulated outliers.

Example 6 (ctd.): Minefield identification

We now return to the problem of minefield identification from an image with a great deal of clutter. We showed the raw data and the true classification in Figure 1.3. This is a challenging problem for a number of reasons. The minefield follows a highly nonlinear pattern, and so could not be represented by a single Gaussian component, since bivariate normal distributions are concentrated around a line in

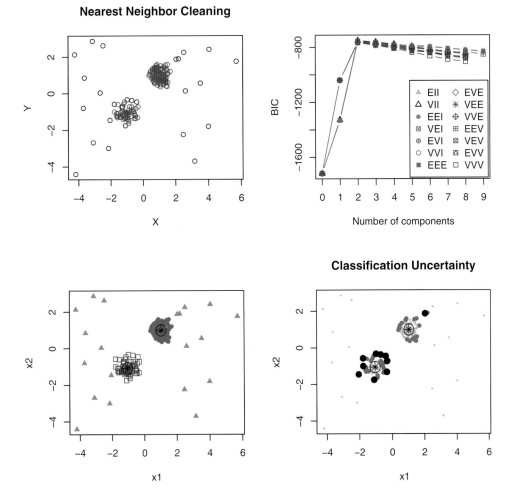

Figure 3.4 Simulated data: Analysis using model-based clustering with uniform noise component. (a) Initialization using nearest neighbor cleaning. The initially identified outliers are shown as black circles. (b) BIC plot. (c) Classification using model-based clustering. The finally identified outliers are shown as green triangles. (d) Uncertainty plot.

two-dimensional space. Also, if one thinks of the clutter as outliers, their number is large, consisting of over two-thirds of the points.

When we apply model-based clustering with no noise component to the data, the best model according to BIC is the varying covariance model VVV with eight clusters. The BIC plot is shown in Figure 3.6, with the BIC values for all models in the left panel, and the BIC values for the VVV model with different numbers of clusters in the right panel.

The resulting classification is shown in Figure 3.7. While the minefield appears

Listing 3.2: Outlier modeling for simulated data using nearest neighbor variance estimation

```
# Loading libraries
library (covRobust)

# Robust covariance estimation and clustering (using the simulated
    data of the previous listing)
nnve.out <- cov.nnve (scale(simdata))
simdata.MclustN.NNVE <- Mclust (simdata, initialization=list(noise
    =(nnve.out$ classification==0)))


# Plotting results
plot (simdata, col=1+nnve.out$classification)
mclust2Dplot(simdata , parameters=simdata.MclustN.NNVE$parameters,
    z=simdata.MclustN.NNVE$z,
              what = "classification", PCH=20)
```

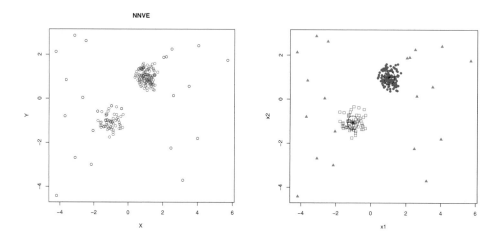

Figure 3.5 Outlier modeling for simulated data with nearest neighbor variance estimation (NNVE): Left: initialization of assignment of outliers via NNVE. Right: classification by model-based clustering initialized with NNVE.

as two clusters, the outliers are represented by six clusters that do not have any reality. Thus, without a noise component, model-based clustering does not give an interpretable result for these data.

The identification of outliers by nearest neighbor cleaning is shown in Figure 3.8. This does identify most of the outliers correctly, but it erroneously identifies some clumps of outliers as part of the minefield. This is because, by chance, their 12th nearest neighbor distances are relatively small. Visually, it is fairly clear from looking at the data as a whole that they do not belong to the main minefield, but nearest neighbor cleaning takes a local rather than a global approach to the data

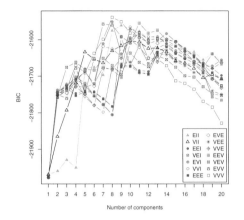
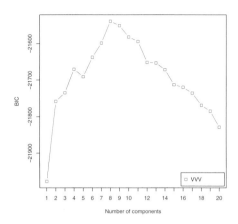

Figure 3.6 BIC values for minefield data for model-based clustering with no noise component. Left: all 14 models. Right: VVV model only.

and so cannot make this kind of inference. It provides a good initialization for model-based clustering, which does take a global view of the data.

The BIC plot for model-based clustering with noise for the minefield data is shown in the left panel of Figure 3.9. The model chosen is the variable-orientation model, EEV, with two clusters, in which the volumes and shapes of the clusters are the same. This means that there are two Gaussian clusters in addition to the noise. The resulting classification is shown in the right panel of Figure 3.9.

The minefield here has a chevron shape, being concentrated around a piecewise linear curve consisting of two lines that connect. Thus it cannot be well represented by any one bivariate Gaussian distribution. This is because the bivariate Gaussian distribution is in general concentrated around a line in two-dimensional space. (When the covariance matrix is proportional to the identity matrix, the distribution is distributed in a circular manner around a point in two-dimensional space.) The model approximates the minefield by two clusters rather than one, each of which is concentrated about a line. This illustrates the fact that a cluster concentrated about a nonlinear curve can be approximated by a piecewise linear curve, which can in turn be represented by several mixture components rather than one. The model was extended to the situation where a single cluster is concentrated around a smooth curve, represented by a principal curve, by Stanford and Raftery (2000).

The noise model gives 63 errors out of 1,104 data points, an error rate of 5.7%. The errors are shown in the left panel of Figure 3.10, and it can be seen that they are concentrated on the edge of the true minefield, which is where one would expect any algorithm to have difficulties. The uncertainties are plotted in the right panel of Figure 3.10, which indicates that the actual errors occurred where the model said that the uncertainty was greatest.

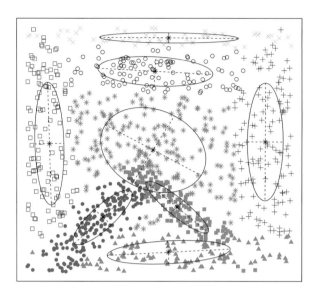

Figure 3.7 Classification of the minefield data by model-based clustering with no noise component.

It is possible to compare the best mixture models without and with noise using BIC to provide an approximate Bayesian test of the presence of uniform noise. For the minefield data, the BIC value for the best model without noise (the unconstrained covariance VVV model with eight clusters) is $-21,538.79$, while for the best model with noise (the varying orientation model EEV with two clusters and noise) it is $-21,127.80$. Thus the model with noise has a BIC value that improves on the model without noise by 411 points, a rather decisive difference.

This analysis also illustrates the fact that with the noise model, one can have a high proportion of outliers, more than 50%, without the analysis breaking down.

3.1.3 Trimming Data with tclust

Another popular technique to deal with noisy data in clustering is the so-called "tclust" method (Gallegos and Ritter, 2005; García-Escudero et al., 2008). tclust can be viewed as a robust version of the EM algorithm for mixture models. The idea of this approach is to exclude observations identified as outliers when estimating the model parameters.

The tclust model assumes that the data at hand contains $n\alpha$ "spurious" observations, $\alpha \in [0, 1]$, that one wants not to include when estimating the mixture

NNclean classification

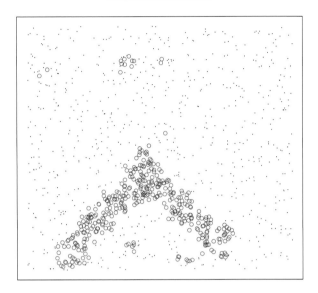

Figure 3.8 Identification of outliers in the minefield data by nearest neighbor cleaning, for use as initialization in model-based clustering with noise. The black dots represent the identified outliers.

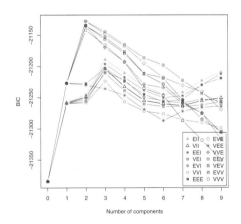
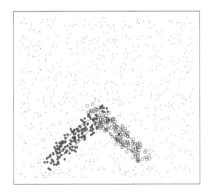

Figure 3.9 Analysis of minefield data by model-based clustering with noise, initialized by nearest neighbor cleaning. Left: BIC plot. Right: classification. The outliers are shown by black dots.

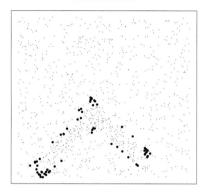 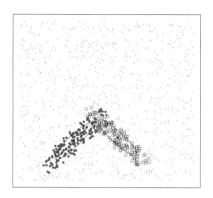

Figure 3.10 Errors and uncertainty in model-based clustering of minefield data with noise. Left: errors; right: uncertainty.

model. To this end, let us introduce a set S of "spurious" observations which is of cardinality $n\alpha$. Then, the spurious mixture model is defined through its likelihood:

$$L(Y; \Theta) = \left[\prod_{y_i \in \bar{S}} \prod_{g=1}^{G} \tau_g \phi(y_i; \mu_g, \Sigma_g) \right] \times \left[\prod_{y_i \in S} f_i(y_i) \right],$$

where the densities f_i are not necessarily Gaussian probability density functions. Under these assumptions, Gallegos and Ritter (2005) and García-Escudero et al. (2008) show that mixture parameters and group memberships can be inferred by just maximizing the "trimmed" log-likelihood:

$$\ell(Y; \Theta, S) = \sum_{y_i \notin S} \sum_{g=1}^{G} \log \left(\tau_g \phi(y_i; \mu_g, \Sigma_g) \right).$$

To do so, García-Escudero et al. (2008) proposed the tclust algorithm for a Gaussian mixture model which can be viewed as an EM-like algorithm with additional trimming step. The tclust algorithm operates as follows at iteration s:

- E-step: for all observations, compute the expectation of the complete data likelihood as usual, and compute and store the quantities:

$$d_i = \max \left\{ D_1(y_i; \Theta), ..., D_G(y_i; \Theta) \right\},$$

  where D_g is a discriminant function defined as $D_g(y_i; \Theta) = \tau_g \phi(y_i; \mu_g, \Sigma_g)$;
- T-step: the $n\alpha$ observations with the smallest d_i values are assigned to the set S of spurious observations;
- M-step: the mixture parameters are updated based only on the non-spurious observations, *i.e.* the observations which do not belong to S.

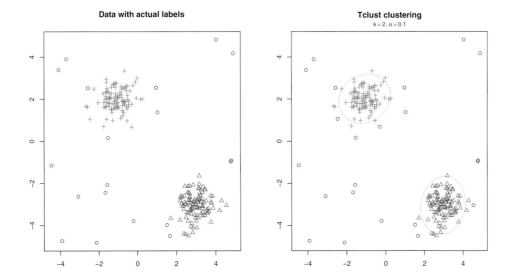

Figure 3.11 tclust for robust clustering. The left panel shows the simulated data with the actual group label (outliers are the black circles). The right panel displays the tclust solution, which gives results similar to the actual group labels. Notice that the true values of G and α were provided to tclust here.

Listing 3.3 shows a simple use of tclust on simulated data, using the `tclust` package for R. Here the data are simulated from a mixture of two Gaussian distributions and a uniform component. The percentage of data in the uniform component (assumed to be the outlier group) is $\alpha = 0.1$. The results are shown in Figure 3.11.

As always, the actual number of groups and the percentage of spurious data are not known in practical situations. It is therefore desirable to be able to automatically choose appropriate values for G and α for the data at hand. Unfortunately, as discussed by García-Escudero et al. (2008), it is not valid to use classical model selection criteria in this context. Instead, García-Escudero et al. (2008) proposed a heuristic which consists of examining the trimmed classification log-likelihood curves and searching for the value of α for which the curves G and $G+1$ are similar for larger values of α. Listing 3.4 presents code which allows the simultaneous choice of G and α with tclust. The plot produced by the last commands is shown in Figure 3.12.

The left panel of Figure 3.12 shows the curves of the trimmed classification log-likelihood for $G = 2$, 3 and 4 against the value of α. The curve for $G = 2$ quickly joins the curves associated with larger numbers of groups. Following the heuristic of García-Escudero et al. (2008), one can conclude that an appropriate value for G is 2, and that α should be between 0.07 and 0.1. The right panel gives a better view of the relative differences between the curves for $G = 2$ and 3. It

Listing 3.3: `tclust` for robust clustering

```
# Loading libraries
library(tclust)
library(MBCbook)

# Data simulation (class 1 is for outliers)
n = 200; alpha = 0.1
m1 = c(-1,2); m2 = 3*c(1,-1)
S1 = 0.25*diag(2); S2 = 0.25 * diag(2)
X = rbind(mvrnorm(n/2,m1,S1),mvrnorm(n/2,m2,S2),
   matrix(runif(alpha*n*2,-5,5),ncol=2))
cls = rep(3:1,c(n/2,n/2,alpha*n))

# tclust with the actual G and percentage of outliers
out = tclust(X,k=2,alpha=alpha)

# Plotting results
par(mfrow=c(1,2))
plot(X,col=cls,pch=cls,xlab='',ylab='',
   main="Data with actual labels")
plot(out,xlab='',ylab='',main='tclust clustering')
```

Listing 3.4: Choice of G and α within tclust

```
# Clustering for various G and alpha (using data of the previous
   example)
out = ctlcurves (X, k=2:4, alpha=seq(0,0.2,len=21))

# Plotting results
par(mfrow=c(1,2))
plot(out)
plot(seq (0, 0.2, len = 21),out$obj[1,] - out$obj[2,], type='b',
   xlab='alpha',ylab='',main='Difference between G=2 and G=3')
abline(h=0,lty=2)
```

appears that the curves join at around $\alpha = 0.09$, which is a reasonable estimate of the actual percentage of outliers (the actual value of α is 0.1).

3.2 Dealing with Degeneracies: Bayesian Regularization

A general problem with estimating Gaussian mixture models is that maximization of the mixture likelihood (2.6) without any constraints could lead to a degenerate solution. This is because as $\mu_g \longrightarrow y_i$ and $|\Sigma_g| \longrightarrow 0$ for any observation i and mixture component g, i.e. as the component mean approaches the observation and the component covariance becomes singular, then the likelihood for that observation becomes infinite and hence so does the whole mixture likelihood (2.6). Thus in general there is at least one path in parameter space per observation along which the likelihood tends to infinity (Day, 1969; Titterington et al., 1985).

The parameter values corresponding to these infinite likelihoods are not inter-

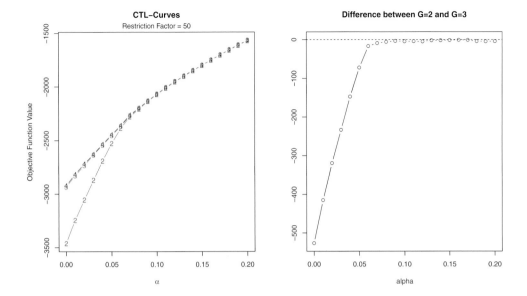

Figure 3.12 Choice of G and α with tclust. The left panel shows the curves of the trimmed classification log-likelihood for $G = 2$, 3 and 4 against the value of α. On the right panel, the difference between the curves for $G = 2$ and 3 is displayed. Following the heuristic of García-Escudero et al. (2008), one can conclude that appropriate values for G and α are respectively 2 and 0.09.

esting in general, and are often spurious or degenerate solutions. They do not have good properties as estimators. Redner and Walker (1984) showed that there is in general a sequence of *interior* maximum likelihood estimates that is consistent, i.e. that approaches the true values in the limit with high probability as the sample size increases. But the spurious estimators do not have this good property.

In practice for estimation, this is not a big problem, because if the model is well specified and the starting values are reasonable, it is unusual for the EM algorithm to get trapped at a spurious solution. However, for model selection it is a bigger problem. This is because model selection inevitably involves fitting models that are bigger than the best fitting model, and so include mixture components that are not well supported by the data. This is precisely the situation where spurious solutions are most likely to be found.

Several solutions have been proposed. Hathaway (1985, 1986b) proposed maximizing the mixture likelihood (2.6) subject to the constraint there exists a constant c such that all the eigenvalues of $\Sigma_g \Sigma_h^{-1}$ are greater than or equal to c for all pairs of mixture components g, h. However, the resulting constrained maximization is difficult.

The geometric constraints described in Section 2.2 provide a partial solution, in that models that impose equal-volume constraints, such as EEE or EVV, do not suffer from this problem. This is not a complete solution, however, because one might be interested in solutions involving models that do have spurious solutions when the number of components is too large, such as the unconstrained covariance model, VVV.

The default solution in `mclust` is to compute the ratio of the smallest to the largest eigenvalue of the estimated covariance matrix for all the mixture components, and not return any solution or BIC value for models for which this value is below a user-specified threshold for any threshold (Fraley and Raftery, 2003). By default this threshold is taken to be the relative machine precision, which is about 2×10^{-16} on IEEE-compliant computers. This avoids spurious solutions, but leaves the problem that log-likelihoods or BIC values are not provided for some models, leaving open the possibility that they might fit the data better than the selected model. This is not likely in practice, because spurious solutions tend to correspond to the presence of components whose separate existence is not well supported by the data.

Nevertheless, any component with fewer than d points will tend to have a singular covariance matrix, and hence produce an infinite likelihood, even if there is a true cluster with fewer than d points, where d is the dimension of the data. Geometric constraints on the component covariance matrices could resolve this, but they may not be justified by the data. Thus the singularities might lead to incorrect model specification if, for example, there are several clusters of which one has fewer than d points, and geometric constraints on the covariance matrices are not justified by the data.

A more general solution that avoids most of these difficulties is to adopt a Bayesian approach, and instead of maximum likelihood, use a maximum a posteriori (MAP) estimate (Fraley and Raftery, 2007a). Unlike what happens with the likelihood, there are generally not paths in parameter space along which the posterior density tends to infinity, with reasonably specified proper prior distributions. This is because the prior distribution is multiplied by the likelihood and tends to smooth or regularize these features, usually eliminating them.

Fraley and Raftery (2007a) used the standard normal-inverse Wishart prior distribution for the component means and covariance matrices. Conditionally on the cluster assignments, this is a conjugate prior, i.e. it yields a posterior distribution with the same functional form (Gelman et al., 2013). They then used the EM algorithm to find the posterior mode rather than the maximum likelihood estimator (Dempster et al., 1977). We modify the definition of BIC in (2.16) so that the maximized likelihood $p(D|\hat{\theta}_{M_k}, M_k)$ is replaced by the likelihood evaluated at the posterior mode.

For the mean μ and covariance matrix Σ of one component, the normal-inverse Wishart prior distribution takes the following form. The prior distribution of the mean given the covariance matrix is normal:

$$\mu|\Sigma \sim MVN(\mu_{\mathcal{P}}, \Sigma/\kappa_{\mathcal{P}}) \tag{3.2}$$

$$\propto |\Sigma|^{-\frac{1}{2}} \exp\left\{-\frac{\kappa_{\mathcal{P}}}{2} \text{tr}\left[(\mu - \mu_{\mathcal{P}})^T \Sigma^{-1} (\mu - \mu_{\mathcal{P}})\right]\right\}. \tag{3.3}$$

The prior distribution of the covariance matrix is inverse Wishart, namely

$$\Sigma \sim \text{inverseWishart}(\nu_{\mathcal{P}}, \Lambda_{\mathcal{P}}) \tag{3.4}$$

$$\propto |\Sigma|^{\frac{\nu_{\mathcal{P}}+d+1}{2}} \exp\left\{-\tfrac{1}{2}\text{tr}\left[\Sigma^{-1}\Lambda_{\mathcal{P}}^{-1}\right]\right\}. \tag{3.5}$$

The parameters of the prior distribution, or hyperparameters, are denoted by a subscript $\mathcal{P}$. The hyperparameter $\kappa_{\mathcal{P}}$ is called the prior sample size, $\nu_{\mathcal{P}}$ is the number of degrees of freedom, and $\Lambda_{\mathcal{P}}$ is the scale matrix of the Wishart distribution. Explicit expressions for the different models were given by Fraley and Raftery (2007a), while detailed derivations of the posterior distributions were provided by Fraley and Raftery (2005).

Default hyperparameters were selected to have two desirable properties. One is that the prior be spread out enough for inference to be almost unchanged from the maximum likelihood inference when the EM algorithm converges to a non-spurious solution. The second property we want is that the prior be concentrated enough for spurious solutions to be avoided in most cases and for the BIC curve to transition smoothly from models with non-spurious solutions, to models where the EM algorithm converges to spurious solutions. The prior mean $\mu_{\mathcal{P}}$ is taken to be equal to the mean of the data. The prior sample size $\kappa_{\mathcal{P}}$ is taken to be 0.01. The prior degrees of freedom $\nu_{\mathcal{P}}$ is taken to be $d+2$, which is the smallest integer value that gives a prior distribution with finite covariance matrix. The prior scale matrix $\Lambda_{\mathcal{P}}$ is taken to be the empirical covariance matrix of the data divided by $G^{2/d}$.

The resulting posterior mode is usually such that when the local maximum of the likelihood found by the EM algorithm is not spurious, the parameter values, maximized likelihood and BIC value are relatively unchanged. When a spurious local maximum is found, however, usually because the number of components is larger than what is supported by the data, the sequence of BIC values tends to decline smoothly, rather than jumping spuriously to infinity (Fraley and Raftery, 2007a).[1]

Example 2 (ctd.): Diabetes diagnosis

We saw in Section 2.5 that this data set is well fit by the VVV model with three components; this is the choice made by BIC. A closer look, however, reveals that for the VVV model, no BIC values were returned by `mclust` for $G \geq 5$, i.e. for five or more mixture components. This is because for these cases the EM algorithm diverged to solutions with infinite likelihoods and so, by default, the results were not reported. It therefore remains possible that a solution with 5 or more groups would fit the data better.

To investigate this further, we removed the constraint forbidding the reporting of results from spurious singular solutions, as shown in Listing 3.5. The control parameter `eps`, which is the reporting threshold on the ratio of smallest to largest eigenvalue of the covariance matrix of any component, was set to zero, so that

---

[1] Currently, as of version 5.4, the `mclust` software implements the Bayesian method with a prior for only 10 of the 14 geometrically constrained models.

Listing 3.5: BIC for `diabetes` data with spurious singular solutions and with prior

```
# Loading libraries and data
library(MBCbook)
data(diabetes)
X = diabetes[,-1]

# Including spurious singular solutions:
Mclust (X, modelNames="VVV", control=emControl(eps=0, tol=c(1e-3,1
    e-3)))

# Including prior:
Mclust (X, prior=priorControl(), modelNames="VVV")
```

spurious singular solutions are reported. The tolerance for convergence of the EM algorithm, specified by the control parameter `tol`, was relaxed to 10^{-3} instead of the default 10^{-5}.

The result is shown in Figure 3.13. The log-likelihoods found by the EM algorithm for five or more components diverged to infinity without regularization, but finite values were found here because the algorithm stopped. These BIC values are shown by dashed red lines and open circles in Figure 3.13; by default they would not be shown. Non-spurious results are found only for four or fewer components, and the corresponding results are shown by solid red circles. Using the full red BIC curve in Figure 3.13 would lead to the erroneous conclusion that there are at least nine components.

When the default prior and posterior mode are used, the result is shown in Figure 3.13 by the solid blue circles and lines. The BIC values are very close to those for the default solution for four or fewer components, when the default solution is non-spurious, as we would wish. For five or more components, the spurious singular solutions are no longer found because of the Bayesian regularization, and the BIC continues to decline in a smooth way. Thus using the prior confirms our earlier conclusion that using three mixture components is best.

To illustrate the issues, we now consider the five-component solution. This is the smallest number of components for which the EM algorithm yields a degenerate solution, shown in the left panel of Figure 3.14. Note that the data do not support five components, so this would not be a good scientific solution, and is shown only for illustration. Cluster 4 has only three points, shown in black in Figure 3.14. The dimension of the data is $d = 3$, and so Cluster 4 has the same number of points as there are dimensions, not enough to estimate the covariance matrix. Thus the estimated covariance matrix for Cluster 4 is singular, with the ratio of the smallest to the largest eigenvalue equal to zero, and determinant equal to zero also. There is zero uncertainty about membership of cluster 4. This is unrealistic, and is another consequence of the degeneracy. In the default `mclust` solution, this result would not be reported.

The right panel of Figure 3.14 shows the classification using the maximum

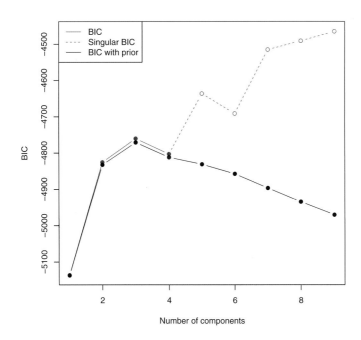

Figure 3.13 Diabetes diagnosis data: BIC values for the unconstrained covariance VVV model calculated in different ways. The red filled circles show the BIC values from the usual EM algorithm for 1 to 4 mixture components; these correspond to non-spurious interior solutions. The red open circles and dashed lines show the spurious computed BIC values for 5 through 9 components; these were obtained by relaxing the convergence tolerance. The blue circles show the BIC values when a prior is used.

a posteriori estimates of the model parameters. Cluster 4 now has only two points (again shown in black). However, the solution is no longer degenerate: the covariance matrix for Cluster 4 has the ratio of smallest to largest eigenvalue equal to 1.6×10^{-4}. While small, this is much larger than the reciprocal machine infinity computed for the unregularized solution (2×10^{-16}). Note that even with the Bayesian regularization, the BIC strongly prefers three components to five, so the five-component solution would not be used in practice.

3.3 Non-Gaussian Mixture Components and Merging

Model-based clustering is based on the idea of a finite mixture of multivariate normal or Gaussian distributions, with each mixture component distribution corresponding to one cluster. Sometimes, however, some of the clusters do not have a Gaussian distribution. In this case, BIC will tend to choose a model that represents a non-Gaussian cluster by a mixture of Gaussian mixture components.

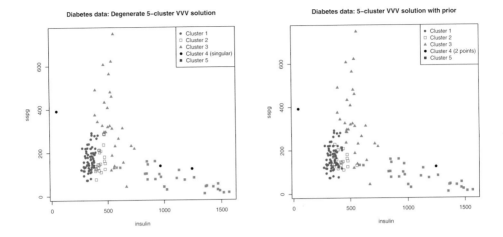

Figure 3.14 Five-cluster solutions for diabetes data using the unconstrained covariance **VVV** model. Left: spurious degenerate solution from EM algorithm. Cluster 4 (shown in black) has only three points and is concentrated on a two-dimensional plane, and so is degenerate. Right: maximum a posteriori solution. Cluster 4 has only two points, but its estimated covariance matrix is not singular because of the Bayesian regularization.

This will give a good estimate of the density, but will overestimate the number of clusters.

On the other hand, ICL will tend to select a number of mixture components that corresponds to the number of clusters. However, in so doing it typically represents a non-Gaussian cluster by a single Gaussian mixture component, which may not be a good fit to the data. The difficulty is illustrated by the Old Faithful data, as shown in Figure 3.15.

As we saw in Section 2.6, BIC chooses three mixture components, as illustrated in the top panels of Figure 3.15. The top left panel suggests that, while the mixture component in red at the bottom left of the plot is cohesive and well-separated from the rest of the data, the two components in blue and green at the top right part of the plot are not very separate from one another, and may in fact make up a single cluster. The density estimate in the top right panel reinforces this. It looks as if there are really two clusters, not three, but the one at the top right has a non-Gaussian distribution and is itself being modeled by a mixture of two Gaussian distributions.

As seen in the bottom left panel of Figure 3.15, ICL chooses two mixture components. This is both good and bad. Good because there do indeed seem to be two clusters, and hence the number of components chosen corresponds to the number of clusters. But bad because the cluster at the top right of the plot is represented by a single Gaussian in the ICL solution, which may not be a very good fit to the data.

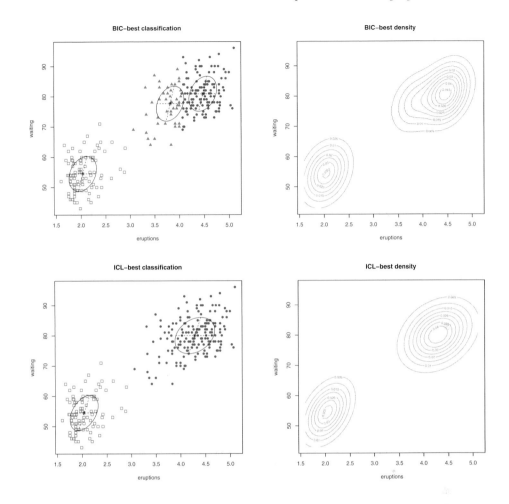

Figure 3.15 Old Faithful data: Top left: classification by the BIC-best model with three components. Top right: corresponding density estimate. Bottom left: classification by the ICL-best model with two components. Bottom right: corresponding density estimate.

A widely used solution to this dilemma is to keep the BIC solution as a model for the density of the data, but to merge mixture components that are close together before clustering. This has the potential to give the best of both worlds: the right number of clusters, but a flexible model for each of them that fits the data well. As summarized by Hennig (2010), methods for merging Gaussian components proceed as follows:

1 Start with all components of the initially estimated Gaussian mixture as current clusters.

2 Find the pair of current clusters most promising to merge.

3 Apply a stopping criterion to decide whether to merge them to form a new current cluster, or to use the current clustering as the final one.

4 If merged, go to 2.

Different choices of initially estimated Gaussian mixture, different methods for finding the most promising pair of components to merge at each stage, and different stopping criteria lead to different merging methods. The model selected by BIC is often used as the basis for the initially estimated Gaussian mixture. Methods proposed differ mainly in the selection of pairs for merging and the stopping criterion.

One approach is to base these on the entropy of the resulting classification (Baudry et al., 2010). Let $\hat{z}_{i,g}^{[G]}$ denote the conditional probability that observation i arises from cluster $g = 1, \ldots, G$, based on the G-cluster solution. The corresponding estimated soft or probabilistic classification, defined by (2.9) at the converged solution for G clusters, has entropy

$$\text{Ent}(G) = -\sum_{g=1}^{G} \sum_{i=1}^{n} \hat{z}_{i,g}^{[G]} \log(\hat{z}_{i,g}^{[G]}) \geq 0. \tag{3.6}$$

For a classification with no uncertainty, i.e. if $\hat{z}_{i,g}^{[G]}$ is equal to 0 or 1 for all i and g, $\text{Ent}(G) = 0$. The entropy is maximized when $\hat{z}_{i,g}^{[G]} = 1/G$ for all i and g, i.e. when the classification is effectively uninformative.

The basic idea is to choose the pair of clusters to merge at each stage that minimize the increase in entropy over all pairs. The calculations are facilitated by the fact that there is a simple relationship between the posterior cluster membership probabilities of a point for any two clusters, and the corresponding value for the merged cluster: the latter is just the sum of the two former. Merging continues as long as the resulting increase in entropy is small, and stops when a merge leads to a large increase in entropy. Using an elbow or change point in the plot of the entropy against the number of observations has worked well for deciding when to stop. This approach is implemented in the `clustCombi` function in the `mclust` package.

If clusters g and g' from the G-cluster solution are combined, the values of $\hat{z}_{i,h}$ remain the same for every group h except g and g'. The new cluster $g \cup g'$ then has the following conditional probability:

$$\hat{z}_{i,g \cup g'}^{[G]} = \hat{z}_{i,g}^{[G]} + \hat{z}_{i,g'}^{[G]}. \tag{3.7}$$

Then the resulting entropy is

$$-\sum_{i=1}^{n} \left\{ \hat{z}_{i,g}^{[G]} + \hat{z}_{i,g'}^{[G]} \log \left(\hat{z}_{i,g}^{[G]} + \hat{z}_{i,g'}^{[G]} \right) + \sum_{h \neq g, g'} \hat{z}_{i,h}^{[G]} \log(\hat{z}_{i,h}^{[G]}) \right\}. \tag{3.8}$$

Thus, the two clusters g and g' to be combined are those that maximize the criterion

$$-\sum_{i=1}^{n} \left\{ \hat{z}_{i,g}^{[G]} \log(\hat{z}_{i,g}^{[G]}) + \hat{z}_{i,g'}^{[G]} \log(\hat{z}_{i,g'}^{[G]}) \right\} + \sum_{i=1}^{n} \hat{z}_{i,g \cup g'}^{[G]} \log(\hat{z}_{i,g \cup g'}^{[G]}) \tag{3.9}$$

among all pairs of clusters (g, g'). Then the conditional classification probabilities, $\hat{z}_{i,g}^{[G-1]}, i = 1, \ldots, n; g = 1, \ldots, G - 1$, can be updated.

At the first step of the combining procedure, G is the number of components selected by BIC, and $\hat{z}_{i,g}^{[G]}$ is the conditional probability that observation i arose from the gth mixture component. But as soon as at least two components are merged into a cluster g, so that G is smaller than the value selected by BIC, $\hat{z}_{i,g}^{[G]}$ is the conditional probability that the ith observation belongs to one of the merged components in cluster g.

Hennig (2010) has proposed several other criteria within the same general framework. These include the Bhattacharyya distance between two cluster densities, directly estimated misclassification probabilities (DEMP), prediction strength (Tibshirani and Walther, 2005), two versions of the ridgeline method of Ray and Lindsay (2005) to prohibit merging if the resulting merged cluster has a density that is not unimodal, and the method of Tantrum et al. (2003) that is based on the dip test of Hartigan and Hartigan (1985) for unimodality. In a simulation study comparing these methods, Hennig (2010) found the Bhattacharyya distance method to have good performance across a wide range of settings. It is also computationally cheap. The prediction strength method also had good performance, but was by far the most computationally expensive of the methods considered. These criteria are implemented in the `fpc` R package (Hennig, 2015a).

We illustrate the methods by applying them to a simulated data set shown in Figure 3.16. These data were used in Baudry et al. (2010, Section 4.1) and are available in the `mclust` package. Visually, this data set appears to contain four well-separated clusters. However, the clusters in the top left and bottom right of the plot are far from Gaussian, and instead appear to follow a cross-like pattern. The BIC solution chooses six mixture components. It divides each of the two cross-like clusters into two mixture components, representing each cluster by a mixture of two Gaussian distributions. The ICL solution (not shown here) chooses four clusters, which is the right number. However, it represents each of the two cross-like clusters by a nearly spherical Gaussian, which clearly is not a very good fit to the data.

The entropy values for different numbers of clusters from the entropy merging method of Baudry et al. (2010) are shown in Figure 3.17. The left panel shows how the entropy changes with the number of clusters, starting with one cluster on the left of the plot (with zero entropy by definition), and going to six clusters on the right of the plot. The entropy is plotted against the cumulative sum of the number of observations merged at each step. On this scale, the plot is expected to increase slowly up to the true number of clusters, and then to increase more quickly and roughly linearly beyond that. The red dotted line comes from a fitted linear change-point model. The change-point clearly occurs at $G = 4$ clusters, which is the number indicated by visual inspection.

The right panel of Figure 3.17 shows the same information in a different form, namely the differences in entropy in going from G to $G + 1$ clusters, normalized by the number of observations in the merged clusters. We see that the difference

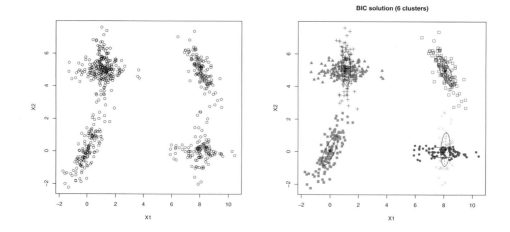

Figure 3.16 Simulated data (left) and BIC-best solution with six mixture components (right).

Listing 3.6: Cluster merging by entropy method for simulated data

```
# Loading libraries and data
library(mclust)
data(Baudry_etal_2010_JCGS_examples)

# Clustering and merging
output <- clustCombi(data = ex4.4.1)

# plots the hierarchy of combined solutions, then some "entropy
    plots" which may help one to select the number of classes
plot(output)
```

in entropy is small in going from one to two clusters, from two to three clusters and from three to four clusters. However, going from four to five clusters involves a big increase in entropy, again suggesting that merging down to four clusters improves the entropy substantially, but merging beyond that does not bring big improvements. Overall, this suggests that four clusters is the best choice.

Figure 3.18 shows the results of merging mixture components, as one goes from the BIC solution of six clusters down to two clusters. The selected four-cluster solution combines the components that make up the cross-like clusters, as desired. The number of clusters is the same as that selected by ICL, but the resulting densities are different. In the ICL solution, each of the two cross-like clusters is represented by a single, nearly spherical Gaussian distribution, which is clearly not fully satisfactory. In the merged four-cluster solution, each of the two cross-like clusters is itself represented by a mixture of two Gaussian distributions, retaining the true, cross-like nature of the resulting clusters. Figures 3.17 and 3.18 were produced by the code in Listing 3.6.

Normalized entropy plot

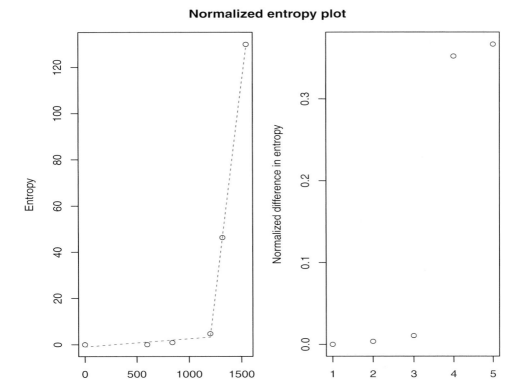

Figure 3.17 Simulated data. Left: entropy values for the G-cluster combined solution, using the entropy merging method of Baudry et al. (2010), plotted against the cumulative sum of the number of observations merged at each step. The dashed line shows the best piecewise linear fit, with a breakpoint at $G = 44$ clusters. Right: rescaled differences between successive entropy values, equal to $(\text{Ent}(G+1)-\text{Ent}(G))/$ Number of merged observations.

We now apply the method of Hennig (2010) where clusters are merged until the Bhattacharyya distance between all remaining pairs of clusters is greater than a user-specified threshold. It is implemented using the code in Listing 3.7 and gives the results shown there. It also selects four clusters, as do the DEMP method, and the dip test method. The ridgeline methods, however, select five clusters, which seems less satisfactory.

We now return to the Old Faithful data, where BIC selected three mixture components but visual inspection suggests two clusters. All the merging methods we have discussed here point to merging the two mixture components at the top right of the top left panel of Figure 3.15, as desired. ICL selects three clusters, each represented by a single Gaussian, as shown in the bottom right panel of Figure 3.15, while the merging solution has the density shown in the top right

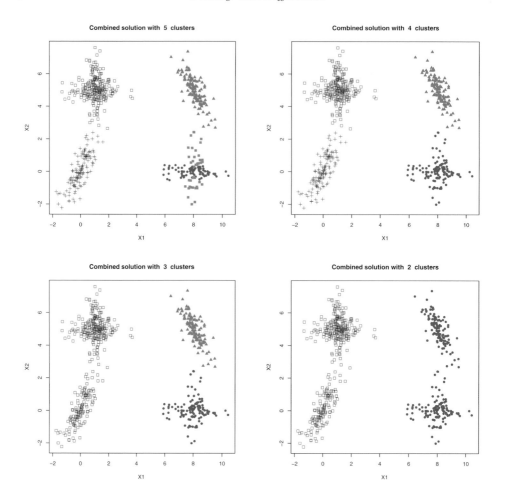

Figure 3.18 Simulated data: results from merging for 5, 4, 3 and 2 clusters.

panel. Thus the merging solution seems to provide a better representation of the apparently non-Gaussian cluster at the top right of the data.

We have found that, even if ICL selects a model that may not be fully satisfactory because it represents each cluster by a single Gaussian distribution, even when the cluster is non-Gaussian, it tends to do well in selecting the right number of *clusters* (as opposed to the right number of mixture components). Thus a hybrid approach in the case where BIC selects more mixture components than ICL would be as follows: use ICL to select the number of clusters, use BIC to select the number of mixture components and the covariance model, and select the final clustering model by merging the components as described here, until the number of clusters selected by ICL is reached.

Listing 3.7: Cluster merging by Hennig–Bhattacharyya method for simulated data

```
# Loading libraries and data
library (mclust)
library (fpc)
data (Baudry_etal_2010_JCGS_examples)

# Clustering and merging
ex4.1.Mclust <- Mclust (ex4.1)
ex4.1.bhat <- mergenormals (ex4.1, method="bhat", clustering=
    ex4.1.Mclust$classification, probs = ex4.1.Mclust$parameters$pro
    , muarray = ex4.1.Mclust$parameters$mean, Sigmaarray =
    ex4.1.Mclust$parameters$variance$sigma, z = ex4.1.Mclust$z)

# Displaying results
summary (ex4.1.bhat)

 Method:  bhat , cutoff value:  0.1
 Original number of components:  6
 Number of clusters after merging:  4
 Values at which clusters were merged:
[1,]    5 0.55978081
[2,]    4 0.55617411
[3,]    3 0.01132068
 Components assigned to clusters:
[1,]    1
[2,]    2
[3,]    3
[4,]    3
[5,]    4
[6,]    1
```

3.4 Bibliographic Notes

Outliers

The model consisting of a mixture of a multivariate normal distribution with a uniform distribution was mentioned and used by Titterington et al. (1985) to model a data set, but not for identifying outliers. The model consisting of a mixture of several multivariate normal distributions with a uniform distribution was proposed by Banfield and Raftery (1993) and Dasgupta and Raftery (1998) for identifying and removing or down-weighting outliers, and further developed by Fraley and Raftery (1998, 2002). The robustness properties of the model were studied by Hennig (2004), in particular its breakdown point (Hampel, 1971). Hennig (2004) showed that the method is theoretically not breakdown-robust, although in practice very extreme outliers are needed to spoil it. It has been used in many applications, including in a large-scale image processing context by Horaud et al. (2011). It was extended to accommodate clusters centered around curves rather than lines in the bivariate case by Stanford and Raftery (2000), using the concept of principal curves.

The nearest neighbor cleaning method for identifying outliers, used here to initialize model-based clustering with noise, was introduced by Byers and Raftery (1998). It was extended to robust covariance estimation by Wang and Raftery (2002), introducing simulated noise points to reduce bias.

Several alternatives to the uniform noise model for outliers have been proposed. Hennig (2004) proposed replacing the uniform distribution here by an improper uniform component, an idea further developed by Hennig and Coretto (2008); Coretto and Hennig (2010, 2011). An approach based on iteratively trimming potential outliers was proposed by García-Escudero et al. (2008, 2010, 2011); it is implemented in the `tclust` R package (Fritz et al., 2012). Evans et al. (2015) have proposed a method for outlier detection in model-based clustering based on leave-one-out re-estimation of cluster parameters.

The normal-uniform mixture model for outliers has also been used in other contexts. For example, Dean and Raftery (2005) used it as the basis for a method for detecting differentially expressed genes in microarray data. There the differentially expressed genes were analogous to the outliers and were modeled with the uniform distribution, while the normal distribution was used to model the non-differentially expressed genes.

An alternative approach is based on replacing the Gaussian mixture components by t-distributions (Peel and McLachlan, 2000). This does not aim to identify and remove outliers and then identify compact clusters, but rather to model mixture components that may include outlying data points, and so its goal is rather different from the other methods discussed in this chapter. We discuss it in more detail along with more recent extensions in Chapter 9.

Degeneracy

The fact that the likelihood for normal mixture models has multiple infinite modes that are not interesting was first pointed out by Day (1969) in the one-dimensional case and further elucidated by Titterington et al. (1985). Redner and Walker (1984) showed that interior local maxima of the likelihood can define estimators that are consistent, but that the spurious infinite maxima on the edge of the parameter space do not.

Methods for dealing with this problem can be classified as constraint methods, Bayesian methods, penalty methods, and others. Hathaway (1985) was the first to propose a constraint method. He proposed maximizing the likelihood subject to the constraints that the eigenvalues of $\Sigma_g \Sigma_h^{-1}$ be greater than or equal to some minimum value $c > 0$, and developed an algorithm for the univariate case (Hathaway, 1986b). Ingrassia and Rocci (2007) showed how the Hathaway constraints can be implemented directly at each iteration of the EM algorithm. Gallegos and Ritter (2009b,a) proposed different, but conceptually similar, constraints based on the Löwner matrix ordering, essentially that no group covariance matrix differs from any other by more than a specified factor.

Fraley and Raftery (2003) proposed computing the condition number for each estimated covariance matrix and not reporting the results for any covariance model and number of components for which any of the condition numbers is zero (or below relative machine precision); this is the default approach in `mclust`.

García-Escudero et al. (2015) proposed a constraint on the ratio of the maximum to the minimum of all the eigenvalues of all the mixture component covariance matrices. This controls differences between groups and singularity of individual components at the same time. They described an algorithm for implementing it. Cerioli et al. (2018) proposed a refinement of this method that modifies the likelihood penalty for model complexity to take account of the higher model complexity that a larger constraint entails.

In addition to Fraley and Raftery (2007a), several other Bayesian approaches have used the EM algorithm to estimate the posterior mode for mixture models. Roberts et al. (1998) used a Dirichlet prior on the mixing proportions and a non-informative prior on the elements of the means and covariances, while Figueiredo and Jain (2002) used non-informative priors for all the parameters, and Brand (1999) proposed an entropic prior on the mixing proportions. These methods work by starting with more components than necessary, and then pruning those for which the mixing proportions are considered negligible.

Chi and Lange (2014) proposed a prior on the covariance matrices that discourages their nuclear norm (sum of their eigenvalues) from being too large or too small. It penalizes large values of a weighted average of the nuclear norm of Σ_g and the nuclear norm of Σ_g^{-1}, leading to a method for covariance regularized model-based clustering. Zhao et al. (2015) proposed a hierarchical BIC using the posterior mode based on an inverse Wishart prior for the covariance matrices and non-informative priors for the component means and mixing proportions.

Bayesian estimation for mixture models can also be done via Markov chain Monte Carlo (MCMC) simulation, typically using conjugate priors on the means and covariances similar to those of Fraley and Raftery (2007a) (Lavine and West, 1992; Bensmail et al., 1997; Dellaportas, 1998; Bensmail and Meulman, 2003; Zhang et al., 2004; Bensmail et al., 2005). These can suffer from label-switching problems (Celeux et al., 2000; Stephens, 2000b) and can be computationally demanding. In our experience, use of the posterior mode is simpler and less computationally expensive, provides most of the benefits of the Bayesian approach, and avoids most of the difficulties.

Ciuperca et al. (2003) and Chen and Tan (2009) proposed maximum penalized likelihood methods to overcome the unboundedness of the likelihood, with purpose-built penalty functions. Ruan et al. (2011) and Bhattacharya and McNicholas (2014) proposed using ℓ_1 penality functions. Halbe et al. (2013) proposed an EM algorithm with Ledoit–Wolf-type shrinkage estimation of the covariance matrices (Ledoit and Wolf, 2003, 2004, 2012).

Rather than dealing with the problem algorithmically, McLachlan and Peel (2000) proposed monitoring the local maximizers of the likelihood function and carefully evaluating them. Seo and Kim (2012) and Kim and Seo (2014) proposed a systematic algorithmic approach to selection of the best local maximizer to avoid spurious solutions.

Non-Gaussian Clusters and Merging

One of the earliest authors to propose merging mixture components for clustering was Li (2005), who assumed that the true number of clusters is known in advance, unlike the other methods discussed here. She used k-means clustering to merge components, which works well when the mixture components are spherical, but may have problems when they are not. This method was compared with the entropy merging method by Baudry et al. (2010). A related method based on finding modes was proposed by Li et al. (2007a).

All the methods discussed here are effectively multilevel clustering methods, in which each cluster can itself be represented by a mixture model. Jörnsten and Keleş (2008) proposed a multilevel mixture model that provides parsimony in modeling both the clusters and the relationships between them, allowing a different number of clusters at each level of an experimental factor of interest.

Various criteria for finding clusters from mixture models have been proposed, other than the ones described here. Longford and Bartošová (2014) proposed a confusion index, Zeng and Cheung (2014) used a minimum message length criterion, while Scrucca (2016b) proposed a method for finding connected components based on the density estimate from the model-based clustering itself. Naim et al. (2014) proposed a merging method based on a Kullback–Leibler divergence criterion, while Mosmann et al. (2014) applied it to flow cytometry data.

The problem of merging mixture components has also arisen in the context of Hidden Markov Models (Ackerson and Fu, 1970; Harrison and Stevens, 1971; West and Harrison, 1989; Volant et al., 2014). There sequential updating using, for example, the multiprocess Kalman filter, leads to a posterior distribution that is a finite mixture, but with a number of components that explodes exponentially as time proceeds. There is a substantial literature on how to approximate the resulting mixture with a more manageable form. One approach is to reduce the number of mixture components, approximating a mixture with many components by one with fewer components. Recent contributions include the merging algorithm of Runnals (2007), the joining algorithm of Salmond (2009), and the variational Bayes approach of Bruneau et al. (2010). Note that this is very different from the goal of clustering that we are discussing here.

A more direct approach to model-based clustering with non-Gaussian clusters would be to use mixture models with non-Gaussian component distributions. The merging approach relies on the fact that one can approximate a continuous non-Gaussian distribution arbitrarily closely by a finite mixture of Gaussian distribution, and that model-based clustering with Gaussian distributions is fairly mature. Model-based clustering with non-Gaussian distributions is developing rapidly, and we discuss it in more detail in Chapter 9. The choice raises philosophical as well as technical issues, and some of these have been discussed by Hennig (2013, 2015b).

4

Model-based Classification

Supervised classification, also known as discriminant analysis, is an important domain of supervised learning which consists of predicting the values of categorical *outputs* from the values of a set of *input* variables. In order to construct a prediction rule, a training data set $(y_1, z_1), \ldots, (y_n, z_n)$ of n iid labeled observations is available, where $y_i \in \mathbf{R}^d$ are the inputs and the labels $z_i \in \{0, 1\}^G$ for $i = 1, \ldots, n$ with G known are the outputs. In particular, $z_{ig} = 1$ if observation belongs to group g, where $g = 1, \ldots, G$, and $z_{ig} = 0$ otherwise. This training data set is used to classify new observations y into one of the G classes.

Thus, supervised classification is the set of methods aiming to predict the output of an individual based on its values for the d predictors. The precise purpose of supervised classification depends on the context. In most cases, good prediction is of the greatest interest, and the aim is to minimize the misclassification error rate. But in some cases, providing a relevant interpretation of the G classes from the predictors is of specific interest. These two goals could be in conflict in some cases.

Examples of popular and successful classification methods not based on a probabilistic model are kernel methods. But here naturally, we concentrate on model-based methods.

4.1 Classification in the Probabilistic Framework

In the probabilistic framework, assumptions are made on the joint distribution $p(y_i, z_i)$ or on the conditional distribution $p(z_i|y_i)$ of the label z_i knowing the d predictors $y_i = (y_{i1}, \ldots, y_{id})$. Then, assuming equal misclassification costs, the optimal Bayes classifier consists of assigning a unit i to the class g by maximizing the conditional probability $P(z_{ig} = 1 \mid y_i)$ that this unit belongs to class g knowing its observed vector y_i.

Different approaches are possible to approximate $P(z_{ig} = 1|y_i)$. *Predictive* or *discriminative* approaches estimate these conditional probabilities directly using a parametric model such as logistic regression (Hosmer et al., 2013), or a nonparametric approach such as classification trees (Breiman et al., 1984), and related methods such as random forests (Breiman, 2001). *Generative* methods derive the conditional probabilities $P(z_{ig} = 1|y_i)$ from the class densities $f_g(y_i)$

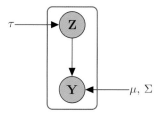

Figure 4.1 Graphical representation of the LDA and QDA models.

through the Bayes formula:

$$P(z_{ig} = 1|y_i) \propto \tau_g f_g(y_i),$$

where τ_g is the prior probability of class g. The class densities can be estimated by a nonparametric method such as the k-nearest neighbors method (Ripley, 1996, Chapter 6), or with a parametric model.

The first documented methods in discriminant analysis were model-based classification methods. These methods use parametric models for which the marginal density of the observations can be written as a finite mixture of parameterized densities, namely

$$p(y_i|\tau, \theta) = \sum_{g=1}^{G} \tau_g f_g(y_i|\theta_g), \qquad (4.1)$$

where θ_g is the parameter of the gth class conditional density, $\tau = (\tau_1, \ldots, \tau_G)$ denotes the mixing proportions, and $\theta = (\theta_1, \ldots, \theta_G)$. Thus, this marginal distribution is a finite mixture of densities with parameters (τ, θ).

The first classification models assumed that the class conditional densities are Gaussian (Anderson, 2003, Chapter 6). When the class covariance matrices are assumed to be equal, it leads to the popular linear discriminant analysis (LDA) which was proposed and studied first by Wald (1944) for $G = 2$ and by Von Mises (1945) for $G \geq 2$. When the class covariance matrices are unequal, it leads to quadratic discriminant analysis (QDA) and the boundaries between the classes are no longer linear, but quadratic.

In this chapter, we restrict attention to model-based classification methods.

4.1.1 Generative or Predictive Approach

The model-based classification methods presented in this chapter are instances of the generative approach supervised classification. This consists of modeling the joint distribution $p(y_i, z_i)$, while the predictive approach directly evaluates the conditional distribution $p(z_i \mid y_i)$ in order to predict z_i for new values of y_i.

The predictive point of view has been much developed recently and has given rise to many popular methods. Most of them, such as k-nearest neighbors, neural networks, support vector machine, random forests are nonparametric and do not assume any model for the joint distribution $p(y_i, z_i)$ (Vapnik, 1998; Bishop, 2006; Hastie et al., 2009). New methods such as deep learning for neural networks are taking advantage of rapidly increasing computer power to address increasingly complex problems involving larger data sets and more complex models.

In this landscape, model-based classification methods provide often slightly higher misclassification error rates than the best predictive methods due to differences between the models and the true distribution of the data. But model-based classification has some advantages: interpretability and simplicity combined with good if not optimal performance, and the ability to deal with partially labeled data. Often, it provides almost optimal misclassification error rates with parsimonious and understandable models. The decision rules of model-based classifiers are also easier to interpret. Hybrid approaches aiming to achieve the best of both worlds have also been proposed (Bouchard and Triggs, 2004; Lasserre et al., 2006).

Moreover, a model-based approach is useful in the semi-supervised classification context where many data are unlabeled (see Chapter 5). By using the information provided by cheap and widespread *unlabeled data*, semi-supervised classification aims to improve the classifiers' performance. Most predictive approaches are unable to take unlabeled data into account without additional assumptions on the marginal distribution $p(y)$ of the predictors.

In a semi-supervised setting it is natural to adopt the generative point of view, since it leads to modeling the joint distribution

$$p(y_i, z_i) = p(z_i)p(y_i|z_i).$$

This in turn leads to writing the marginal distribution $p(y_i)$ as a mixture model

$$p(y_i) = \sum_{z_i} p(z_i)p(y_i|z_i).$$

Thus, the maximum likelihood estimate of the model parameters can be simply computed using the EM algorithm. The straightforward formulae of EM in the semi-supervised context are given in Chapter 5 and details can be found in Miller and Browning (2003).

4.1.2 An Introductory Example

We illustrate the use of a nonparametric method and a parametric model-based classification method on a simple data set, originally analyzed by Habbema et al. (1974). The objective is to find a procedure for detecting potential haemophilia A carriers on the basis of two measured variables, $Y_1 = \log_{10}(\text{AHF activity})$ and $Y_2 = \log_{10}(\text{AHF-like antigen})$. The first class of $n_1 = 30$ women consists of known non-carriers (normal class) and the second class of $n_2 = 45$ women is selected from known haemophilia A carriers (obligatory carriers). This data set is depicted in Figures 4.2 and 4.3.

The nonparametric classification method we consider is a k-nearest neighbors (NN) method which allocates an observation to the majority class among its

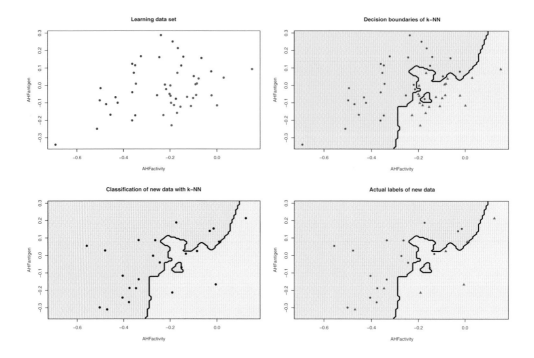

Figure 4.2 The panel on the top left displays the training data set for the haemophilia data set (no carriers in blue and carriers in red); the panel on the top right displays the boundary of the 1NN decision rule on the training data set; the bottom left panel displays the test data set and the bottom right panel its assignment.

k-nearest neighbors in the training set. In this illustration, we took $k = 1$. The parametric model-based classification method is the quadratic discriminant analysis (QDA) which consists of assuming that each class follows a Gaussian distribution with no restrictions on its parameters.

In order to assess the predictive ability of the classifiers, the whole data set was divided at random into a training data set (75%) and a test data set (25%). The training data set was used to design the classifiers while the test data set was used to estimate their misclassification rates (see Section 4.7). Figure 4.2 displays the classification with the 1NN method. It appears that the 1NN classifier overfitted the training data set. It produced no error on the training data set while it led to 5 misclassified observations on the test data set (4 for the non carrier class on 8 observations and 1 for the carrier class on 16 observations).

Figure 4.3 displays the classification with the QDA method. Contrary to the 1NN classifier, there is no sensitive difference between the misclassification error rates on the training and the test data sets with QDA. On the test data set, it leads to 4 misclassified observations (3 for the noncarrier class on 8 observations and 1 for the carrier class on 16 observations).

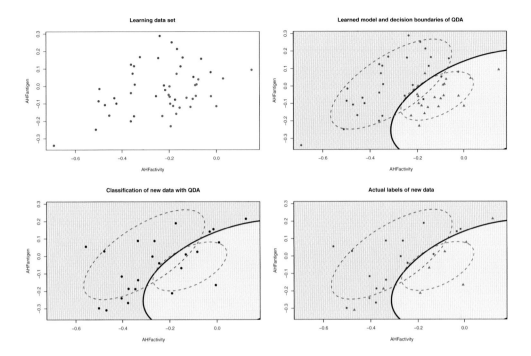

Figure 4.3 The panel on the top left displays the training data set for the haemophilia data set (non carriers in blue and carriers in red); the panel on the top right displays the boundary of the QDA decision rule on the training data set; the bottom left panel displays the test data set and the bottom right panel its assignment.

4.2 Parameter Estimation

In the model-based classification context, the estimation task is easier than in the clustering context because both the number G of classes and the labels $z_1, \ldots, z_n$ are known. In this context, the n labeled observations $(y_1, z_1), \ldots, (y_n, z_n)$ are identically and independently distributed random observations with distribution

$$p(y_i, z_i | \tau, \theta) = \prod_{g=1}^{G} \tau_g^{z_{ig}} [f_g(y_i | \theta_g)]^{z_{ig}}. \tag{4.2}$$

The way the whole vector parameter (τ, θ) is estimated depends on the way the training data set is sampled.

Sampling Schemes

In the *mixture* sampling, the training data set arises from a random sample from the whole population under study, while in *retrospective* sampling, the training data set is the concatenation of G independent sub-samples with fixed sizes n_g and distribution $f_g(y_i | \theta_g)$ for $k = 1, \ldots, G$.

With mixture sampling, the y_i are sampled from the distribution

$$p(y_i|\tau,\theta) = \sum_{g=1}^{G} \tau_g f_g(y_i|\theta_g).$$

Mixture sampling is more natural and the maximum likelihood estimate of τ_g is the proportion n_g/n for $g = 1,\ldots,G$, where n_g denotes the number of objects arising from the gth class in the training data set. However, when rare classes are present, retrospective sampling is preferred. But it requires the class proportions to be known since they cannot be estimated from the fixed proportions n_g and, in such cases, the vector parameter to be estimated reduces to $\theta = (\theta_1,\ldots,\theta_G)$.

An estimate $\hat{\theta}$ of θ is obtained by the maximum likelihood method

$$\hat{\theta} = \arg\max_{\theta} L(\tau,\theta|y,z), \tag{4.3}$$

where the log-likelihood function $L(\tau,\theta|y,z)$ is defined by

$$L(\tau,\theta|y,z) = \sum_{i=1}^{n}\sum_{g=1}^{G} z_{ig}\log[\tau_g f_g(y_i|\theta_g)]. \tag{4.4}$$

The maximum likelihood estimate of θ depends on the parameterized model being used. For some models, $\hat{\theta}$ is not available in closed form. The computation needed to get $\hat{\theta}$ is equivalent to a single M-step of the EM algorithm in a model-based clustering context.

As soon as $\hat{\theta}$ is derived, it is possible to assign any new point y_{n+1} with unknown label to one of the G classes by the maximum a posteriori (MAP) procedure. The estimated conditional probability, $\hat{z}_{(n+1)g}$, that y_{n+1} arises from the gth class is

$$\hat{z}_{(n+1)g} = \frac{\hat{\tau}_g f_g(y_{n+1}|\hat{\theta}_g)}{\sum_{\ell=1}^{G} \hat{\tau}_\ell f_g(y_{n+1}|\hat{\theta}_\ell)}. \tag{4.5}$$

The MAP procedure consists of assigning y_{n+1} to the class maximizing this conditional probability, namely

$$z^*_{(n+1)\,g} = \begin{cases} 1 \text{ if } g = \arg\max_{\ell=1\ldots,G} \hat{z}_{(n+1)g} \\ 0 \text{ otherwise.} \end{cases} \tag{4.6}$$

4.3 Parsimonious Classification Models

4.3.1 Gaussian Classification with EDDA

When the variables are continuous, it is possible to take advantage of the numerous Gaussian models proposed in the model-based clustering context to get a collection of classification models of different complexity and with simple geometric interpretation. This leads to the so-called EDDA (*Eigenvalue Decomposition Discriminant Analysis*) method which makes use of the eigenvector decomposition of the class variance matrices

$$\Sigma_g = \lambda_g D_g A_g D_g^T, \tag{4.7}$$

where $\lambda_g = |\Sigma_g|^{1/d}$, D_g is the eigenvector matrix of Σ_g and A_g is a diagonal matrix with determinant 1, whose diagonal contains the normalized eigenvalues of Σ_g in decreasing order (see Bensmail and Celeux, 1996).

Recall that the parameter λ_g determines the volume of class g, D_g its orientation and A_g its shape. By allowing some of these elements to vary or not between classes, we get different, more or less parsimonious, and easy to interpret models. Moreover, by considering particular situations (diagonal or proportional to identity variance matrices), we get other interesting and highly parsimonious models. Overall, there are 14 competing models, and EDDA chooses the model that minimizes the cross-validated error rate, the parameters being estimated by maximum likelihood.

In many cases, EDDA provides relevant parsimonious classifiers with simple geometric interpretation and the interest of EDDA increases when the number of classes is large. For instance, the standard linear (LDA) and quadratic (QDA) classifiers (McLachlan, 1992) are special cases of the EDDA methodology. The linear classifier is obtained by assuming equal volumes, shapes and orientations for the covariance matrices Σ_g and the quadratic classifier is obtained when the Σ_g are allowed to vary freely.

As an illustration, some models available with EDDA are applied in Listing 4.1 on the diabetes data set, using the `Rmixmod` package. The classes to be recovered are the three classes of the clinical classification. For this classification problem, we assumed that the data have been drawn according to the mixture sampling and thus the class proportions are free. Figure 4.4 displays the classification boundaries of the quadratic classifier ($p_g L_g C_g$ in the terminology of the `mixmodLearn` function), the classifier allowing free class volumes ($p_g L_g C$) and the linear classifier with free proportions ($p_g L C$) in the plane of the first two variables.

In the standard version of EDDA (Bensmail and Celeux, 1996), one of the available models is selected by minimizing the cross-validated error rate (see Section 4.7). This criterion provides usually good results especially in a small sample setting, but other model selection criteria such as BIC could also be used. For the diabetes data set, we compare the behavior of BIC and the cross-validated error rate (CV). Note that in Figure 4.5 and Listing 4.2 we consider the criterion $-\text{BIC}$ because BIC is to be maximized in `Rmixmod`. For this data set, there is little difference between the two criteria (see Figure 4.5) and both of them select the most complex model QDA.

4.3.2 Regularized Discriminant Analysis

A standard way to design classifier estimates with reduced variances, when the size of the training data set is small, consists of replacing plug-in estimates with shrunk estimates. In the Gaussian framework, Friedman (1989) proposed regularized discriminant analysis (RDA) based on the following regularization of the class covariance matrices

$$\tilde{\Sigma}_g(\lambda, \gamma) = (1 - \gamma)\hat{\Sigma}_g(\lambda) + \gamma \left(\frac{\text{tr}(\hat{\Sigma}_g(\lambda))}{d} \right) I_d,$$

Listing 4.1: Classification boundaries for EDDA models

```
# Loading libraries and data
library(MBCbook)
data(diabetes)
diabetes.data = diabetes[2:4]
diabetes.class = unlist(diabetes[1])

# Learning the classifier with Mixmod
# LDA = 'Gaussian_pk_L_C', QDA = 'Gaussian_pk_Lk_Ck'
res = mixmodLearn(diabetes.data[,2:3],diabetes.class,models=
    mixmodGaussianModel(listModels='Gaussian_pk_Lk_C'))

# Drawing the classification boundaries
prec = 150; Z = diabetes.data[,2:3]
x1 = seq(-200,1800,length=prec)
x2 = seq(-200,900,length=prec)
s = expand.grid(x1,x2); s <- as.data.frame(s)
P = mixmodPredict(as.data.frame(s),res@bestResult)@proba
plot(diabetes.data[,2:3],type='n',xlim=c(-100,1700),ylim=c(-100
    ,800))
pastel <- .9
points(s,type='p',pch=16,col=c(rgb(1,pastel,pastel),rgb(pastel,1,
    pastel),rgb(pastel,pastel,1))[max.col(P)])
points(diabetes.data[,2:3],col=as.numeric(diabetes.class)+1,pch=19,
    xlim=c(-100,1700),ylim=c(-100,800))
for (i in 1:3){
  T = P[,i] - apply(P[,-i,drop = FALSE],1, max)
  contour(x1,x2,matrix(T,prec,prec),level=0,add=1,col="black",lwd
      =3,drawlabels=0);
}
title(main="EDDA with model 'pk_Lk_Ck'"); box()
```

where

$$\hat{\Sigma}_g(\lambda) = \frac{(1-\lambda)(n_g-1)\hat{\Sigma}_g + \lambda(n-G)\hat{\Sigma}}{(1-\lambda)(n_g-1) + \lambda(n-G)},$$

$\hat{\Sigma}_g$ are the estimates of the covariance matrices for the QDA model for $g = 1, \ldots, G$, that is

$$\hat{\Sigma}_g = \frac{\sum_{i=1}^{n} z_{ig}(y_i - \hat{\mu}_g)(y_i - \hat{\mu}_g)^T}{n_g},$$

where

$$\hat{\mu}_g = \frac{\sum_{i=1}^{n} z_{ig} y_i}{n_g},$$

and $\hat{\Sigma}$ is the common covariance matrix estimate for the LDA model, namely

$$\hat{\Sigma} = \frac{\sum_{g=1}^{G} \sum_{i=1}^{n} z_{ig}(y_i - \hat{\mu}_g)(y_i - \hat{\mu}_g)^T}{n}.$$

The complexity parameter λ ($0 \leq \lambda \leq 1$) controls the contribution of QDA and LDA, while the regularization parameter γ ($0 \leq \gamma \leq 1$) controls the amount of

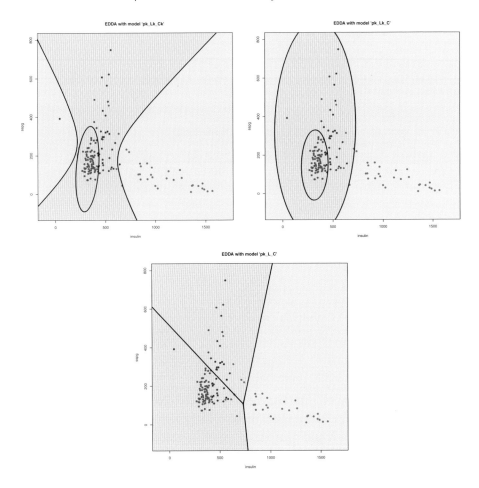

Figure 4.4 Classification boundaries for EDDA models $p_gL_gC_g$, p_gL_gC and p_gLC on the diabetes data set.

shrinkage of the eigenvalues towards equality, since $\text{tr}(\hat{\Sigma}_g(\lambda))/d$ is the mean of eigenvalues of $\hat{\Sigma}_g(\lambda)$.

Thus, with fixed $\gamma = 0$, varying λ leads to an intermediate method between LDA and QDA. Now, with fixed $\lambda = 0$, increasing γ leads to more biased estimates of the eigenvalues of the covariance matrices. For $\lambda = 1$ and $\gamma = 1$, we get a quite simple classifier which consists of assigning any vector to the class whose center is the nearest.

Figure 4.6 illustrates the role of the parameters λ and γ on simulated data. Finally, note that it is also possible to use the Moore–Penrose pseudo-inverse of $\hat{\Sigma}$ instead of the usual inverse $\hat{\Sigma}^{-1}$. The reader can also refer to Mkhadri et al. (1997) who provides a comprehensive overview of regularization techniques in discriminant analysis.

Listing 4.2: Model selection with BIC and CV

```
# List of models and learning with mixmodLearn
models = mixmodGaussianModel(family='all')
res = mixmodLearn(diabetes.data,diabetes.class,models=models,
    criterion=c('CV','BIC'))

# Collecting the BIC and CV values
BIC = CV = rep(NA,length(models@listModels))
for (i in 1:length(res@results)){
  ind = which(res@results[[i]]@model == models@listModels)
  CV[ind] = res@results[[i]]@criterionValue[1]
  BIC[ind] = - res@results[[i]]@criterionValue[2]
}

# Plot of the results
par(mfrow=c(2,1))
plot(BIC,type='b',xlab='',xaxt='n',col=2); axis(1,at=1:length(
    res@results),labels=substr(models@listModels,10,30),cex.axis=0.8
    ,las=2)
abline(v=which.max(-BIC), col=1, lty=2)
plot(CV,type='b',xlab='',xaxt='n',col=3); axis(1,at=1:length(
    res@results),labels=substr(models@listModels,10,30),cex.axis=0.8
    ,las=2)
abline(v=which.max(CV), col=1, lty=2)
```

The solution based on regularization does not have the same drawbacks as dimension reduction and can be used more widely (see Chapter 8).

The tuning parameters λ and γ are computed by minimizing the cross-validated error rate. In practice, this method performs well compared with LDA and QDA for small sample sizes. The drawbacks of RDA are that it provides classifiers that can be difficult to interpret and the methods are not very sensitive to the parameters λ and γ; Figure 4.7 provides an illustration of the similar performance for large ranges of values of the parameters.

The next illustration allows us to compare the EDDA and RDA behaviors on the haemophilia data. The data set presented in Section 4.1.2 consists of a population of 75 women (45 were haemophilia A carriers and 30 were not) described by two variables $y_1 = 100 \log(AHF$ activity$)$ and $y_2 = 100 \log(AHF$ like antigen$)$. We use this data set to illustrate the ability of EDDA to suggest a plausible geometric model. We applied RDA (see Listing 4.3), LDA, QDA and EDDA to this data set. The prior probabilities of the two classes were assumed to be equal.

The cross-validated misclassification risks were .14, .16, .17 and .13 for the four methods. For RDA, the complexity parameter was $\lambda = 0.15$ and the shrinkage parameter was $\gamma = 0.8$. Thus, RDA proposed a shrunk version of a quadratic classifier. But, as shown in Figure 4.7, the two selected tuning parameters vary greatly with the cross-validated samples and so are difficult to interpret. Conversely, the selected model with EDDA was $[\lambda_g DA_g D']$ and the boundaries of this classifier are depicted in Figure 4.8.

Thus EDDA suggests that the best model for this data set assumes different

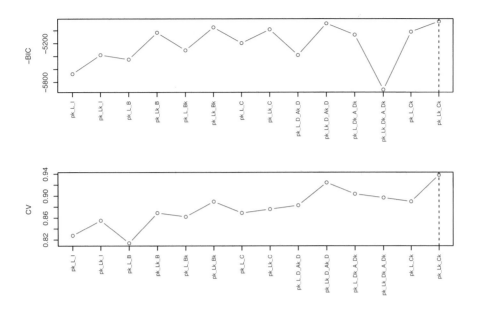

Figure 4.5 Model selection with BIC and CV on the diabetes data.

volumes and shapes and the same orientations for the two classes. Its boundaries
are depicted in Figure 4.8 and show that this proposal seems quite reasonable.
EDDA performs well and provides a clear representation of the class distributions.
For this example, we do not claim that EDDA performs better than RDA or LDA.
The .632 bootstrap, (Efron and Tibshirani, 1997), with 100 bootstrap replications
gives the same estimated misclassification risk (.08) for RDA, LDA and EDDA.
But, we think that, for this data set, EDDA provides a clearer and more relevant
representation of the class distributions than either RDA or LDA.

4.4 Multinomial Classification

When the variables are categorical, the empirical model is simple and natural.
It consists of assuming that the variables follow a multivariate multinomial
distribution within each class. Unfortunately the number of parameters to be
estimated grows rapidly with data dimension. For instance, with d binary variables,
the number of parameters of the empirical model is 2^d per class. Clearly, this
model suffers from the curse of dimensionality (see Chapter 8) and there is the
need to define far simpler models with fewer parameters.

4.4.1 The Conditional Independence Model

A simple way to address the dimensionality problem is to assume that the
variables within an observation are independent conditional on knowing the class
membership of the observation. This leads to the *conditional* or *local independence*

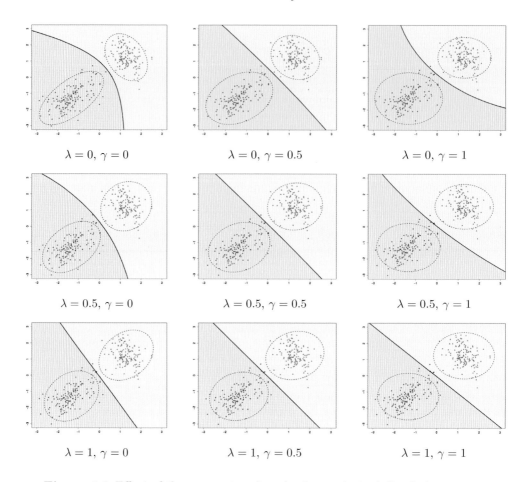

$\lambda = 0, \gamma = 0$ $\lambda = 0, \gamma = 0.5$ $\lambda = 0, \gamma = 1$

$\lambda = 0.5, \gamma = 0$ $\lambda = 0.5, \gamma = 0.5$ $\lambda = 0.5, \gamma = 1$

$\lambda = 1, \gamma = 0$ $\lambda = 1, \gamma = 0.5$ $\lambda = 1, \gamma = 1$

Figure 4.6 Effect of the parameters λ and γ in regularized discriminant analysis (RDA) on simulated data. The parameter λ allows the classifier to vary from QDA to LDA whereas γ controls the ridge regularization.

model which can be regarded as a reference generative classification model for categorical data. This model assumes that the d descriptors are independent conditionally on knowing the classes and that the class densities are multivariate multinomial distributions. This is similar to the latent class model described in Chapter 6 for model-based clustering.

Assuming that variable j has m_j response levels, the data are $(y_1, \ldots, y_n)$ where y_i is coded by the vector $(y_i^j; j = 1, \ldots, d)$ of response level h for each variable j or, equivalently, by the binary vector $(y_i^{jh}; j = 1, \ldots, d; h = 1, \ldots, m_j)$ with

$$\begin{cases} y_i^{jh} = 1 & \text{if } y_i^j = h \\ y_i^{jh} = 0 & \text{otherwise.} \end{cases}$$

Then, the class conditional densities are multivariate multinomial distributions,

Listing 4.3: Regularized discriminant analysis on the `hemophilia` data

```
# Loading libraries and data
library(klaR); library(rrcov)
library(MBCbook)
data(hemophilia)
X = hemophilia[,-3]
cls = as.factor(hemophilia[,3])

# Learning the classifier with RDA
out = rda(X,cls,prior = 1,crossval = TRUE,fold = 10)
out$regularization
out$error.rate[2]

# Analysis of the parameter sensitivity
OUT = matrix(NA,25,3)
for (i in 1:25){
   out = rda(X,cls,prior = 1,crossval = TRUE,fold = 10)
   OUT[i,1:2] = out$regularization
   OUT[i,3] = out$error.rate[2]
}
boxplot(OUT,names=c('Gamma','Lambda','CV error rate'),col=2:4)

# Learning the classifier with EDDA
# (LDA = 'Gaussian_p_L_C', QDA = 'Gaussian_pk_Lk_Ck')
res = mixmodLearn(X,cls,models=mixmodGaussianModel(listModels='
    Gaussian_p_Lk_D_Ak_D'))
```

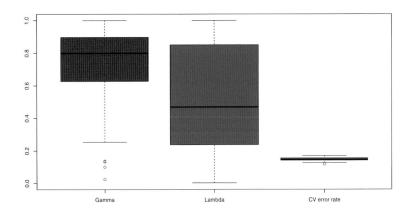

Figure 4.7 Boxplot of the tuning parameters γ and λ of RDA found using repeated cross-validated samples.

for $g = 1, \ldots, G$,

$$f_g(y_i|\boldsymbol{\alpha}_g) = \prod_{j,h}(\alpha_g^{jh})^{y_i^{jh}}, \tag{4.8}$$

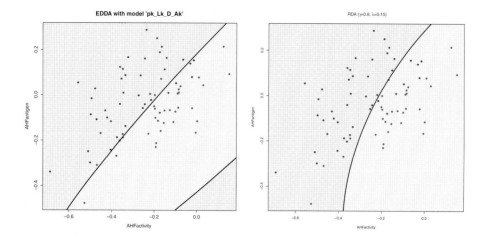

Figure 4.8 Classification boundaries for the EDDA model $[\lambda_g D A_g D^T]$ (left) and for the RDA method with tuning parameters $\lambda = 0.15$ and $\gamma = 0.80$ on the haemophilia data set (right).

where α_g^{jh} denotes the probability that variable j has level h in class g, and $\boldsymbol{\alpha}_g = (\alpha_g^{jh}; j = 1, \ldots, d; h = 1, \ldots, m_j)$. The maximum likelihood estimates of the $\boldsymbol{\alpha}_g$ parameters are

$$(\alpha_g^{jh}) = \frac{(u_g^{jh})}{n_g}, j = 1, \ldots, d; h = 1, \ldots, m_j; g = 1, \ldots, G,$$

where $n_g = \sum_{i=1}^{n} z_{ig}$ and $(u_g^{jh}) = \sum_{i=1}^{n} z_{ig} y_i^{jh}$.

But maximum likelihood estimation with multivariate categorical data remains problematic because of the curse of dimensionality. This conditional independence model requires $(G - 1) + G \sum_j (m_j - 1)$ parameters to be estimated. The model is more parsimonious than the empirical model which requires $\prod_j m_j$ parameters, but it can still be too complex with regard to the sample size n and it can even involve numerical difficulties ("divide by zero" occurrences or parameters α_g^{jh} estimated to zero).

One way of avoiding such drawbacks is to use the regularized maximum likelihood estimator:

$$(\alpha_g^{jh}) = \frac{(u_g^{jh}) + c - 1}{n_g^{(r)} + m_j(c - 1)},$$

where c is a fixed positive number, for updating the estimation of the α_g^{jh}. The estimates can be sensitive to the value of c and considering this regularization issue in a Bayesian setting could be beneficial (see Section 6.2.7).

With this in mind, the parsimonious models presented in Chapter 6 in the clustering context could be thought of as desirable. These models use a reparameterization of $\boldsymbol{\alpha}_g$ in two parts. For any variable j, let a_g^j denote the most frequent response level in class g, the parameter $\boldsymbol{\alpha}_g$ can be replaced by $(\mathbf{a}_g, \boldsymbol{\varepsilon}_g)$

with $\mathbf{a}_g = (a_g^1, \ldots, a_g^d)$ and $\boldsymbol{\varepsilon}_g = (\varepsilon_g^{11}, \ldots, \varepsilon_g^{dm_d})$ where

$$a_g^j = \arg\max_h \alpha_g^{jh} \quad \text{and} \quad \varepsilon_g^{jh} = \begin{cases} 1 - \alpha_g^{jh} & \text{if} \quad h = a_g^j \\ \alpha_g^{jh} & \text{otherwise.} \end{cases}$$

Vector $\mathbf{a}_g$ provides the modal levels in class g for the variables and the elements of vector $\boldsymbol{\varepsilon}_g$ can be regarded as scattering values.

Using this form, it is possible to impose various constraints on the scattering parameters ε_g^{jh}. The models we consider are the following.

- the standard conditional independence model $[\varepsilon_g^{jh}]$: the scattering depends upon classes, variables and levels.
- $[\varepsilon_g^j]$: the scattering depends upon classes and variables but not upon levels.
- $[\varepsilon_g]$: the scattering depends upon classes, but not upon variables.
- $[\varepsilon^j]$: the scattering depends upon variables, but not upon classes and levels.
- $[\varepsilon]$: the scattering is constant over variables and classes.

We do not detail here the maximum likelihood estimation of these models in the classification context. This can be straightforwardly derived from the M-step formulae given in Chapter 6.

As an example, the maximum likelihood formulae for the model $[\varepsilon_g^j]$, are

$$(a_g^j) = \arg\max_h (u_g^{jh})$$

and

$$(\varepsilon_g^j) = \frac{n_g - v_g^j}{n_g},$$

for $g = 1, \ldots, G$ and $j = 1, \ldots, d$, where $v_g^j = u_g^{jh^*}$, h^* is the modal level a_g^j (Aitchison and Aitken, 1976). Notice that ε_g^j is simply the percentage of times the level of an object in class g is different from its modal level for variable j.

By considering this particular representation of the multivariate multinomial distribution, it is possible to get parsimonious and simple classifiers. The parameters of these classifiers are estimated using the formulae presented in the latent class model context. In particular, it could be beneficial to consider regularized maximum likelihood or Bayesian estimates to avoid the numerical problems occurring when dealing with many categorical variables.

4.4.2 An Illustration

To illustrate this family of models for supervised classification with categorical data, a data set describing the morphology of birds (puffins) is used. This data set was first used by Bretagnolle (2007). Each bird is described by five categorical variables. One variable for the gender and four variables providing a morphological description of the birds: the eyebrow stripe (4 levels), the collar description (4 levels), the sub-caudal description (4 levels) and the border description (3 levels). There are 69 puffins from two species: *lherminieri* (34) and *subalaris* (35). The purpose is to recover the puffin species with the five categorical variables. Assuming a mixture sampling scheme, the model minimizing the cross-validated error rate (leave-one-out) is the standard model $[\varepsilon_g^{jh}]$ and the estimated error rate is 1.5%. Listing 4.4 presents the code for learning the classifier on the **puffin** data set with Rmixmod.

```
****************************************************************
* Number of samples    =  69
* Problem dimension    =  5
****************************************************************
* Number of cluster =   2
* Model Type =  Binary_pk_Ekjh
* Criterion =   CV(0.9855)
* Parameters =  list by cluster
* Cluster  1 :
  Proportion =  0.4928
  Center =  1.0000 2.0000 3.0000 1.0000 1.0000
  Scatter = |      0.4714        0.4714 |
            |      0.1786        0.3929       0.2071       0.0071 |
            |      0.1500        0.3786       0.5929       0.0643 |
            |      0.4500        0.3214       0.0929       0.0357 |
            |      0.0762        0.0667       0.0095 |
* Cluster  2 :
  Proportion =  0.5072
  Center =  2.0000 3.0000 1.0000 1.0000 1.0000
  Scatter = |      0.4306        0.4306 |
            |      0.0069        0.0069       0.1319       0.1181 |
            |      0.0208        0.0069       0.0069       0.0069 |
            |      0.0486        0.0069       0.0069       0.0347 |
            |      0.0741        0.0370       0.0370 |
*    Log-likelihood =   -237.3621
****************************************************************
```

Figure 4.9 displays the distribution of the two classes for the five descriptors. The discriminative variables are eyebrow stripe, collar description and sub-caudal description, while gender is (not surprisingly) useless as border description.

4.5 Variable Selection

In the model-based classification setting, selecting the relevant variables is often of primary interest to get an efficient classifier. Each of the d predictive variables brings discriminant information (its own ability to separate the classes), and noise (its sampling variance). It is important to select the variables bringing more discriminant information than noise. In practice there are three kinds of variables:

Listing 4.4: Multinomial classification for the `puffin` data set

```
# Loading libraries and data
library(MBCbook)
data(puffin)
X = puffin[,-1]
cls = puffin$class

# Learning the classifier with Mixmod
out = mixmodLearn(X,as.factor(cls),dataType="qualitative")
barplot(out)

# Computing and displaying the alphas
p = ncol(X)
prms = out@bestResult@parameters
M = ncol(prms@scatter[[1]])
alpha = array(NA,c(max(cls),p,M))
for (c in 1:max(cls)){
  center = prms@center[c,]
  a = matrix(0,p,M)
  for (j in 1:M) a[center==j,j] = 1
  alpha[c,,] = abs(a - prms@scatter[[1]])
}
par(mfrow=c(2,3))
for(j in 1:p){
  alph = alpha[,j,]
  alph = alph[,colSums(alph) != 0]
  barplot(alph,beside=TRUE,names.arg=levels(X[,j]),legend.text=c("
      Borealis","Diomedea"),col=c(1,10),ylab='Alpha')
}
summary(out)
```

- Useful variables bringing strong discriminant information. For instance, to classify men and women, the height is clearly a useful variable.

- Redundant variables for which the discriminant information is essentially provided by other variables: for the same example, weight does not give information on the gender knowing the height of a person.

- Irrelevant or noisy variables. Eye color does not contain any information on the gender of a person.

Thus variable selection is in general a crucial step to get a reliable classifier. In model-based classification, it is possible to define the roles of the predictive variables in a proper way, as precisely described in Chapter 7 and to derive algorithms to assign a relevant status (useful, redundant or independent) to the predictors. In the model-based context, it is possible to design efficient variable selection procedures that work well. This possibility could enhance the efficiency of complex nonlinear generative classification methods such as quadratic discriminant analysis (QDA) in high dimensional contexts and make them competitive in many situations with popular classifiers such as LDA, logistic regression, k-nearest neighbors classifier or support vector machine classifiers.

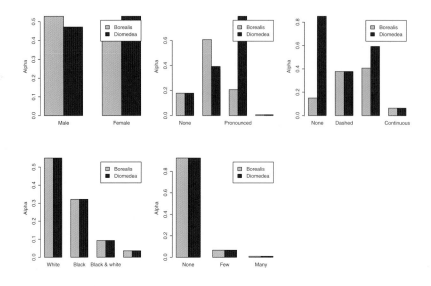

Figure 4.9 The distribution of the two puffin species on the five descriptors: gender, eyebrow stripe, collar description, sub-caudal description and border description (from left to right and top to bottom).

4.6 Mixture Discriminant Analysis

Contrary to situations where regularization is desirable, there exist situations, with a reasonable training sample size, where the Gaussian assumption for the class-conditional densities is too restrictive. Thus, as proposed by Hastie and Tibshirani (1996), it is possible to model the class-conditional densities with a mixture of distributions. In what follows, we restrict attention to mixtures of Gaussian distributions on continuous predictors. Mixture discriminant analysis (MDA) assumes that, for $g = 1, \ldots, G$, the class density is a mixture density

$$f_g(y_i) = \sum_{\ell=1}^{r_g} \pi_{g\ell} \Phi(y_i | m_{g\ell}, \Sigma_{g\ell}),$$

where the mixture proportions $\pi_{g\ell}$ satisfy $\sum_{\ell=1}^{r_g} \pi_{g\ell} = 1$.

The parameters of the mixture distributions for each class can be estimated using the EM algorithm described in Chapter 2. All the models proposed in EDDA can be used when defining an MDA classifier (see Fraley and Raftery, 2002). But such a general method would lead to a large number of parameters to be estimated. Thus, MDA would need a large sample size for each class to be relevant. For this reason, in Hastie and Tibshirani (1996), the authors restricted attention to a mixture model with a common covariance matrix for all mixture components of each class: $\Sigma_{g\ell} = \Sigma$, $(1 \le \ell \le r_g \,;\, 1 \le g \le G)$, the number of mixture components r_g being chosen a priori by the user. This allows for a dramatic reduction in the number of parameters to be estimated, but the price to be paid is a great loss of

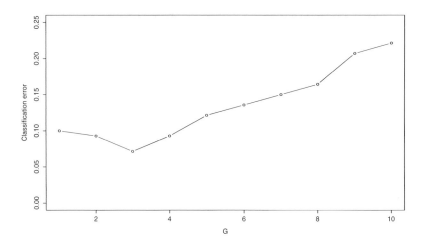

Figure 4.10 Choice of the common number of components per class for MDA with the diabetes data set.

the flexibility of MDA. An illustration of MDA with the most general Gaussian mixture with $r_g = r$ for $g = 1, \ldots, G$ is given in Figures 4.10 and 4.11 for the diabetes data set. The best cross-validated error rate is achieved when $r = 3$. The more and more complex boundaries of the classifications with $r = 1, 3$ and 10 are displayed in Figure 4.11.

An alternative solution is to consider covariance matrices of the form $\Sigma_{g\ell} = \sigma_{g\ell}^2 I$, I being the identity matrix. Acting in such a way, the class conditional densities $f_g(y_i)$ are modeled by a union of Gaussian balls with free volumes, where a Gaussian ball is a multivariate Gaussian distribution with covariance matrix proportional to the identity matrix. It is a flexible model since (numerous) Gaussian balls could be expected to fit the cloud of the n_g points of class g, and it remains parsimonious since each covariance matrix is parameterized with a single parameter.

As an illustration, we consider in Listing 4.5 MDA with Gaussian balls for learning a supervised classifier on the diabetes data set. For each class, the number of Gaussian balls was chosen by the BIC criterion (3, 5 and 3 components for the Type 1 diabetic, Type 2 diabetic and non-diabetic classes, respectively). The boundaries of the classification rule are displayed in Figure 4.13. It can be seen that they are somewhat less complex than the boundaries displayed in Figure 4.11 but remain flexible.

4.7 Model Assessment and Selection

A classifier aims at providing a decision function to predict the unknown labels of future observations. Thus, it is important to assess the expected misclassification error rate of a classifier designed on a finite training set, but supposed to be used on an independent sample from a population of potentially infinite size. The most

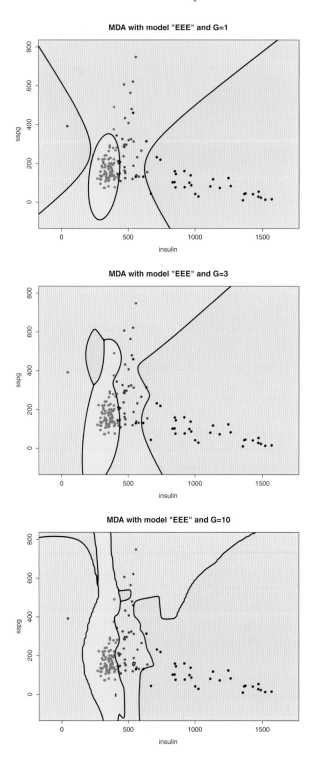

Figure 4.11 Classification boundaries for MDA with $r_g = r = 1, 3, 10$ components per class for the diabetes data set.

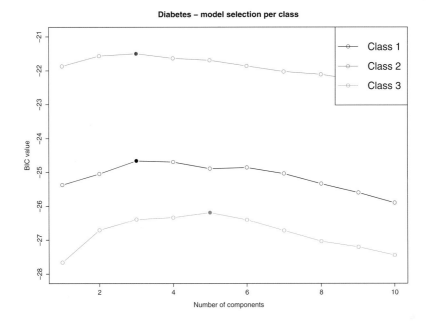

Figure 4.12 Choice of the number of components per class for MDA with the diabetes data set.

widely used method to estimate the misclassification error rate of a classifier is *cross-validation*.

4.7.1 The Cross-validated Error Rate

Cross-validation in its most common version is known as *leave-one-out* cross-validation, and is as follows. For $i = 1, \dots, n$ the classifier is designed from the training set A without the ith observation and the ith observation is assigned to one of the classes with this classifier. The error rate is then estimated by the frequency of misclassified points by this procedure. The resulting estimated error rate is almost unbiased. But its variability is underestimated for large n since the different classifiers designed with $n - 2$ common observations have a strong tendency to be identical. Moreover, leave-one-out is costly for large n even if it is often possible to derive the $(n - 1)$-classifier from the n-classifier by simplified calculations.

So-called v-fold cross-validation is now often preferred to leave-one-out. It consists of dividing at random the training data set A into v parts of approximatively equal sizes. For $\ell = 1, \dots, v$, a classifier is designed from A without the ℓth part, then the observations of the ℓth part are assigned to one of the classes with this classifier. Finally, the misclassification error rate is estimated by the mean of the corresponding v obtained error rates. If $v = n$, v-fold cross-validation is *leave-one-out*. Conversely if $v = 2$ v-fold cross-validation is known as *half sampling*

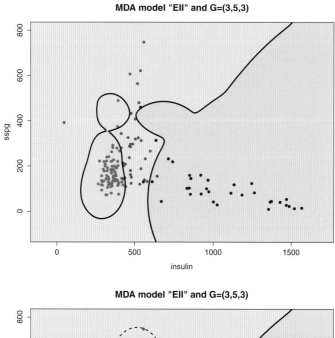

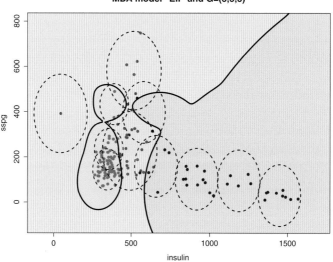

Figure 4.13 Classification boundaries for MDA with 3 Gaussian balls for the Type 1 diabetic class (red), 5 for the Type 2 diabetic class (blue) and 3 for the non-diabetic class (green) for the diabetes data set.

where the whole sample is separated into two equal parts used alternatively as training and test data sets.

Results can be sensitive to the choice of v. The larger the value of v, the smaller the estimation bias, but the larger the variance. Empirical studies show that when n is small, $v = n$ could be preferred. When n is large, $v = 2$ could be preferred. In many cases where A is of reasonable size, choosing v around five or ten is

Listing 4.5: Mixture discriminant analysis on the diabetes data

```
# Loading libraries and data
library(MBCbook)
data(diabetes)
X <- diabetes[3:4]
class <- unlist(diabetes[1])

# Choice of G by cross-validation
G = 10; V = 20; ind = sample(nrow(X))
B = round(nrow(X)/V); err = matrix(NA,G,V)
for (g in 1:G){
  for (v in 1:V){
    test = ind[(B*(v-1)+1):(v*B)]
    res = MclustDA(X[-test,],class[-test],G=g,modelNames='VVV')
    err[g,v] = sum(predict(res,X[test,])$class != class[test]) / B
  }
}

# Plotting the class. error according to G
plot(1:G,rowMeans(err),type='b',ylab='Classification error',xlab='G
    ',ylim=c(0,0.25))
```

sensible. Moreover for small values of v it is possible to use a Monte Carlo version of the v-fold procedure, namely repeat it a large number L of times, to get stabler estimation of the misclassification error rate. In practice, cross-validation produces almost unbiased estimation of error rates, but its variability can be large (Hastie et al., 2009, Section 7.10).

A possible alternative with large data sets is to use a *test data set* drawn at random from the whole data set according to the sampling scheme at hand (mixture sampling or retrospective sampling) and independent of the resulting training data set. Typically, a test sample is often 20% of the whole data set, but this percentage may vary greatly. The error rate estimated on a test sample is unbiased but its variance remains unknown. The main interest of this methodology to assess the error rate of a classifier is its simplicity and rapidity.

4.7.2 Model Selection and Assessing the Error Rate

In most situations designing a classifier involves a selection step: choosing the predictors for LDA or QDA models, such as selecting the best model among many models when using the EDDA methodology, or choosing the number of mixture components through the MDA approach. Denote by $\mathcal{M}$ the family of models in which we want to select a "best" model $M^\star \in \mathcal{M}$ minimizing the misclassification error rate according to a selection step.

The performance of a classifier is highly dependent on this selection step. It is crucial to properly take into account the variability of this selection step when assessing the performances of a classifier. As stated by Hastie et al. (2009), Section 7.10.2, computing the cross-validated error rate on the whole training sample while ignoring the variability of the selection step is an incorrect use of cross-validation and can lead to overoptimistic estimation of the error rate.

Denoting by $A_1, \ldots, A_v$ the v equal random parts of the training data set A, the following procedure is a correct way to use cross-validation:

For $\ell = 1, \ldots, v$:

(i) **Selection step:** run the selection step on the sample $A^{-\ell} = A - A_\ell$: it leads to the selection of a model $M^\ell \in \mathcal{M}$.

(ii) **Estimating the misclassification error rate:**

(a) Consider the classifier on the sample $A^{-\ell}$ with the model M^ℓ selected in (i).

(b) Classify A_ℓ with the classifier considered in (a). This leads to an estimated misclassification error rate, say E_ℓ.

Then the estimated cross-validated misclassification error rate is $E_{\text{CV}} = \frac{1}{v} \sum_{\ell=1}^{v} E_\ell$.

This procedure aims to estimate the misclassification error rate of the classifier associated with the model $M^\star$, which is based on the whole training data set A. Moreover, as in EDDA or RDA, the cross-validated error rate could be used as a model selection criterion in the selection step. This suggests that a double cross-validation procedure can be used in the previous procedure (see Stone, 1974), as follows:

Divide the training sample A at random into v parts of equal sizes $A_1, \ldots, A_v$ and divide each $A^{-\ell} = A - A_\ell$ into v parts of equal sizes $A_1^\ell, \ldots, A_v^\ell$.

For $\ell = 1, \ldots, v$

(i) **Selection step:** for each model $M \in \mathcal{M}$, compute its cross-validated misclassification error rate E_ℓ^M on the data set $A^{-\ell}$. Computing this misclassification error rate requires making use of the data sets $A_1^\ell, \ldots, A_v^\ell$. It is given by $E_\ell^M = \frac{1}{v} \sum_{k=1}^{v} E_{\ell k}^M$ where $E_{\ell k}^M$ is the misclassification error of the classifier associated with M on data set $A_k^{-\ell}$. Then select the model $M^\ell = \arg \min_M E_\ell^M$.

(ii) **Estimating the misclassification error rate:**
     For $\ell = 1, \ldots, v$:

(a) Consider the classifier on the sample $A^{-\ell}$ with the model selected in (i).

(b) Classify A_ℓ with the classifier considered in (a). This leads to a misclassification error rate E_ℓ.

Then the cross-validated misclassification error rate is $E_{\text{CV}} = \frac{1}{v} \sum_{\ell=1}^{v} E_\ell$.

In the same way as before, this procedure aims to estimate the misclassification error rate of the classifier associated with the model $M^\star$, computed for the whole training data set A.

4.7.3 Penalized Log-likelihood Criteria

In the model-based classification setting, penalized log-likelihood criteria such as the Bayesian Information Criterion (BIC) (Schwarz, 1978) or the Akaike criterion (AIC) (Akaike, 1974) are available. However, BIC and AIC appear sometimes to yield disappointing results in a discriminant analysis context (see Bouchard and Celeux, 2006). The reason for this disappointing behavior could be related to the fact that those criteria do not take into account the classification purpose. They simply measure the fit of the models to the training set and implicitly assume that the best model fits the training set well. Thus, the v-fold cross-validated error rate might be expected to be a better model selection criterion in model-based classificaition, despite the sensitivity of the choice of v and the expensive cost of the double cross-validation procedure it requires.

For very large data sizes, it is possible to randomly divide the data set into three parts: a training part A (say about 60%) to fit the competing models, a selection part S (say about 20%) to select the model minimizing the classification error rate on S, and a test set T (say 20%) to assess the actual classification error rate of the selected classifier. It is important not to confuse S with T.

For instance for the diabetes data set, the selected model for EDDA is the model EVV with equal proportions and free shapes and orientations. The 10-fold cross-validated error rate evaluated without taking account of the selection uncertainty, i.e. in an incorrect way, was 7% with a standard error of 6.7%. The 10-fold cross-validated error rate evaluated in a fair way was 8.6% with a standard error of 6.6%. Thus the optimism of the wrong evaluation of the error rate was about 1.6%. The standard error can also be sensitive to the method used.

5

Semi-supervised Clustering and Classification

Although clustering (Chapter 2) and classification (Chapter 4) are well-established and important fields, several intermediate situations are of interest in practical applications.

On one hand, semi-supervised classification considers the situation of learning a supervised classifier from both labeled and unlabeled data. This is of particular interest in applications where the learning data are available in large quantities (such as texts or images) and for which the labeling of each observation is either difficult or expensive. On the other hand, semi-supervised clustering aims to improve the quality of a data clustering by exploiting pairwise relations, such as must-link or cannot-link relations.

Aside from those two main semi-supervised situations, several other intermediate cases also arise. First, in the supervised classification framework, human supervision is often required for labeling the learning data, which are then used for building the classifier. However, human labeling can be imprecise, difficult and/or expensive, depending on the complexity of the data. In such cases, it is therefore important to take into account the possible uncertainty about the data labels when learning the supervised classifier.

Another important issue usually not taken into account by supervised classification methods is that a class represented in the test set may not have been encountered earlier in the learning phase. Adaptive supervised classification methods then allow the discovery of new classes in the test data.

5.1 Semi-supervised Classification

By using the information provided by cheap and widespread *unlabeled data*, semi-supervised classification aims to improve the classifier's performance. Predictive models are unable to take into account unlabeled data without additional assumptions about the marginal distribution $p(y_i)$ of the predictors. However, some extensions of these methods to the semi-supervised setting have been proposed that make implicit assumptions regarding $p(y_i)$ in order to take unlabeled data into account. Among these the low density separation assumption is the most commonly used (Joachims, 1999; Grandvalet and Bengio, 2004).

In a semi-supervised setting, it is natural to adopt the generative point of view,

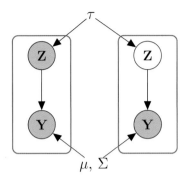

Figure 5.1 Graphical view of the semi-supervised classification problem.

since it leads to modeling the joint distribution

$$p(y_i, z_i) = p(z_i)p(y_i|z_i)$$

and to writing the marginal distribution $p(y_i)$ as a mixture model:

$$p(y_i) = \sum_{z_i} p(z_i)p(y_i|z_i).$$

Thus, the maximum likelihood estimate of the model parameters can be simply derived through the EM algorithm. The formulae describing the EM algorithm for semi-supervised classification are straightforward, and are given in the next section.

5.1.1 Estimating the Model Parameters through the EM Algorithm

In the semi-supervised setting, assuming that the labels of the first n observations are known while the labels of the m next observations are unknown, the complete-data log-likelihood of the mixture model is:

$$L(\theta|y, z) = \sum_{i=1}^{n}\sum_{g=1}^{G} z_{ig} \log\left(\tau_g p(y_i|\theta_g)\right) + \sum_{i=n+1}^{n+m}\sum_{g=1}^{G} z_{ig} \log\left(\tau_g p(y_i|\theta_g)\right),$$

where $\{z_{n+1}, ..., z_{n+m}\}$ are the missing labels and $\theta = (\tau_1, \ldots, \tau_G, \theta_1, \ldots, \theta_G)$. The expectation of the completed log-likelihood $Q(\theta|\theta^*)$, involved in the EM algorithm, is then:

$$Q(\theta|\theta^*) = \sum_{i=1}^{n}\sum_{g=1}^{G} z_{ig} \log\left(\tau_g p(y_i|\theta_g)\right) + \sum_{i=n+1}^{n+m}\sum_{g=1}^{G} t_g(y_i|\eta) \log\left(\tau_g p(y_i|\theta_g)\right),$$

where θ^* is the current estimate of the model parameters.

The EM algorithm then takes the following form at iteration s:

- E-step: the conditional posterior probabilities $\tau_g(y_i|\theta^{(s)})$ of the unlabeled sample set $\{y_{n+1}, ..., y_{n+m}\}$ are updated

$$\tau_g(y_i|\theta^{(s)}) = \frac{\hat{\tau}_g^{(s-1)}p(y_i|\hat{\theta}_g^{(s-1)})}{p(y_i|\hat{\theta}^{(s-1)})}, \; i = n+1, ..., n+m, \; g = 1, ..., G,$$

where $\hat{\tau}_g^{(s-1)}$ and $\hat{\theta}_g^{(s-1)}$ are the mixture parameters estimated in the M-step at the iteration $(s-1)$.

- M-step: updates for the model parameters are obtained by maximizing the conditional expectation of the complete likelihood $Q(\theta|\theta^*)$.

As an example, for the standard Gaussian mixture with free covariance matrices, the M-step of the EM algorithm consists of the following updates:

$$\hat{\tau}_g^{(s)} = \frac{(n_g + m_g^{(s)})}{n+m},$$

$$\hat{\mu}_g^{(s)} = \frac{1}{n}\left(\sum_{i=1}^{n} z_{ig}y_i + \sum_{i=n+1}^{n+m} t_g(y_i|\theta^{(s)})y_i \right),$$

$$\hat{\Sigma}_g = \frac{1}{n}\left(S_g + S_g^{(s)} \right),$$

where $n_g = \sum_{i=1}^{n} z_{ig}$, $m_g^{(s)} = \sum_{i=n+1}^{n+m} t_g(y_i|\theta^{(s)})$, $S_g = \sum_{i=1}^{n} z_{ig}(y_i - \hat{\mu}_g^{(s)})^T(y_i - \hat{\mu}_g^{(s)})$ and $S_g^{(s)} = \sum_{i=n+1}^{n+m} t_g(y_i|\theta^{(s)})(y_i - \hat{\mu}_g^{(s)})^T(y_i - \hat{\mu}_g^{(s)})$.

This EM algorithm is not expected to suffer from the well-documented initialization variability of EM since it can be initialized in a natural way by the maximum likelihood estimate of θ provided by the labeled sample $\{(y_1, z_1), \ldots, (y_n, z_n)\}$. Denoting by $\hat{\theta}_{y,z}$ the maximum likelihood estimator of θ obtained by the EM algorithm at convergence, a new observation y_{n+1} is assigned to one of the classes with the maximum a posteriori (MAP) classification rule

$$\hat{z}_{n+1} = \arg\max_z p(z|y_{n+1}, \hat{\theta}_{y,z}). \tag{5.1}$$

5.1.2 A First Experimental Comparison

Here we show how the use of unlabeled data can improve a classification rule. The `mixmodCluster` function in the `Rmixmod` package for R allows the user to provide known labels for some of the observations. Another option would be to use the `upclass` package (Russell et al., 2014), which is a companion package for `mclust`. Listing 5.1 shows a simple call of the `mixmodCluster` function on the `wine27` data with only 30% of the labels assumed to be known (70% of unobserved labels). The actual and estimated class proportions are also displayed to show the ability of the semi-supervised classification approach to recover the model parameters.

Figure 5.2 allows one to compare the influence of the percentage of labeled data on supervised and semi-supervised classifiers. The left panel of each row

Listing 5.1: Semi-supervised classification with `Rmixmod`

```
# Loading libraries and data
library(MBCbook)
data(wine27)
X = wine27[,1:27]
cls = wine27$Type

# Displaying actual class proportions
summary(cls) / length(cls)

    Barbera      Barolo Grignolino
   0.2696629   0.3314607   0.3988764

# 70% of the labels are assumed not to be known
ind = sample(1:nrow(X),round(0.7*nrow(X)))
cls[ind] = NA

# Semi-supervised learning
res = mixmodCluster(X,3,knownLabels=cls)

# Displaying estimated class proportions
res@bestResult@parameters@proportions

   [1] 0.2697579 0.3266665 0.4035756
```

shows the classification boundaries of QDA (model 'Gaussian_pk_Lk_Ck' within Rmixmod) when learned with all labeled data. The center and right panels of each row respectively show the boundaries generated by QDA with only a given percentage of labeled data and by a semi-supervised classifier (also with the model 'Gaussian_pk_Lk_Ck') with in addition unlabeled data. As expected, the semi-supervised classifier turns out to provide decision boundaries closer to the ones of the supervised classifier learned with all data than those of the classifier learned with only a few labeled data. The use of unlabeled data appears to be helpful for designing robust classifiers in situations where only few labeled data are available.

The effect of the use of unlabeled data for robustifying the classifier can also be observed in the prediction phase. Listing 5.2 provides an experiment which aims to highlight this effect when predicting for new (test) observations. The whole data set is first split into a learning set of 150 observations and a test set with the remaining 28 observations. Within the learning phase, three different classifiers are learned: a supervised classifier learned on all labeled data (res0), a semi-supervised classifier (res1) learned with 80% of unknown labels and a supervised classifier learned on the 20% of labeled data (res2). On the one hand, the classifier learned with all labeled data makes 7 errors for the test data. On the other hand, res2 misclassifies 13 test observations whereas the semi-supervised, with the help of the unlabeled data, makes only 9 misclassifications.

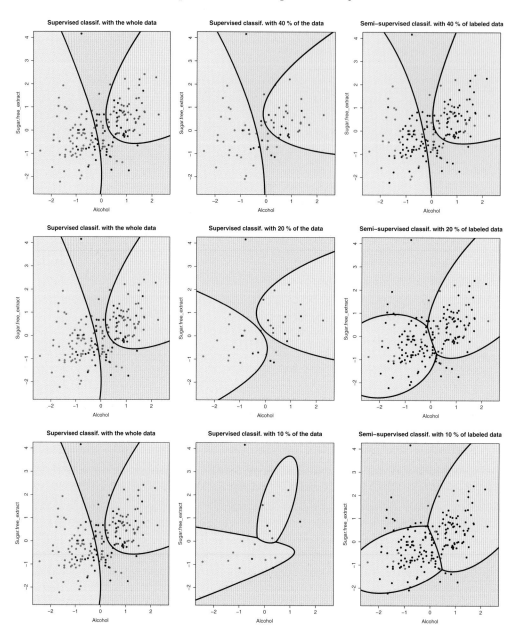

Figure 5.2 Decision boundaries of QDA with all labeled data (left panels), QDA with only a few labeled data (center panels) and a semi-supervised classifier with additional unlabeled data (right panels) on the `Wine` data. From top to bottom, the percentage of labeled data varies from 40% to 10%.

5.1.3 Model Selection Criteria for Semi-supervised Classification

In the semi-supervised context, an important question is to what extent can the unlabeled data be useful in guiding the choice of a relevant generative model. This

Listing 5.2: Semi-supervised classification on the `wine27` data

```
# Loading libraries and data
library(MBCbook)
data(wine27)

# Splitting the data into learning and test sets
sel = sample(nrow(wine27),150)
X = wine27[sel,1:27]; labels = X.labels = wine27$Type[sel]
Y = wine27[-sel,1:27]; Y.labels = wine27$Type[-sel]

# Learning the classifiers
per = 0.8 # percentage of unlabeled data
labels[1:round(per*nrow(X))] = NA
model = mixmodGaussianModel(listModels='Gaussian_pk_Lk_Ck')
res0 = mixmodLearn(X,X.labels,models=model)
res1 = mixmodCluster(X,3,knownLabels=labels,models=model)
res2 = mixmodLearn(X[!is.na(labels),],labels[!is.na(labels)],models
    =model)

# Predicting for test data (results may vary due to sampling)
table(mixmodPredict(Y,res0@bestResult)@partition,Y.labels)
   Y.labels
    1 2 3
  1 9 0 3
  2 0 8 4
  3 0 0 4

table(mixmodPredict(Y,res1@bestResult)@partition,Y.labels)
   Y.labels
    1 2 3
  1 9 0 4
  2 0 8 5
  3 0 0 2

table(mixmodPredict(Y,res2@bestResult)@partition,Y.labels)
   Y.labels
    1 2 3
  1 7 3 4
  2 0 5 4
  3 2 0 3
```

choice often consists of selecting the discriminant variables for any model, or the form of the class conditional variance matrix for EDDA, or the number of mixture components for MDA. For this model selection task, well-known model selection criteria such as the information criteria BIC or AIC are available. However, those criteria do not take into account the classification purpose and thus can give disappointing performance. The cross-validated error rate could be regarded as a more relevant criterion, but cross-validation is highly time-consuming in the semi-supervised context since the EM algorithm is involved in the estimation process. Alternative model selection criteria specific to the semi-supervised classification context are now presented.

A Bayesian Entropic Criterion

A good approximation of the conditional distribution $p(z_i|y_i)$ produces a good classifier (see Equation (5.1)). Consequently, it makes sense to choose a generative classification model m with parameter $\theta^{(m)}$ that gives the largest conditional integrated likelihood $p(z|y, m)$. In this Bayesian perspective, the Bayesian entropic criterion (BEC), to be maximized, is a BIC-like approximation of $\log p(z|y, m)$:

$$\mathrm{BEC}(m) = \log p(y, z|\hat{\theta}^m_{y,z}) - \log p(y|\hat{\theta}^m_y), \tag{5.2}$$

where $\hat{\theta}^m_y$ is the maximum likelihood estimate of θ^m derived from y. The computational cost of BEC is approximately twice as large as the computational cost of AIC or BIC, since both $\hat{\theta}^m_{y,z}$ and $\hat{\theta}^m_y$ have to be estimated through an EM algorithm, but it remains significantly cheaper than cross-validation.

From a theoretical point of view, if the sampling distribution belongs to a single model of the model collection, this model will be asymptotically selected by BEC (Bouchard and Celeux, 2006). However, when there are several nested true models, BEC can select arbitrarily complex models among them.

From a practical point of view, BEC has been found to perform better than AIC and BIC for many classification problems, though it often selects more complex generative classifiers than the cross-validated error rate criterion (Bouchard and Celeux, 2006).

A Predictive Deviance Criterion

From a frequentist perspective, it is desirable to choose the model m minimizing the predictive deviance of the classification model, which is related to the conditional likelihood of the model knowing the predictor variables. The aim is to find the model that minimizes the expected Kullback–Leibler (KL) divergence between the estimated conditional distribution of $z_i|y_i$ and the true conditional distribution. Following computations analogous to those on which AIC is based (see Ripley, 1996, pp. 30–35), it leads to the so-called predictive deviance criterion (AIC_{cond}) (Vandewalle et al., 2013), namely

$$\mathrm{AIC}_{cond}(m) = 2\log p(z|y, \hat{\theta}^m_{y,z}) - 4\log \frac{p(y|\hat{\theta}^m_y)}{p(y|\hat{\theta}^m_{y,z})}. \tag{5.3}$$

Note that AIC_{cond} is different from the standard AIC criterion in the predictive setting, even in the absence of additional unlabeled data. The AIC_{cond} criterion can be viewed as an over-penalized BEC criterion, since

$$\mathrm{AIC}_{cond}(m) = 2\mathrm{BEC}(m) - 2\log \frac{p(y|\hat{\theta}^m_y)}{p(y|\hat{\theta}^m_{y,z})}. \tag{5.4}$$

Formally, the additional penalty of AIC_{cond} ensures consistency over nested models. In other words, AIC_{cond} tends to prefer the less complex model among two nested true models (see Vandewalle et al., 2013). AIC_{cond} is thus able to overcome BEC's tendency to select too complex models when different nested models are in competition. Numerical experiments have shown that both criteria outperform standard criteria such as AIC and BIC to select a generative classification model (Bouchard and Celeux, 2006; Vandewalle et al., 2013). Moreover, there is evidence for the superiority of AIC_{cond} over BEC: in most situations they perform the same, but when they give different answers, AIC_{cond} outperforms BEC (Vandewalle et al., 2013).

An Illustration

The behavior of these model selection criteria is illustrated with a reduced version of the wine data set where we keep only the first two variables from the 27 measurements. Twenty-five semi-supervised data sets have been considered by removing 80% of the labels from the 178 observations at random. Equal proportions for the three regions (labels) were assumed; this setup was chosen in order to exhibit a situation where AIC_{cond} outperforms BEC. Figure 5.3 displays from left to right the cross-validated correct classification rate for the learning data set and the average values of the BEC and AIC_{cond} criteria for the 25 semi-supervised samples.

From this figure it is apparent that AIC_{cond}, which selects the same model as the "pseudo-oracle" classifier for this data set, based on the completely labeled data set, outperforms BEC. The values of BIC are also displayed. The comparison highlights the usefulness of BEC and AIC_{cond} which are designed specifically for semi-supervised classification.

5.2 Semi-supervised Clustering

In some practical situations such as document clustering, image analysis or bioinformatics, it may be possible to ask some experts whether a pair of observations should be assigned to the same group or not. In bioinformatics, when clustering genes based on expression data, information of this kind can be obtained directly from a protein–protein interaction network for instance. This kind of clustering can be called semi-supervised clustering or constrained clustering. In this case we do indeed have additional information about a set of pairs of observations. The information on the pairs is usually binary: the two observations should be assigned to the same group (must-link) or they should be assigned to two different groups (cannot-link). This approach is quite attractive since it does not require that the experts examine all the observation pairs and know the number of groups to search for. It can be therefore applied in a wide variety of contexts.

Shental et al. (2003) considered must-link and cannot-link constraints in the context of clustering with a mixture of Gaussian distributions. Before going further, it is important to highlight that both constraints are different from the transitivity point of view. Indeed, must-link constraints are transitive whereas cannot-link ones are not. This implies that the latter constraints are more difficult than the former ones to take into account when inferring the mixture model. We now explain how must-link and cannot-link constraints can be taken into account within the EM algorithm for model-based clustering.

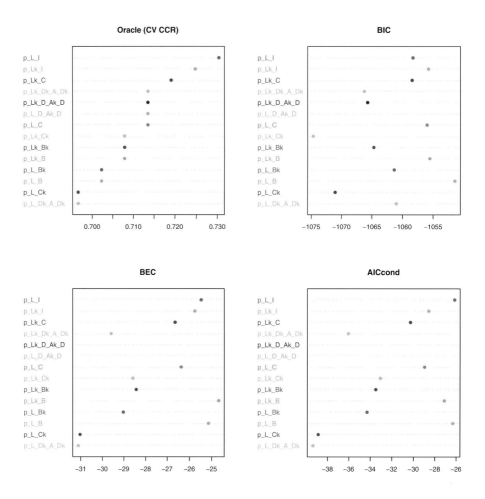

Figure 5.3 Selection of the most appropriate model by cross-validation on the whole sample (left panel, leave-one-out correct classification rate) and with BIC, BEC and AIC$_{cond}$ on semi-supervised samples for the wine data set.

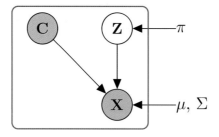

Figure 5.4 Graphical view of the semi-supervised clustering problem with must-link constraints.

5.2.1 Incorporating Must-link Constraints

We refer to the elements of the set $C = \{C_1, ..., C_L\}$, $L \leq N$ as "chunklets," where C_ℓ is such that $z_i = z_j$ if $y_i \in C_\ell$ and $y_j \in C_\ell$. The chunklets are also assumed to contain all observations, i.e. $\bigcup_{\ell=1}^{L} C_\ell = \{y_1, ..., y_n\}$. In particular this means that unconstrained points have their own chunklet. With this notation, we can write the conditional expectation of the complete-data log-likelihood:

$$Q(\theta|\theta^*) = E\left[\log\left(p(Y, Z|\theta, C)\right)|Y, \theta^*, C\right] = \sum_{i=1}^{n}\sum_{g=1}^{G} p(Z|Y, \theta^*, C)\log\left(p(Y, Z|\theta, C)\right),$$

and it is possible to derive an EM algorithm for inference. Thus, the EM algorithm taking into account the constraints encoded by the chunklets has the following form at iteration s:

- E-step: it reduces in computing the chunklet conditional probabilities $\tau_{\ell g} = P(Z_\ell = g|C_\ell, \theta^{(s-1)})$, for $\ell = 1, \ldots, L$ and $g = 1, \ldots, G$:

$$\tau_{\ell g} = \frac{1}{\xi}\tau_g^{(s-1)}\prod_{y_i \in C_\ell} p(y_i|Z_\ell = g, \theta_g^{(s-1)}),$$

  where $\xi = \sum_{j=1}^{G} \tau_j^{(s-1)}\prod_{y_i \in C_\ell} p(y_i|Z_\ell = j, \theta_j^{(s-1)})$ is the normalization constant.
- M-step: the update equations for model parameters which maximize the conditional expectation of the complete-data log-likelihood are given by:

$$\tau_g^{(s)} = \frac{1}{L}\sum_{\ell=1}^{L}\tau_{\ell g},$$

$$\mu_g^{(s)} = \frac{1}{n_g}\sum_{\ell=1}^{L}\tau_{\ell g}\bar{y}_\ell,$$

$$\Sigma_g^{(s)} = \frac{1}{n_g}\sum_{\ell=1}^{L}\tau_{\ell g}\Sigma_{\ell g},$$

  where $n_\ell = |C_\ell|$ is the cardinality of chunklet C_ℓ, $n_g = \sum_{\ell=1}^{L} n_\ell\tau_{\ell g}$ is the expectation of the number of observations in group g and $\Sigma_{\ell g}$ is the sample covariance matrix of the ℓth chunklet of the gth group.

An implementation of the constrained EM algorithm for must-link constraints is available in the MBCbook package. The function is simply called constrEM.

Listing 5.3 presents a basic use of the function for a single must-link constraint. It simulates a data set of 600 observations, generated from a mixture of three Gaussians, in which observations 398 and 430 are forced to belong to the same cluster (whereas they in fact come from different Gaussians). The left panel of Figure 5.5 shows the data and the must-link constraint (red line). If one applies for instance mclust on this data set, the resulting clustering should be the one presented on the center panel of the figure. As expected, the mclust solution consists of three components and recovers the simulated pattern almost perfectly,

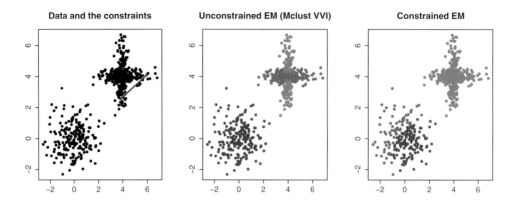

Figure 5.5 The constrained EM algorithm for must-link constraints: simulated data made of three Gaussians and the must-link constraint (left), clustering results of an unconstrained EM algorithm (Mclust, selected model "VVI", center) and clustering results of the constrained EM algorithm (right). The number G of groups has been set to 3 in both cases.

without the constraint. If, however, one wants to take into account the must-link constraint with `constrEM`, the solution will be quite different and will merge the two crossing components and divides the other group into two parts. The right panel of Figure 5.5 shows the latter solution.

5.2.2 Incorporating Cannot-link Constraints

The incorporation of cannot-link constraints is significantly more difficult than the previous case. Indeed, the fact that two data points should not be in the same group does not provide any clear information about their group membership. All possible cases have to be explored and this clearly makes the clustering task more difficult. In Shental et al. (2003), the authors introduce a new latent variable to indicate whether the observation complies with the constraints. We refer to Shental et al. (2003) for further detail on the way to incorporate cannot-link constraints in clustering. Melnykov et al. (2015) recently proposed a unified approach for both must-link and cannot-link constraints.

5.3 Supervised Classification with Uncertain Labels

In the supervised classification framework, human supervision is required to associate labels with a set of learning observations in order to construct a classifier. However, in many applications, this kind of supervision is imprecise, difficult or expensive. For instance, in biomedical applications, domain experts are asked to manually label a sample of learning data (MRI images, DNA micro-array, ...) which is then used for building a supervised classifier. The cost of the supervision

Listing 5.3: The constrained EM algorithm for must-link constraints

```
library(MBCbook)

# Data simulation
set.seed(123)
n = 200
m1 = c(0,0); m2 = 4*c(1,1); m3 = 4*c(1,1)
S1 = diag(2); S2 = rbind(c(1,0),c(0,0.05))
S3 = rbind(c(0.05,0),c(0,1))
X = rbind(mvrnorm(n,m1,S1),mvrnorm(n,m2,S2),
    mvrnorm(n,m3,S3))
cls = rep(1:3,c(n,n,n))

# Encoding the constraints through chunklets
# Observations 397 and 408 are in the same chunklet
a = 398
b = 430
C = c(1:(b-1),a,b:(nrow(X)-1))

# Clustering with Mclust
res0 = Mclust(X,G=3)
t(res0$parameters$mean)
      [,1]    [,2]
[1,] -0.040 -0.007
[2,]  3.982  4.011
[3,]  3.975  4.033

# Clustering with constrEM
res1 = constrEM(X,K=3,C)
res1$mu
       [,1]    [,2]
[1,]  0.259 -0.224
[2,] -0.419  0.273
[3,]  3.979  4.022
```

phase is usually high due to the difficulty of labeling complex data. Furthermore, a human error is always possible in such a difficult task and an error in the supervision phase could have big effects on the decision phase, particularly if the size of the learning sample is small. It is therefore important to provide supervised classifiers robust enough to deal with data with uncertain labels.

5.3.1 The Label Noise Problem

In statistical learning, it is very common to assume that the data are noisy. Two types of noise can be considered in supervised learning: the noise on the explanatory variables and the noise on the response variable. Noise on explanatory variables has been widely studied in the literature whereas the problem of noise on the response variable has received less attention in some supervised situations. While almost all approaches model the noise on the response variable in regression analysis (see Chap. 3 of Hastie et al., 2009, for details), label noise remains an important and unsolved problem in supervised classification. Brodley and Friedl (1999) summarized the main reasons for which label noise can occur. Since the main assumption of supervised classification is that the labels of learning samples are true, existing methods giving a full confidence to the labels of the learning data naturally provide disappointing classification results when the learning data set contains some wrong labels. Particularly, model-based discriminant analysis methods such as Linear Discriminant Analysis (LDA, see Chap. 3 of McLachlan, 1992) or Mixture Discriminant Analysis (MDA, see Hastie and Tibshirani, 1996) are sensitive to label noise. Learning a supervised classifier from data with uncertain labels can be achieved using three main strategies: cleaning the data, using robust estimations of model parameters and finally modeling the label noise. We consider the latter approach here.

5.3.2 A Model-based Approach for the Binary Case

Among all these solutions, the model proposed in Lawrence and Schölkopf (2001) has the advantage of explicitly including the label noise in the model with a sound theoretical foundation in the binary classification case. Denoting respectively by z and $\tilde{z}$ the actual and the observed class labels of an observation x, it is assumed that their joint distribution can be factorized:

$$p(y, z, \tilde{z}) = P(z|\tilde{z})p(y|z)P(\tilde{z}).$$

The class conditional densities $p(y|z)$ are modeled by Gaussian distributions while the probabilistic relationship $P(z|\tilde{z})$ between noisy and observed class labels is specified by a 2×2 probability table (Table 5.1).

Table 5.1 *A table of the mislabeling probabilities. The rows indicate the true label and the columns give the observed labels.*

		z	
		0	1
$\tilde{z}$	0	$1 - \gamma_0$	γ_0
	1	γ_1	$1 - \gamma_1$

The prior probability of each class is modeled as

$$P(z = 1) = \tau,$$
$$P(z = 0) = 1 - \tau,$$

leading to:

$$P(\tilde{z} = 1) = (1 - \tau)\gamma_0 + \tau(1 - \gamma_1) = \tilde{\tau},$$
$$P(\tilde{z} = 0) = 1 - \tilde{\tau}.$$

In this model the actual labels of the data, denoted by z, are viewed as latent variables. Lawrence and Schölkopf (2001) have proposed a variational EM

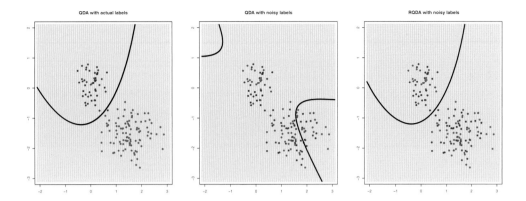

Figure 5.6 Classification boundaries under label noise: QDA with the actual labels (left), QDA with the noisy labels (center) and the robust version RQDA of Lawrence and Schölkopf (2001) on the noisy labels (right).

algorithm for the model inference. The E-step of their inference algorithm aims to compute:

$$\tau_{ig} = P(z = g | y_i, \tilde{z}_i, \hat{\theta}) \propto p(y_i, z = g | \tilde{z}_i, \hat{\theta}) = p(y_i | z = g, \hat{\theta}) P(z = g | \tilde{z}_i), \; g \in \{0, 1\}$$

where $p(y | z = g, \hat{\theta}) = \mathcal{N}(y; \hat{\mu}_g, \hat{\Sigma}_g)$. The M-step updates the model parameters by maximizing the log-likelihood:

$$\hat{\mu}_g = \frac{1}{\nu_g} \sum_{i=1}^{n} \tau_{ig} y_i,$$

$$\hat{\Sigma}_g = \frac{1}{\nu_g} \sum_{i=1}^{n} \tau_{ig} (y_i - \tilde{\mu}_y)(y_i - \tilde{\mu}_y)^T,$$

$$\hat{\gamma}_g = \frac{1}{\nu_g} \sum_{i=1}^{n} \tau_{ig} 1_{\{\tilde{z}_i = g\}},$$

where $\nu_g = \sum_{i=1}^{n} \tau_{ig}$.

In the original paper, Lawrence and Schölkopf (2001) combined this model with a kernelized version of Fisher discriminant analysis to be able to deal with a complicated classification task on real-world images. This work has also been extended by Li et al. (2007b) who proposed an extension of the noise model in the classifier and relaxed the distribution assumption by allowing each class density $p(x | z)$ to be modeled by a mixture of several Gaussians.

The method of Lawrence and Schölkopf (2001), which we call hereafter robust quadratic discriminant analysis (RQDA), is implemented in the MBCbook package. The methodology can also be implemented with more parsimonious models than the full Gaussian model. Listing 5.4 presents an application example of the method to simulated data with label noise. The data considered here come from two classes with unbalanced proportions (respectively 50 and 100 observations per class).

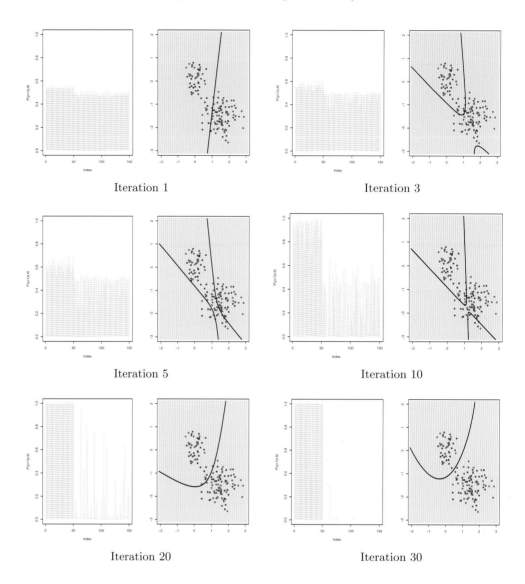

Figure 5.7 Behavior of RQDA at different iterations of the EM algorithm: the left plot of each panel shows the posterior probabilities $P(z = g|y_i, \tilde{z}_i, \hat{\theta})$ for each observation while the right plot shows the estimated decision boundary.

The label noise is important here: 40% of the observations are incorrectly labeled. However, as one can observe in Figure 5.6, the structure of the data is clear enough to recover the true classification boundary. The method performs very well in this case and the estimated classifier can be used with confidence on new data.

In order to better understand how the methodology works, Figure 5.7 shows the posterior probabilities $\tau_{ig} = P(z = g|y_i, \tilde{z}_i, \hat{\theta})$ and the estimated decision

Listing 5.4: Robust quadratic discriminant analysis

```
# Loading libraries
library(MBCbook)

# Data simulation
n = 50
m1 = c(0,0); m2 = 1.5*c(1,-1)
S1 = 0.1*diag(2); S2 = 0.25 * diag(2)
X = rbind(mvrnorm(n,m1,S1),mvrnorm(2*n,m2,S2))
cls = rep(1:2,c(n,2*n))

# Label perturbation
ind = rbinom(3*n,1,0.4); lb = cls
lb[ind==1 & cls==1] = 2
lb[ind==1 & cls==2] = 1

# Classification with RQDA
res = rqda(X,lb,X)
table(cls,res$cls)
     cls
        1    2
   1   49    0
   2    1  100
```

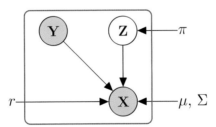

Figure 5.8 Graphical view of the model of robust mixture discriminant analysis (RMDA).

boundary at different iterations of the EM algorithm. In this context where the label noise is large (40%), the model parameters computed at the first iteration are poorly estimated and so the decision boundary is far away from the true one (left panel of Figure 5.6). Nevertheless, as the algorithm iterates, the conditional probabilities τ_{ig} give a better picture of the actual labels of the observations and the estimation of the model is improved. At convergence, the estimated conditional probabilities τ_{ig} are very clear about the class memberships of the data, except for some observations close to the boundary. Notice that the experiment can be replicated on two-dimensional data with the function **rqda** by activating the optional argument **disp**.

5.3.3 A Model-based Approach for the Multi-class Case

The approach presented in the previous paragraph, though very efficient, is unfortunately limited to binary classification cases. It is of great interest to be able to deal with label noise in a multi-class context. A simple model-based approach to take into account the noise label was proposed by Bouveyron and Girard (2009). It considers a mixture model in which two different structures coexist: an unsupervised structure of G clusters (represented by the random discrete variable Z) and a supervised structure, provided by the learning data, of C classes (represented by the random discrete variable W). As in the standard mixture model, we assume that the data $(y_1, ..., y_n)$ are independent realizations of a random vector $Y \in \mathcal{Y}$ with density function:

$$p(y) = \sum_{g=1}^{G} P(Z = g)p(y|Z = g), \qquad (5.5)$$

where $P(Z = g)$ is the prior probability of the gth cluster and $p(y|Z = g)$ is the corresponding conditional density. Let us now introduce the supervised information carried by the learning data. Since $\sum_{c=1}^{C} P(W = c|Z = g) = 1$ for all $g = 1, ..., G$, we can plug this quantity in (5.5) to obtain:

$$p(y) = \sum_{c=1}^{C}\sum_{g=1}^{G} P(W = c|Z = g)P(Z = g)p(y|Z = g), \qquad (5.6)$$

where $P(W = c|Z = g)$ can be interpreted as the probability that the gth cluster belongs to the cth class. Thus, $P(W = c|Z = g)$ measures the consistency between classes and clusters. Using the classical notations of parametric mixture models and introducing the notation $r_{cg} = P(W = c|Z = g)$, we can reformulate (5.6) as follows:

$$p(y) = \sum_{c=1}^{C}\sum_{g=1}^{G} r_{cg}\tau_g p(y|Z = g), \qquad (5.7)$$

where $\tau_g = P(Z = g)$. Therefore, Equation (5.7) exhibits both the "modeling" part of our approach, based on the mixture model, and the "supervision" part through the parameters r_{cg}. Since the modeling is based on the mixture model, we can use any conditional density to model each cluster.

Due to the nature of the model, learning the classifier consists of two steps corresponding respectively to the unsupervised and to the supervised part of the comparison. The first step aims at estimating the parameters of the mixture model in an unsupervised way, leading to the clustering probabilities $P(Z = g|Y = y)$. This first step then consists of fitting the chosen mixture model (the Gaussian mixture model will be used below in the numerical experiments) using for instance the EM algorithm. The choice of the number K of mixture components can be done using model selection criteria such as BIC.

In this second step of the procedure, the labels of the data are introduced to estimate the $C \times G$ matrix of parameters $R = (r_{cg})$ and we use the parameters

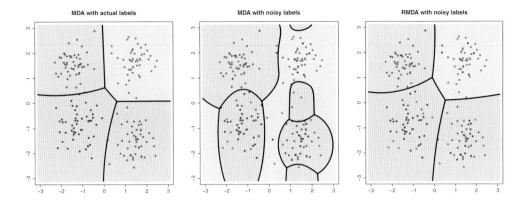

Figure 5.9 Classification boundaries under label noise in a 3-class context: MDA with the actual labels (left), MDA with the noisy labels (center) and RMDA (Bouveyron and Girard, 2009) with the noisy labels (right).

learned in the previous step as the mixture parameters. The parameters r_{cg} modify the unsupervised model taking account of the label information. They thus indicate the consistency of the unsupervised modeling of the data with the supervised information carried by the labels of the learning data. Since in this section we are considering supervised problems, the labels $w_1, \ldots, w_n$ of the learning data $y_1, \ldots, y_n$ are known, and we can therefore introduce $S_c = \{y_i, i = 1, \ldots, n \,|\, w_i = c\}$.

The log-likelihood associated with the model is

$$\ell(R) = \sum_{c=1}^{C} \sum_{y \in S_c} \log P(Y = y, W = c)$$

$$= \sum_{c=1}^{C} \sum_{y \in S_c} \log \left(\sum_{g=1}^{G} r_{cg} P(Z = g | Y = y) \right) + \xi,$$

where, in view of Equation (5.5), $\xi = \sum_{c=1}^{C} \sum_{y \in S_c} \log p(y)$ does not depend on R. This relation can be rewritten in matrix form as:

$$\ell(R) = \sum_{c=1}^{C} \sum_{y \in S_c} \log \left(R_c \Psi(y) \right) + \xi, \qquad (5.8)$$

with the $\mathbb{R}^G$-vector $\Psi(y) = (P(Z = 1 | Y = y), \ldots, P(Z = G | Y = y))^T$ and where R_c is the cth row of R. Consequently, we end up with a constrained optimization problem:

$$\begin{cases} \text{maximize} & \sum_{c=1}^{C} \sum_{y \in S_c} \log \left(R_c \Psi(y) \right), \\ \text{with respect to} & r_{cg} \in [0, 1], \ \forall c = 1, \ldots, C, \ \forall g = 1, \ldots, G, \\ \text{and} & \sum_{c=1}^{C} r_{cg} = 1, \ \forall g = 1, \ldots, G. \end{cases}$$

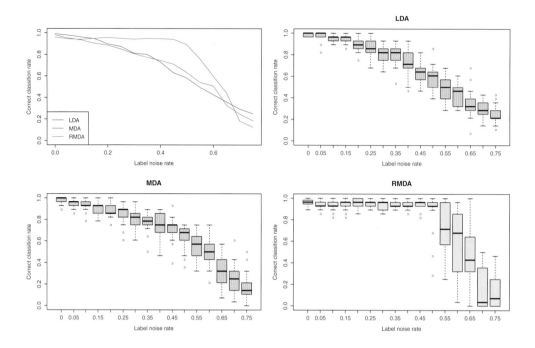

Figure 5.10 Effect of the label noise proportion on the classification performance of LDA, MDA and RMDA. The performance is measured by the correct classification rate (CCR) on a test set and results are averaged on 25 replications of the experiment. The left top panel shows the CCR according to the noise proportion. The three other panels detail the behavior of each method through boxplots.

Unfortunately, this optimization problem does not have an explicit solution and an iterative optimization procedure has to be used to compute the maximum likelihood estimators of the parameters r_{cg}. Such an optimization problem can be handled by sequential quadratic programming (SQP) solvers.

This method has been called robust mixture discriminant analysis (RMDA). RMDA is strongly related to mixture discriminant analysis (MDA, Hastie and Tibshirani, 1996). RMDA is indeed a generalization of MDA to situations with label noise and MDA can be recovered from RMDA when each group corresponds to a unique class (only one 1 per column in the matrix R).

The RMDA method is implemented in the R package `robustDA`. Listing 5.5 presents an example of the use of RMDA on a three-class classification problem with label noise.

Here, some data are simulated in a two-dimensional space. One of the three classes is made of two Gaussian distributions while the other two are made of a single Gaussian. A significant noise is randomly added to the labels: 40% of the labels are incorrect. The RMDA classifier is implemented in the `rmda` function. It is here applied with a possible number K of components ranging from 3 to 8. BIC is used here to pick up the most appropriate number of groups and it chooses

Listing 5.5: Robust mixture discriminant analysis

```
# Loading libraries
library(robustDA)

# Data simulation
n = 50
m1 = c(-1,-1); m2 = 1.5*c(1,-1)
m3 = 1.5*c(-1,1); m4 = 1.5*c(1,1)
S1 = 0.1*diag(2); S2 = 0.25 * diag(2)
X = rbind(mvrnorm(n,m1,S2),mvrnorm(n,m2,S2),mvrnorm(n,m3,S2),
    mvrnorm(n,m4,S2))
cls = rep(1:3,c(n,2*n,n))

# Adding label noise
tau = 0.4
lbl = cls; sel = rbinom(nrow(X),1,tau)
lbl[sel & cls==1] = sample(c(2,3),sum(sel & cls==1),replace=TRUE)
lbl[sel & cls==2] = sample(c(1,3),sum(sel & cls==2),replace=TRUE)
lbl[sel & cls==3] = sample(c(1,2),sum(sel & cls==3),replace=TRUE)

# Learning the RMDA classifier
res = rmda(X,lbl,K=3:8)
round(res$R,4)
        [,1]    [,2]    [,3]    [,4]
[1,] 0.9998 0.0052 0.0000 0.0013
[2,] 0.0000 0.9948 0.9994 0.0000
[3,] 0.0002 0.0000 0.0006 0.9987
```

$K = 4$. The matrix R of the estimated probabilities $r_{cg} = P(W = c|Z = g)$ can be accessed through the R component of the learned classifier. Here, RMDA correctly recognizes the two Gaussians of class 2 and chooses only one Gaussian for classes 1 and 3.

Figure 5.9 shows the classification boundaries of MDA with the actual labels (without label noise, left panel), MDA with the noisy labels (center panel) and RMDA with the noisy labels (right panel). It appears that, even in this specific case where the label noise is important, RMDA is able to recover the true classification boundaries. Learning a supervised classifier without taking into account the possibility that labels are corrupted leads here to a classifier (center panel of Figure 5.9) significantly different from the true one.

As a final experiment, we focus on the robustness of RMDA to label noise and compare it to non-robust methods such as LDA and MDA. Figure 5.10 shows the effect of the label noise proportion on the classification performance of LDA, MDA and RMDA. The performance is measured by the correct classification rate (CCR) on a test set and results are averaged on 25 replications of the experiment. The left top panel shows the CCR according to the noise proportion for the three methods while the other panels detail the behavior of each method through boxplots. Unsurprisingly, the performances of LDA and MDA decrease linearly regarding the label noise. Conversely, RMDA turns out to be highly robust to label corruption: its performance breaks down only after 45% of corrupted labels.

5.4 Novelty Detection: Supervised Classification with Unobserved Classes

The usual framework of supervised classification assumes that all existing classes in the data have been observed during the learning phase and does not take into account the possibility of having observations in the test data belonging to a class which has not been observed in the learning phase. In particular, such a situation could occur in the case of rare classes or in the case of an evolving population. For instance, an important problem in biology is the detection of novel species which could appear at any time resulting from structural or physiological modifications. In the same manner, the detection of unobserved communities is a major issue in social network analysis for security or commercial reasons. Unfortunately, classical supervised algorithms will automatically label observations from a novel class as belonging to one of the known classes in the training set and will not be able to detect new classes. It is therefore important to allow supervised classification methods to detect unobserved situations and to adapt to the new configuration.

As an introductory example, let us consider the iris data set, made famous by its use by Fisher (1936) as an example for discriminant analysis. This data set, in fact collected by Anderson (1935) in the Gaspé Peninsula (Canada), involves three classes corresponding to different species of iris (*setosa*, *versicolor* and *virginica*) among which the classes *versicolor* and *virginica* are known to be difficult to discriminate. Let us now suppose that botanists are studying iris species and have only observed the two species *setosa* and *versicolor*. For this experiment, the data set has been randomly split into a learning data set without *virginica* examples and a test data set with several *virginica* examples. The left panel of Figure 5.11 shows what the botanists observe in the learning phase. The second left panel of the same figure presents a sample of new observations of iris to be classified. However, as the third panel indicates, this new sample contains individuals from a class which has not been observed by the botanists in the learning phase and the iris experts will likely classify all these new observations as belonging to either the class *setosa* or the class *versicolor*. The right panel of Figure 5.11 shows the result of such a scenario, using quadratic discriminant analysis (QDA), which yields a classification of all of the *virginica* observations into the class *versicolor*. Note that even though this result is disappointing from our point of view, it is understandable both for a human expert and a classification method since the classes *versicolor* and *virginica* are indeed difficult to discriminate.

The most related topic to supervised classification with unobserved classes is called novelty detection. Novelty detection focuses on the identification of new or unknown structure for which the learned classifier was not aware during the learning phase. Even though many methods are able to detect new or unobserved data points, none of them is able to recognize homogeneous groups of unobserved

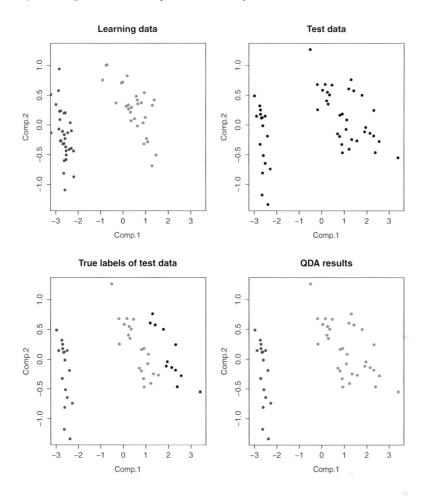

Figure 5.11 Supervised classification on the iris data set with one unobserved class (*virginica*) in the learning set.

points and to adapt the classifier to the new situation for classifying future observations. We focus on such methods in this section.

5.4.1 A Transductive Model-based Approach

A first work in the line of unobserved class detection is due to Miller and Browning (2003). They consider a mixture model with observed and unobserved classes where the observations can be labeled or unlabeled. Such a situation, which mixes labeled and unlabeled data for predicting the class membership of unlabeled data, is called transductive learning.

The model of Miller and Browning (2003) assumes a mixture model with G

components which they confront with the C known classes as follows:

$$p(y) = \sum_{g=1}^{G} \tau_g \beta_{cg} f(y; \theta_g),$$

where θ_g and $\tau_g \geq 0$ are respectively the density parameter and the prior probability of component g, with $\sum_{g=1}^{G} \tau_g = 1$, and $\beta_{cg} = P(c|\theta_g)$, such that $\sum_{g=1}^{G} \beta_{cg} = 1$ for $c \in \{1, ..., C, u\}$, parameterizes the conditional label generator. Notice that the specific label, u, is devoted to the meta-class of the unlabeled data and allows to confront the unsupervised partition with the supervised one. It also worth noticing that this mixture model is quite similar to the one used by Bouveyron and Girard (2009) for robust classification.

Given a set of $n = n_\ell + n_u$ observations, this model is therefore associated with the following log-likelihood function:

$$\ell(y_1, ..., y_n; \Theta) = \sum_{i=1}^{n} \log \sum_{g=1}^{G} \tau_g \beta_{c_i g} f(y_i; \theta_g)$$

$$= \sum_{i=1}^{n_\ell} \log \sum_{g=1}^{G} \tau_g \beta_{c_i g} f(y_i; \theta_g) + \sum_{i=n_\ell+1}^{n} \log \sum_{g=1}^{G} \tau_g \beta_{ug} f(y_i; \theta_g),$$

where $\Theta = \{\{\tau_g\}, \{\beta_{cg}\}, \{\theta_g\}\}$ is the whole set of model parameters, n_ℓ is the number of labeled observations and n_u is the number of unlabeled data. Miller and Browning (2003) propose to use the EM algorithm for inferring this model. The update equations for the EM algorithm are in this case as follows:

- E-step: This step aims to compute the expected complete data log-likelihood given the current parameter estimates and reduces here to computing the following quantities, first for the labeled observations:

$$t_{ig} = \begin{cases} \dfrac{\hat{\tau}_g^{(s-1)} \hat{\beta}_{c_i g}^{(s-1)} f(y_i; \hat{\theta}_g^{(s-1)})}{\sum_{\ell=1}^{C} \hat{\tau}_\ell^{(s-1)} \hat{\beta}_{c_i \ell}^{(s-1)} f(y_i; \hat{\theta}_\ell^{(s-1)})} & \text{if } g \in \{1, ..., C\} \\ 0 & \text{otherwise} \end{cases},$$

for $i = 1, ..., n_\ell$, and for the unlabeled observations with $i = n_\ell + 1, ..., n$, we have:

$$t_{ig} = \begin{cases} \dfrac{\hat{\tau}_g^{(s-1)} \hat{\beta}_{ug}^{(s-1)} f(y_i; \hat{\theta}_g^{(s-1)})}{\sum_{\ell=1}^{C} \hat{\tau}_\ell^{(s-1)} \hat{\beta}_{u\ell}^{(s-1)} f(y_i; \hat{\theta}_\ell^{(s-1)}) + \sum_{\ell=C+1}^{G} \hat{\tau}_\ell^{(s-1)} f(y_i; \hat{\theta}_\ell^{(s-1)})} & \text{if } g \in \{1, ..., C\} \\ \dfrac{\hat{\tau}_g^{(s-1)} f(y_i; \hat{\theta}_g^{(s-1)})}{\sum_{\ell=1}^{C} \hat{\tau}_\ell^{(s-1)} \hat{\beta}_{u\ell}^{(s-1)} f(y_i; \hat{\theta}_\ell^{(s-1)}) + \sum_{\ell=C+1}^{G} \hat{\tau}_\ell^{(s-1)} f(y_i; \hat{\theta}_\ell^{(s-1)})} & \text{otherwise.} \end{cases} \quad (5.9)$$

- M-step: this step aims to maximize the expected complete data log-likelihood, computed in the previous step. The update formulae for model parameters are fairly standard and depend on the chosen density $f(\cdot)$. The only specific update is the one of β_{cg}:

$$\hat{\beta}_{cg}^{(s)} = \frac{\sum_{y_i | c_i = c} t_{ig}}{n_g}, \quad \forall c \in \{1, ..., C\} \text{ and } \forall g \in \{1, ..., G\},$$

where $n_g = \sum_{i=1}^{n} t_{ig}$.

The classification is made as usual using the Bayes rule:

$$P(C = c|y_i; \theta) = \frac{\sum_{g=1}^{C} \tau_g \beta_{cg} f(y_i; \theta_g)}{\sum_{g=1}^{C} \tau_g f(y_i; \theta_g)}.$$

The a posteriori probability that an observation belongs to a new class is given by:

$$P(C = u|y_i) = 1 - \sum_{g=1}^{C} P(G = g|y_i; \theta),$$

where $P(G = g|y_i; \theta)$ is given by (5.9).

Notice that this model does not explicitly discover new classes and it is necessary to use a model selection criterion to perform the class discovery; Miller and Browning (2003) propose using a BIC criterion. Let us also remark that another model was proposed in Miller and Browning (2003), simply called Model II, which allows a class-conditional generating scheme for missing labels. This model turned out to be too flexible and was outperformed by the model which we have presented here.

5.4.2 An Inductive Model-based Approach

More recently, Bouveyron (2014) introduced a novel model-based approach for detecting unobserved classes and proposed two inference procedures for it. The author proposed in particular an inductive approach which is closer to classical supervised classification, in contrast to the transductive situation. Indeed, in such a situation, it is not required to have access to the learning data for the class prediction on a set of new data and for the discovery of potential new classes. In particular, the inductive approach is the only extant tenable approach for large data set classification and real-time dynamic classification.

The model is a simple parametric mixture model with G components, $G \geq C$: the observations $\mathcal{Y} = \{y_1, ..., y_n\} \in \mathbb{R}^p$ are assumed to be independent realizations of a random vector $Y \in \mathbb{R}^p$ with density:

$$f(y; \Theta) = \sum_{g=1}^{G} \tau_g f_g(y; \theta_g), \tag{5.10}$$

where $\tau_g \geq 0$ for $g = 1, ..., G$ are the mixing proportions (with the constraint $\sum_{g=1}^{G} \tau_g = 1$), $f_g(y; \theta_g)$ is the density of the gth component of the mixture parameterized by θ_g and finally $\Theta = \{\tau_1, ..., \tau_G, \theta_1, ..., \theta_G\}$. We refer to the previous chapters regarding the choice of the mixture densities.

The proposed approach involves two stages: a learning phase and a discovery phase.

The Learning Phase

In this first phase, only learning observations are considered and, since the data of the learning set are complete, i.e. a label z_i, where $z_{ic} = 1$ if observation i comes from class c, is associated to each observation in the learning set $(i = 1, ..., n)$, we fall into the classical estimation framework of model-based discriminant analysis. In such a case, the maximization of the likelihood reduces to separately estimating the parameters of each class density by maximizing the associated conditional log-likelihood

$$\ell_g(\mathcal{Y}; \Theta) = \sum_{i=1}^{n} z_{ig} \log\left(\tau_g f_g(y_i; \theta_g)\right),$$

for $g = 1, ..., C$, and this leads to estimating τ_g by $\hat{\tau}_g = n_g/n$ where $n_g = \sum_{i=1}^{n} z_{ig}$ is the number of observations of the kth class, and to estimating θ_g by $\hat{\theta}_g$, which depends on the chosen component density. For instance, in the case of a Gaussian density, the maximization of $\ell_g(\mathcal{Y}; \Theta)$ leads to an estimation of $\theta_g = \{\mu_g, \Sigma_g\}$ by $\hat{\mu}_g = \frac{1}{n_g} \sum_{i=1}^{n} z_{ig} y_i$ and $\hat{\Sigma}_g = \frac{1}{n_g} \sum_{i=1}^{n} z_{ig}(y_i - \hat{\mu}_g)(y_i - \hat{\mu}_g)^T$, for $g = 1, ..., C$.

The Discovery Phase

Usually, in discriminant analysis, the classification phase consists only in assigning new unlabeled observations to one of several known classes. However, here, it is assumed that some classes could not have been observed during the learning phase. It is therefore necessary to search for new classes before classifying the new observations in order to avoid the misclassification of observations from an unobserved class (by assigning them to one of the observed classes).

Using the model and the notation introduced above, it remains to find $G-C$ new classes in the set of n^* new unlabeled observations $\mathcal{Y}^* = \{y_1^*, ..., y_{n^*}^*\}$. Since these new observations are unlabeled, we have to fit the mixture model in a partially unsupervised way. In this case, the complete log-likelihood has the following form:

$$\ell_c(\mathcal{Y}^*; \Theta) = \sum_{i=1}^{n^*} \left(\sum_{g=1}^{C} z_{ig}^* \log\left(\tau_g f_g(y_i^*; \theta_g)\right) + \sum_{g=C+1}^{G} z_{ig}^* \log\left(\tau_g f_g(y_i^*; \theta_g)\right) \right),$$

where the parameters θ_g for $g = 1, ..., C$ have been estimated in the learning phase and the parameters θ_g for $g = C + 1, ..., G$ remain to be estimated. Due to the constraint on the parameters τ_g, i.e. $\sum_{g=1}^{G} \tau_g = 1$, the mixture proportions of the C known classes have to be re-normalized according to the proportions of the $G - C$ new classes which will be estimated on the new sample $\{y_1^*, ..., y_{n^*}^*\}$. However, the test set $\{y_1^*, ..., y_{n^*}^*\}$ is an incomplete data set since the labels z_i^* are missing and the z_{ig}^* are consequently unknown for all observations of this set.

In such a situation, direct maximization of the likelihood is an intractable problem and Bouveyron (2014) proposed a constrained EM algorithm for estimating the parameters of the $G - C$ unobserved classes which alternates between the following E- and M-steps at each iteration q:

- **E-step**: the conditional probabilities $t_{ig}^{*(s)} = P(Z = g | Y = y_i^*)$, for $i = 1, ..., n^*$

and $g = 1, ..., G$, are updated according to the mixture parameters as follows:

$$t_{ig}^{*(s)} = \frac{\hat{\tau}_g^{(s-1)} f_g(y_i^*; \hat{\theta}_g^{(s-1)})}{f(y_i; \hat{\Theta}^{(s-1)})},$$

where $\hat{\tau}_g^{(s-1)}$ and $\hat{\theta}_g^{(s-1)}$ are the mixture parameters estimated in the M-step at step $(s-1)$.

- **M-step**: the parameters of the $G - C$ unobserved classes are estimated by maximizing the conditional expectation of the completed likelihood whereas the estimated parameters of the observed classes remain fixed to the values obtained in the learning phase except for the proportions which are re-estimated. Therefore, this step only updates the estimates of parameters τ_g for $g = 1, ..., G$ and θ_g for $g = C + 1, ..., G$. In the case of the Gaussian mixture, the update formulae for the parameter estimates are:

$$\begin{cases} \text{for } g = 1, ..., C & \hat{\tau}_g^{(s)} = \left(1 - \sum_{\ell=C+1}^{G} \frac{n_\ell^{*(s)}}{n^*}\right) \frac{n_g}{n}, \\ \text{for } g = C + 1, ..., G & \hat{\tau}_g^{(s)} = \frac{n_g^{*(s)}}{n^*} \end{cases}$$

where $n_g^{*(s)} = \sum_{i=1}^{n^*} t_{ig}^{*(s)}$ and for $g = C + 1, ..., G$:

$$\hat{\mu}_g^{(s)} = \frac{1}{n_g^{*(s)}} \sum_{i=1}^{n^*} t_{ig}^{*(s)} y_i^*, \quad \hat{\Sigma}_g^{(s)} = \frac{1}{n_g^{*(s)}} \sum_{i=1}^{n^*} t_{ig}^{*(s)} (y_i^* - \hat{\mu}_g^{(s)})(y_i^* - \hat{\mu}_g^{(s)})^T.$$

This procedure has been called adaptive mixture discriminant analysis (AMDA) by Bouveyron (2014). Notice that, unlike in the transductive situation, it is only necessary in this case to initialize the parameters of the $G - C$ new mixture components; any initialization strategy discussed in Chapter 2.4 can be used to do this.

Class Discovery through Model Selection

As in the work of Miller and Browning (2003), AMDA does not explicitly discover new classes and it relies on model selection to effectively do it. In Bouveyron (2014), three model selection criteria (AIC, BIC and ICL) have been assessed in the context of class discovery and AIC turns out to be the most efficient one to detect small new classes. Note that the model complexity term involved in those criteria has to be adapted to the inductive context. For instance, if the classical Gaussian model is used within the transductive approach, $\nu(\mathcal{M})$ is equal to $(G-1)+Gd+Gd(d+1)/2$ whereas it is equal to $(G - C) + (G - C)d + (G - C)d(d+1)/2$ with the inductive approach.

We now briefly illustrate the use of AMDA, implemented in the `adaptDA` package, in combination with AIC on the iris data to detect some possible unobserved class. Let us consider once again the situation exposed at the beginning of this section: only the classes *setosa* and *versicolor* are assumed to have been observed and the *virginica* class is missing in the learning set. Listing 5.6 simulates such a situation from the original iris data set.

Figure 5.12 shows the classification results using QDA and AMDA. The bottom

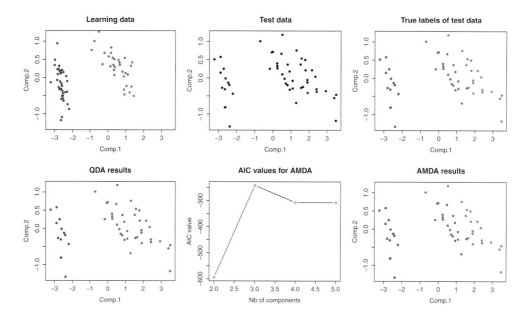

Figure 5.12 Discovery of one unobserved class (*virginica*) using the inductive approach AMDA (Bouveyron, 2014) on the iris data set. The model selection criterion used here is AIC. The data have been resampled such that the learning set does not contain *virginica* examples (see text for details).

middle panel allows to see the evolution of the AIC value regarding the number of components. It clearly peaks at $G = 3$, meaning that the test set contains one new class which has not been observed previously. Thus, unlike QDA which misclassifies all *virginica* examples (bottom left panel), AMDA succeeds in detecting and recovering the *virginica* class.

5.5 Bibliographic Notes

Semi-supervised Clustering and Classification

Semi-supervised classification has been, quite naturally, considered in different contexts such as skewed distributions (Vrbik and McNicholas, 2014). Regarding the constrained clustering problem, Li et al. (2015) have extended it to spectral clustering and Yi et al. (2015) considered sequential constraints in online clustering. Finally, classification under label noise is probably the topic which has received the most attention in recent years. Frénay and Verleysen (2014) have recently written a review on the subject which gathers generative and discriminative solutions to this problem.

Supervised Classification with Uncertain Labels

Early approaches that do not model label noise, tried to clean the data by removing misclassified instances using some kind of nearest neighbor algorithm (Gates, 1972; Dasarathy, 1980; Wilson and Martinez, 1997). Other works treat the noisy data using the C4.5 algorithm (John, 1995; Zhu et al., 2003), neural networks (Zeng and Martinez, 2003) or a saturation filter (Gamberger et al., 1999). Hawkins and McLachlan (1997) identified as outliers the data subset whose deletion gives the smallest value of the determinant of the within-group covariance matrix. Guyon et al. (1996) proposed to remove noisy observations with a cumulative information criterion and further checking by human experts. Other researchers proposed not to remove any learning instance and to build instead supervised classifiers robust to label noise. Bashir and Carter (2005) focused on robust estimation of model parameters in the mixture model context. Maximum likelihood estimators of the mixture model parameters are replaced by the corresponding S-estimators (see Rousseeuw and Leroy (1987) for a general account of robust estimation) but the authors only observed a slight reduction in the average probability of misclassification. Similarly, Mingers (1989), Sakakibara (1993) and Vannoorenbergue and Denoeux (2002) proposed noise-tolerant approaches to make decision tree classifiers robust to label noise. Boosting (Schapire, 1990; Quinlan, 1996) can also be used to limit the sensitivity of the built classifier to label noise.

Novelty Detection

Novelty detection has become popular in several application fields such as fault detection (Dasgupta and Nino, 2000), medical imaging (mass detection in mammograms, Tarassenko et al., 1995) or e-commerce (Manikopoulos and Papavassiliou, 2002). In recent years, many methods have been proposed to deal with this problem. An excellent review on novelty detection methods can be found in Markou and Singh (2003a,b) which splits novelty detection methods into two main categories: statistical and neural network based approaches. In addition to the methods presented in this chapter, among parametric techniques, Chow (1970) was the first to propose a threshold for outlier rejection which has been improved in Hansen et al. (1997) by introducing the classification confidence in the rejection. Gaussian densities were also used in Roberts and Tarassenko (1994) for modeling the learning data and detecting outliers using a measure based on the Mahalanobis distance. Extreme value theory was used in Roberts (1999) for novelty detection by searching for low or high values in the tails of data distributions. Nonparametric approaches include k-NN based techniques (Hellman, 1970; Odin and Addison, 2000) or Parzen windows (Yeung and Chow, 2002) for estimating the distribution of the data. More recently, Tax and Duin (1999) and Schölkopf et al. (1999) used support vector machines (SVM) for distinguishing between known and unknown objects. The problem has also been considered in the context of deep learning, but with limited success so far.

Listing 5.6: Class discovery with AMDA

```
# Loading libraries and setting up a seed
library(adaptDA)
set.seed(12345)

# Data simulation
data(iris); N = nrow(iris)
Z.data = iris[,-5]
Z.cls = as.numeric(iris[,5])
Z.cls[as.numeric(iris[,5]==2)] = 3
Z.cls[as.numeric(iris[,5]==3)] = 2

# Sampling
ind = sample(1:N,N)
X.data = Z.data[ind[1:(2*N/3)],]
X.cls = Z.cls[ind[1:(2*N/3)]]
X.data = X.data[X.cls!=3,]
X.cls = X.cls[X.cls!=3]
Y.data = Z.data[ind[(2*N/3+1):N],]
Y.cls = Z.cls[ind[(2*N/3+1):N]]

# Usual classification with QDA
c1 = qda(X.data,X.cls)
res1 = predict(c1,Y.data)

# Classification with AMDA
c2 = amdai(X.data,X.cls,model='qda')
B = rep(c(-Inf),5)
myPRMS <- vector(mode='list', length=7)
for (i in 2:5){
  myPRMS[[i]] = predict(c2,Y.data,K=i)
  B[i] = myPRMS[[i]]$crit$bic
}
res2 = myPRMS[[which.max(B)]]

# Classification results
cat("\tQDA:\t",sum(res1$class == Y.cls) / length(Y.cls),"\n")
print(table(res1$class,Y.cls))
   QDA:    0.64
       Y.cls
         1  2  3
     1 13  0  0
     2  0 19 18

cat("\tAMDAi:\t",sum(res2$cls == Y.cls) / length(Y.cls),"\n")
print(table(res2$cls,Y.cls))
   AMDAi:    1
       Y.cls
         1  2  3
     1 13  0  0
     2  0 19  0
     3  0  0 18
```

6

Discrete Data Clustering

This chapter is about mixture models for clustering when the data are categorical or count data. As we have seen in Chapter 2, the Gaussian distribution is the reference distribution for model-based clustering with continuous data. However, the Gaussian distribution is not well adapted for dealing with categorical or count data, because it is designed for continuous-valued data, not discrete data.

For categorical data the group conditional distributions are typically assumed to be multinomial, whereas they are most often assumed to be Poisson distributions for count data. We will first introduce the most commonly used form of model-based clustering for categorical data, namely the Latent Class Model (LCM). We will then present mixture models for contingency tables.

6.1 Example Data

In order to illustrate the presented models, several data sets will be considered and we will now introduce them. Two data sets (the carcinoma data set and the credit data set) concern categorical variables and one data set (the Vélib data set) involves count data. In Section 6.3, a prostate cancer data set will be briefly presented and used to illustrate a specific method to deal with data that include continuous, ordinal and categorical variables.

The Carcinoma Data Set

This data set consists of seven dichotomous variables providing the ratings by seven pathologists of 118 slides on the presence or absence of carcinoma in the uterine cervix. The purpose of Agresti (2002) when studying these data was to model inter-observer agreement. Here, the context is different. It is to analyze how subjects could be divided into groups depending upon the consistency of their diagnoses. In this chapter we use this small data set to exemplify the behavior of the models presented in Section 6.2.2 and the model selection criteria presented in Section 6.2.4.

The Credit Data Set

These data consist of 66 customers who took out loans from a credit company described with the following 11 categorical or ordinal variables:

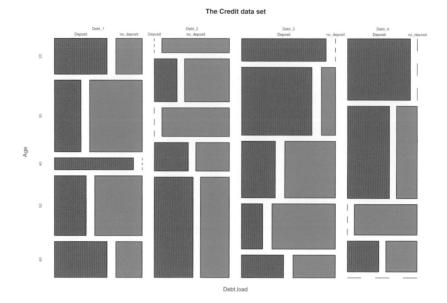

Figure 6.1 Credit data set: joint representation of the variables Age and Debt load. The surfaces of the rectangles are proportional to the frequency of the corresponding cells. Blue (resp. green) rectangles are associated with the level deposit (resp. no deposit).

- Loan: Renovation, Car, Scooter, Motorbike, Furnishings, Sidecar. This is the item for which customers took out a loan.
- Deposit: Yes, No. This indicates whether or not clients paid a deposit before taking out the loan.
- Unpaid: 0, 1 or 2, 3, more. This indicates the number of unpaid loan repayments.
- Debt load: 1 (low), 2, 3, 4 (high). This indicates the customer debt load.
- Insurance: No insurance, DI (Disability Insurance), TPD (Total Permanent Disability), Senior (for people over 60).
- Family: Common-law, Married, Widowed, Single, Divorced.
- Children: 0, 1, 2, 3, 4, more.
- Accommodation: Home owner, First-time buyers, Tenant, Lodged by family, Lodged by employer.
- Profession: Technician, Manual Laborer, Retired, Management, Senior management.
- Title: Mr., Mrs., Miss.
- Age: 20 (18 to 29 years old), 30 (30 to 39), 40 (40 to 49), 50 (50 to 59), 60 and over.

Figure 6.1 gives a joint representation of two important variables, Age and Debt Load. The aim of the study is to characterize the different banking behavior profiles.

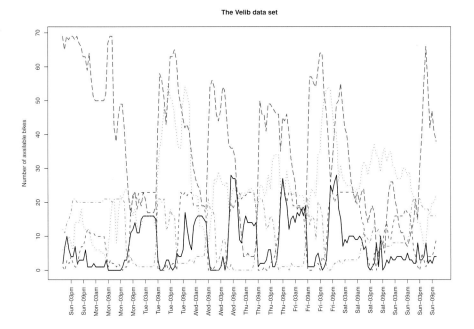

The Velib data set

Figure 6.2 Vélib data: an example of the load of four stations during a week.

The Vélib Data Set

This data set has been extracted from the Vélib large-scale bike sharing system of Paris. The data we used were studied by another method by Bouveyron et al. (2015). They concern station occupancy data collected on the Paris bike sharing system over five weeks, between February 24 and March 30, 2014. We focused on count data grouped by hour for the working days (Monday-Friday) of the week for the 18,000 available bikes at 1,230 stations. We restrict attention to those five days because the customers' behavior is expected to be different during the weekend. Figure 6.2 shows the load of several of the stations over the course of one week.

6.2 The Latent Class Model for Categorical Data

The observations to be clustered are described by d categorical variables. The jth variable has m_j response levels. The set of responses for observation i is denoted by y_i, coded by the vector $(y_i^j; j = 1, \ldots, d)$ for response level h for each variable j or, equivalently, by the binary vector $(y_i^{jh}; j = 1, \ldots, d; h = 1, \ldots, m_j)$ with

$$\begin{cases} y_i^{jh} = 1 & \text{if } y_i^j = h \\ y_i^{jh} = 0 & \text{otherwise.} \end{cases}$$

Latent class analysis was first introduced by Lazarsfeld (1950b) as a way of explaining respondent heterogeneity in survey response patterns involving

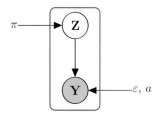

Figure 6.3 Graphical model of a latent class model (LCM).

dichotomous items. The standard LCM was formulated in its modern form by
Goodman (1974) and is as follows. (See its graphical sketch in Figure 6.3.) Data
are assumed to arise from a mixture of G multivariate multinomial distributions
with probability distribution function

$$f(y_i; \theta) = \sum_{g=1}^{G} \tau_g M_g(y_i; \alpha_g) = \sum_g \tau_g \prod_{j,h} (\alpha_g^{jh})^{y_i^{jh}},$$

where α_g^{jh} denotes the probability that variable j has level h if object i is in cluster
g, $\alpha_g = (\alpha_g^{jh}; j = 1, \ldots, d; h = 1, \ldots, m_j)$, $\mathbf{p} = (\tau_1, \ldots, \tau_G)$ denotes the vector of
mixing proportions of the G latent clusters, and $\theta = (\tau_g, \alpha_g, g = 1, \ldots, G)$ is the
vector parameter of the latent class model to be estimated.

The LCM assumes that the variables are *conditionally* independent knowing
the latent clusters. This is known as the *local independence* assumption. This
local independence assumption basically decomposes the dependence between the
variables into mixture components. It seems incongruous that this assumption is
widely made for latent class models, but would be regarded as rather restrictive
for mixture models of continuous data. The main reason is probably that there is
no single simple measure of dependence between discrete variables analogous to
the linear correlation coefficient for continuous data. Some efforts have been made
to avoid the local independence assumption. For instance, Bock (1986) proposed
a mixture of loglinear models. But more flexible models along these lines require
many more parameters to be estimated and have not been widely adopted.

In the same spirit, mixtures of latent trait models, which are particular cases
of the Rasch model (Agresti, 2002), have been proposed (Uebersax and Grove,
1993; Gollini and Murphy, 2014). For instance, the latent trait model used by
Gollini and Murphy (2014) assumes that there is a D-dimensional continuous
latent variable u underlying the behavior of the binary categorical variables y. It
assumes that

$$f(y) = \int f(y|u)du,$$

where the conditional distribution of y given u is

$$f(y|u) = \prod_{j=1}^{d} \pi_j(u)^{y_j} (1 - \pi_j(u))^{1-y_j},$$

and π_j is the logistic function

$$\pi_j(u) = f(y_j = 1|u) = \frac{1}{1 + \exp[-(b_j + w_j^T u)]}, 0 \le \pi(y) \le 1,$$

where b_j and w_j are the intercept and slope parameters in the logistic function. Estimating the parameters of the mixture in this latent trait model involves difficulties that Gollini and Murphy (2014) circumvent using a variational EM approximation. These authors show illustrations where this model, which avoids the local independence assumption, provides good and interpretable clusterings.

Ad hoc practices, such as pooling variables for which the local independence assumption should be avoided, could be used. But experience in many applications suggests that models that assume local independence can work well for bringing similar observations together, which one typically aims at in cluster analysis (Anderlucci and Hennig, 2012).

Identifiability of the LCM

A necessary condition for an LCM to be identifiable is that the number of possible values of the categorical variable table be greater than the number of parameters in the LCM (Goodman, 1974). For instance, for a standard LCM with G clusters on a binary table with d variables, this necessary condition is $2^d - 1 \ge (G-1) + dG$.

Strictly speaking, this necessary condition is not sufficient to ensure the identifiability of the LCM (Gyllenberg et al., 1994). Examples of simple binary mixtures leading to the same distribution were given by Carreira-Perpiñán and Renals (2000). Those authors argue that this lack of identifiability of the LCM is not an issue from the practical point of view. This claim has been confirmed by the work of Allman et al. (2009) who gave sufficient conditions under which the LCM is generically identifiable, meaning that the set of non-identifiable parameters has measure zero. This implies that the statistical inference remains meaningful even if the LCM is not strictly identifiable, as conjectured by Carreira-Perpiñán and Renals (2000). For binary mixtures with d variables and G clusters, the condition of Allman et al. (2009) is $d \ge 2\lceil \log_2 G \rceil + 1$, where $\lceil a \rceil$ is the smallest integer at least as large as a. For a mixture of d variables with m levels each, their condition is $d \ge 2\lceil \log_m G \rceil + 1$.

6.2.1 Maximum Likelihood Estimation

The log-likelihood of the LCM is

$$\mathcal{L}(\theta) = \mathcal{L}(\theta; y) = \sum_{i=1}^{n} \log \left(\sum_{g=1}^{G} \tau_g \prod_{j=1}^{d} \prod_{h=1}^{m_j} (\alpha_g^{jh})^{y_i^{jh}} \right).$$

We let $z = (z_1, \ldots, z_G)$ with $z_g = (z_{1g}, \ldots, z_{ng})$, and we define $z_{ig} = 1$ if y_i arose from cluster g and $z_{ig} = 0$ otherwise, where the z_{ig} are the unknown indicator vectors of the clusters. Then the complete-data log-likelihood is

$$\mathcal{L}_C(\theta) = \mathcal{L}(\theta; y, z) = \sum_{i=1}^{n} \sum_{g=1}^{G} z_{ig} \log \left(\tau_g \prod_{j=1}^{d} \prod_{h=1}^{m_j} (\alpha_g^{jh})^{y_i^{jh}} \right).$$

The updating formulae of the EM algorithm for finding the maximum likelihood (ML) estimates of this standard model by maximizing the conditional expectation of the complete log-likelihood are straightforward. Starting from an initial value of the parameters, $\theta^{(0)} = (\mathbf{p}^{(0)}, \alpha^{(0)})$, the two steps of the EM algorithm are as follows:

- E-step: calculation of $(z_{ig}^{(r)}, i = 1, \ldots, n, g = 1, \ldots, G)$ where $z_{ig}^{(r)}$ is the conditional probability that y_i arose from cluster g, namely

$$z_{ig}^{(r)} = \frac{\tau_g^{(r)} M_g(y_i; \alpha_g^{(r)})}{\sum_{\ell=1}^{G} \tau_\ell^{(r)} M_\ell(y_i; \alpha_\ell^{(r)})}.$$

- M-step: updating of the mixture parameter estimates,

$$\tau_g^{(r+1)} = \frac{n_g^{(r)}}{n}, g = 1, \ldots, G,$$

$$(\alpha_g^{jh})^{(r+1)} = \frac{(u_g^{jh})^{(r)}}{n_g^{(r)}}, j = 1, \ldots, d; h = 1, \ldots, m_j; g = 1, \ldots, G,$$

where $n_g^{(r)} = \sum_{i=1}^{n} z_{ig}^{(r)}$ and $(u_g^{jh})^{(r)} = \sum_{i=1}^{n} z_{ig}^{(r)} y_i^{jh}$.

Analyzing multivariate categorical data can be difficult because of the curse of dimensionality. The standard latent class model requires $(G-1) + G \sum_j (m_j - 1)$ parameters to be estimated. It is more parsimonious than the saturated loglinear model which requires $G \prod_j m_j$ parameters; this is simply the empirical multinomial model. For example, with $G = 5$, $d = 10$, $m_j = 4$ for all variables, the latent class model is defined by 154 parameters whereas the saturated loglinear model requires about one million parameters. However, the standard LCM can still be too complex in relation to the sample size n, and can involve numerical difficulties, such as "divide by zero" occurrences, or parameters α_g^{jh} estimated to be zero. To avoid such difficulties, using the regularized formula:

$$(\alpha_g^{jh})^{(r+1)} = \frac{(u_g^{jh})^{(r)} + c - 1}{n_g^{(r)} + m_j(c-1)},$$

c being some fixed positive number, can be recommended for updating the estimation of the α_g^{jh}s. The results can be sensitive to the choice of c, and considering this regularization issue in a Bayesian setting could be beneficial (see Section 6.2.7). In this perspective, more parsimonious LCMs could be thought of as desirable. Such models have been proposed by Celeux and Govaert (1991) and are now presented.

6.2.2 Parsimonious Latent Class Models

We first describe a model reparameterization which will allow us to define a family of parsimonious LCMs.

The Reparameterization

Denoting, for any variable j, a_g^j the most frequent response level in cluster g, the parameter α_g can be replaced by (a_g, ε_g) with $a_g = (a_g^1, \ldots, a_g^d)$ and with $\varepsilon_g = (\varepsilon_g^{11}, \ldots, \varepsilon_g^{dm_d})$ where

$$a_g^j = \arg\max_h \alpha_g^{jh} \quad \text{and} \quad \varepsilon_g^{jh} = \begin{cases} 1 - \alpha_g^{jh} & \text{if } h = a_g^j \\ \alpha_g^{jh} & \text{otherwise.} \end{cases}$$

For instance, for two variables with $m_1 = 3$ and $m_2 = 2$, if $\alpha_g = (0.7, 0.2, 0.1; 0.2, 0.8)$, the new parameters are $a_g = (1, 2)$ and $\varepsilon_g = (0.3, 0.2, 0.1; 0.2, 0.2)$. Vector a_g provides the modal levels in cluster g for the variables and the elements of vector ε_g can be regarded as scattering values.

Five Latent Class Models

Using this form, it is possible to impose various constraints on the scattering parameters ε_g^{jh}. We consider the following models.

- The standard LCM $[\varepsilon_g^{jh}]$: the scattering depends upon clusters, variables and levels.
- $[\varepsilon_g^j]$: the scattering depends upon clusters and variables but not upon levels.
- $[\varepsilon_g]$: the scattering depends upon clusters, but not upon variables.
- $[\varepsilon^j]$: the scattering depends upon variables, but not upon clusters and levels.
- $[\varepsilon]$: the scattering is constant over variables and clusters.

Some of these constrained models are unrealistic in many contexts. In particular, the model $[\varepsilon]$ appears simplistic. However, it could be useful when the number of observations is much smaller than the number of variables. Moreover, this simple model provides a probabilistic interpretation of the k-medoids clustering with the Manhattan distance (Hennig and Liao, 2013). From our experience, the model $[\varepsilon_g^j]$, which is equivalent to the general model $[\varepsilon_g^{jh}]$ when all the variables are binary, is of particular interest. It often offers a good compromise between parsimony and different within-cluster variances for different variables and clusters.

The $[\varepsilon_g^j]$ Model

In this model, the scattering vector parameter is characterized by the scattering value ε_g^{jh} of the modal level $h = a_g^j$. Denoting this value by ε_g^j, we have:

$$\varepsilon_g^{jh} = \begin{cases} \varepsilon_g^j & \text{if } h = a_g^j \\ \frac{\varepsilon_g^j}{m_j - 1} & \text{otherwise.} \end{cases}$$

This model has $(G-1) + Gd$ parameters, compared to the $(G-1) + G\sum_j(m_j - 1)$ parameters of the standard LCM, $[\varepsilon_g^{jh}]$. The model $[\varepsilon_g^j]$ was proposed by Aitchison and Aitken (1976) in a supervised classification context.

The model can be written

$$f(y_i; \theta) = \sum_{g=1}^{G} \tau_g \prod_{j=1}^{m_j} (1 - \varepsilon_g^j) \left(\frac{\varepsilon_g^j}{(m_j - 1)(1 - \varepsilon_g^j)} \right)^{1 - \delta(y_i^j, a_g^j)},$$

where δ is the Kronecker function. The complete-data log-likelihood is

$$\mathcal{L}_C(z, \theta) = \sum_{g=1}^{G} n_g \log \tau_g$$

$$+ \sum_{g=1}^{G} \sum_{j=1}^{d} \log \left(\frac{\varepsilon_g^j}{(m_j - 1)(1 - \varepsilon_g^j)} \right) \left(n_g - \sum_{i=1}^{n} z_{ig} \delta(x_i^j, a_g^j) \right)$$

$$+ \sum_{g=1}^{G} n_g \sum_{j=1}^{d} \log(1 - \varepsilon_g^j).$$

The maximum likelihood estimation of these parsimonious model parameters with the EM algorithm does not involve difficulties. The E-step remains unchanged and the M-step is as follows.

M-step

Updating the mixing proportions is unchanged and updating the a_g^j does not differ over the four parsimonious models.

$$(a_g^j)^{(r+1)} = \arg \max_h (u_g^{jh})^{(r)}.$$

Changes occur when updating the scattering parameters ε.

We let $u^{jh} = \sum_{g=1}^{G} u_g^{jh}$, and we distinguish the modal level a_g^j by denoting $v_g^j = u_g^{jh}$, $v_g = \sum_{j=1}^{d} v_g^j$, $v^j = \sum_{g=1}^{G} v_g^j$ and $v = \sum_{j=1}^{d} \sum_{g=1}^{G} v_g^j$ when h is the modal level a_g^j.

Updating the Scattering Estimate for the $[\varepsilon_g^j]$ Model

For this model we obtain

$$(\varepsilon_g^j)^{(r+1)} = \frac{n_g^{(r)} - (v_g^j)^{(r)}}{n_g^{(r)}}.$$

This means that $(\varepsilon_g^j)^{(r)}$ is the proportion of the time that the level of an object in cluster g is different from its modal level for variable j.

Scattering Estimate for Models $[\varepsilon_g]$, $[\varepsilon^j]$ and $[\varepsilon]$

- $(\varepsilon_g)^{(r+1)} = \frac{n_g^{(r)} d - v_g^{(r)}}{n_g^{(r)} d}$: this is the proportion of the time that the level of an object in cluster g is different from its modal level.
- $\varepsilon^j = \frac{n - (v^j)^{(r)}}{n}$: this is the proportion of the time that the level of an object is different from its modal level for variable j.
- $\varepsilon = \frac{nd - v^{(r)}}{nd}$: this is the proportion of the time the level of an object is different from its modal level.

Limitations

As noted by Vandewalle (2009, chapter 5, page 124), when the categorical variables do not have the same number of levels, the models $[\varepsilon_g]$ and $[\varepsilon]$ suffer from a possible inconsistency. When the scattering does not depend on the levels, the estimated probability of the modal level for a variable with few levels can be smaller than the estimated probability of the minority levels. Vandewalle (2009) proposed an alternative parameterization that avoids such an inconsistency. Unfortunately, the M-step of the resulting EM algorithm is no longer in closed form. In our opinion, when the categorical variables have different numbers of levels, models with scattering parameters that do not depend on the variables, such as models $[\varepsilon_g]$ and $[\varepsilon]$, are not relevant. In the following, we will assume m_j to be constant for these two models.

6.2.3 The Latent Class Model as a Cluster Analysis Tool

The LCM is a parsimonious hidden structure model using a conditional indepen- dence assumption to analyze relations between categorical variables. This model can be used for a variety of purposes, and we now discuss how it can be used for cluster analysis.

A latent class model, estimated for instance using the EM algorithm, leads to a clustering of the data in a simple way, using a classification step. This consists of assigning each object y_i to the cluster maximizing the conditional probability τ_{ig} that y_i arose from cluster g, $1 \leq g \leq G$, based on the maximum a posteriori (MAP) principle. Alternatively, a Classification EM, or CEM algorithm can be used to estimate the LCM parameters and the clustering z simultaneously (Celeux and Govaert, 1995). We now describe the CEM algorithm. The quantities n_g, u_g^{jh} and u^{jh} are as previously defined except that the conditional probabilities τ_{ig} are replaced by the labels z_{ig}.

CEM Algorithm

- E-step: as in the EM algorithm, calculate the current conditional probabilities $\tau_{ig}^{(r)}$ that y_i arose from cluster g, $(i = 1, \ldots, n, g = 1, \ldots, G)$.
- C-step: assign each observation to one of the g clusters using the MAP (maximum a posterior) operator:

$$z_{ig}^{(r+1)} = 1 \quad \text{if} \quad g = \arg\max_g \tau_{ig}^{(r)} \quad \text{and} \quad 0 \quad \text{otherwise.}$$

- M-step: update the parameter estimate by maximizing the complete-data log- likelihood $\mathcal{L}(\theta; y, z^{r+1})$. For the standard LCM, this leads to

$$p_g^{(r+1)} = \frac{n_g^{(r+1)}}{n} \quad \text{and} \quad (\alpha_g^{jh})^{(r+1)} = \frac{(u_g^{jh})^{(r+1)}}{n_g^{(r+1)}}.$$

Note that α_g^{jh} is simply the current relative frequency of level h for variable j in cluster g.

The CEM algorithm maximizes the complete-data log-likelihood $\mathcal{L}_C(\theta)$ rather than the actual log-likelihood $\mathcal{L}(\theta)$. Therefore, it produces biased estimates of the LCM parameters. On the other hand, it converges in a finite and typically small number of iterations. When the mixture components are well separated with similar proportions, the CEM algorithm is expected to provide a relevant clustering.

Clustering using the LCM allows one to highlight connections between the LCM and distance-based methods. Thus, Celeux and Govaert (1991) proved that maximizing the complete-data log-likelihood of the standard LCM is equivalent to maximizing the information criterion

$$H(z) = \sum_{g=1}^{g} \sum_{j=1}^{d} \sum_{h=1}^{m_j} u_g^{jh} \log u_g^{jh} - d \sum_{g=1}^{G} n_g \log n_g.$$

Moreover, several authors, including Benzecri (1973) and Govaert (1983), have shown that maximizing this information criterion $H(z)$ is practically equivalent to minimizing the popular χ^2 criterion

$$W(z) = \sum_{g=1}^{G} \sum_{j=1}^{d} \sum_{h=1}^{m_j} \frac{(n u_g^{jh} - n_g u^{jh})^2}{n n_g u^{jh}}.$$

Minimizing this χ^2 criterion can be performed with a k-means-like algorithm where the distance between observations is the χ^2 distance.

Comparing the LCM and distance-based methods in clustering is a subject of interest. Anderlucci and Hennig (2012) conducted an extensive simulation study for comparing the standard LCM estimated with the EM algorithm and a k-means type algorithm where the distance between the objects was the Manhattan distance. When the clusters were well separated with similar proportions, the two approaches performed similarly. In such cases, the distance-based approach could be preferred since it is faster. In other situations, such as when the clusters overlap, or the cluster proportions are unequal, the LCM gave more satisfying results. In particular, the adjusted Rand index (ARI) of Hubert and Arabie (1985) between the true clustering and the clustering derived with the EM algorithm applied to the LCM was better. See Anderlucci (2012) for details.

6.2.4 Model Selection

In the mixture context, the criteria BIC (Schwarz, 1978) and ICL (Biernacki et al., 2000) are popular for choosing a mixture model and especially the number of mixture components. If BIC can be recommended to select a model for density estimation, ICL could be preferred when the focus of the mixture analysis is clustering. Both criteria are asymptotic approximations of integrated likelihoods: BIC approximates the integrated observed-data likelihood while ICL-BIC approximates the complete-data integrated likelihood.

The integrated observed-data likelihood of an LCM is

$$p(y) = \int_{\Theta} f(y; \theta) p(\theta) d\theta, \tag{6.1}$$

where Θ is the parameter space and $p(\theta)$ is the prior distribution of the parameter θ. The asymptotic approximation of $\log p(y)$ is given by

$$\log p(y) = \log f(y; \hat{\theta}) - \frac{\nu}{2} \log n + O_{\mathrm{p}}(1), \tag{6.2}$$

where $\hat{\theta}$ is the maximum likelihood estimator of the model parameter and ν is the number of free parameters of the LCM at hand. This leads to the BIC criterion to be maximized:

$$\mathrm{BIC} = \log f(y; \hat{\theta}) - \frac{\nu}{2} \log n. \tag{6.3}$$

If BIC is motivated from a Bayesian point of view, it does not depend explicitly on the prior distribution, although in the context of other regular models it corresponds approximately to a unit-information prior (Raftery, 1995).

The integrated complete-data likelihood of an LCM is defined by

$$p(y, z) = \int_{\Theta} p(y, z; \theta) p(\theta) d\theta, \tag{6.4}$$

and the asymptotic BIC-like approximation of $\log p(y, z)$ is

$$\log p(y, z) = \log p(y, z; \hat{\theta}) - \frac{\nu}{2} \log n + O_{\mathrm{p}}(1). \tag{6.5}$$

Replacing the missing labels z by their MAP values $\hat{z}$ for $\hat{\theta}$ defined as

$$\hat{z}_{ig} = \begin{cases} 1 & \text{if } \arg\max_\ell t_{i\ell}(\hat{\theta}) = g \\ 0 & \text{otherwise,} \end{cases} \tag{6.6}$$

leads to the following ICL criterion (Biernacki et al., 2000):

$$\mathrm{ICL} = \log p(x, \hat{z}; \hat{\theta}) - \frac{\nu}{2} \log n. \tag{6.7}$$

This criterion aims at favoring mixture situations giving rise to a clear partitioning of the data and, as a result, it appears to be relatively insensitive to model misspecification (Biernacki et al., 2000).

For the standard LCM, using the conjugate priors defined in Section 6.2.7, namely Dirichlet priors with hyperparameter b for the mixing proportions and c for the α parameters, the integrated complete-data likelihood (6.4) can be derived in closed form (Biernacki et al., 2010). Replacing the missing labels z by $\hat{z}$, it is

$$
\text{ICL} = \quad \log \Gamma(bG) - G \log \Gamma(b) + \sum_{g=1}^{G} \log \Gamma(\hat{n}_g + b) - \log \Gamma(n + bG)
$$

$$
+ g \sum_{j=1}^{d} \left\{ \log \Gamma(cm_j) - m_j \log \Gamma(c) \right\}
$$

$$
+ \sum_{g=1}^{G} \sum_{j=1}^{d} \left\{ \sum_{h=1}^{m_j} \log \Gamma\left(\hat{u}_g^{jh} + c\right) - \log \Gamma(\hat{n}_g + cm_j) \right\}, \tag{6.8}
$$

where $\hat{n}_g = \#\{i : \hat{z}_{ig} = 1\}$ and $\hat{u}_g^{jh} = \#\{i : \hat{z}_{ig} = 1, y_i^{jh} = 1\}$.

For the parsimonious models, ICL can also be derived in closed form with the conjugate prior distributions given in Section 6.2.7, but it involves an increasing number of combinatorial sums to be calculated as the number of scattering parameters to be estimated decreases.

The AIC criterion,

$$
\text{AIC} = \log f(y; \hat{\theta}) - \nu, \tag{6.9}
$$

is known to tend to select mixture models that are too complex (McLachlan and Peel, 2000; Frühwirth-Schnatter, 2006). However, a slight modification of this criterion, namely

$$
\text{AIC3} = \log f(y; \hat{\theta}) - \frac{3}{2}\nu, \tag{6.10}
$$

can, surprisingly, outperform BIC for the LCM (Nadif and Govaert, 1998; Nadolski and Viele, 2004). All these criteria are compared in the next numerical experiments section.

6.2.5 Illustration on the Carcinoma Data Set

This data set includes 118 observations described by seven binary variables. In addition to the four models of Section 6.2.2 (for binary data, the models $[\varepsilon_g^{jh}]$ and $[\varepsilon_g^j]$ are equivalent), we define additional models by assuming the mixing proportions to either vary freely or be constrained to be equal. Therefore eight models are considered, ranging from the most complex model with free proportions and free scattering parameters $[\tau_g, \varepsilon_g^j]$, to the most parsimonious model with equal proportions and scattering parameter $[p, \varepsilon]$.

The maximum likelihood (ML) and regularized maximum likelihood (RML) estimates have been computed using the EM and EM-Bayes algorithms respectively for the eight models with a number of classes ranging from 1 to 10. For the RML estimates, the Bayesian hyperparameters have been set to $b = c = 2$. To select one of the estimated models, the criteria BIC, ICLexact, ICL, AIC and AIC3, described in Section 6.2.4, have been computed. The results obtained with these criteria, except AIC3, have been reported in Figure 6.4 for the maximum likelihood estimates and in Figure 6.5 for the RML estimates. The results for AIC3 are not reported here since in this simulation study the behavior of AIC3 was quite similar to that of AIC.

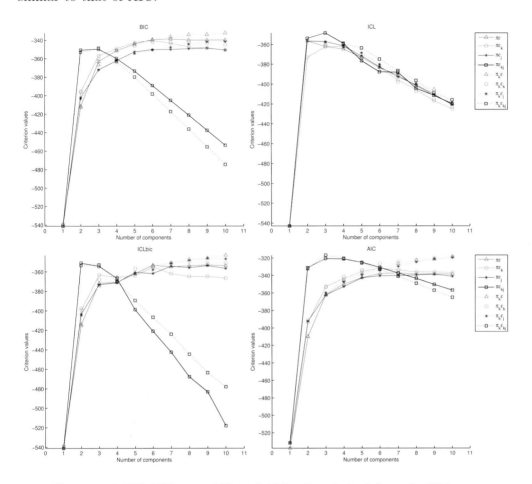

Figure 6.4 BIC, ICLexact, ICL and AIC values derived from the EM algorithm for the eight latent class models as a function of the number of classes.

Running Conditions of the Algorithms

In practical situations, finding the maximum likelihood estimates with an EM-like algorithm is often not easy. This is in part because the results can be sensitive to the starting values used, and finding good starting values is not always easy. There is no gold standard way to get optimal solutions.

For these experiments we used the em-EM strategy (Biernacki et al., 2003). This consists of repeating the following procedure N times:

1 carry out r short runs of the EM algorithm without waiting for its convergence by using loose criteria of convergence (the 'em' pass);

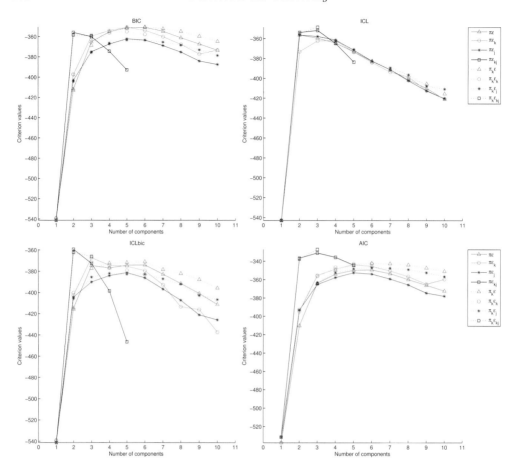

Figure 6.5 BIC, ICLexact, ICL and AIC values derived from the
EM-Bayes algorithm for the eight latent class models as a function of the
number of classes.

2 choose as initial position the solution providing the highest likelihood value
among these r short runs;

3 run the EM algorithm from this initial position with tight convergence criteria
(the 'EM' pass).

   With this kind of strategy, the choice of the criterion for stopping the EM
algorithm is crucial. Since EM may have a painfully slow convergence rate, we do
not recommend basing a stopping rule on the log-likelihood. In these experiments,
we stopped the algorithm as soon as one of the criteria $\|\varepsilon^{(r+1)} - \varepsilon^{(r)}\| < s$ or the
number of iterations greater than a maximum value ITMAX was satisfied.

   The following values have been considered for the em-EM strategy described
above: $N = 20$, $r = 50$, $s = 10^{-2}$ (resp. $s = 10^{-6}$) and ITMAX $= 100$ (resp.
ITMAX $= 2000$) for the em pass (resp. for the EM pass). Although we do not
claim that this em-EM strategy is optimal, this strategy does generally lead

to dramatically improved behavior of the EM algorithm. For instance, using a standard random initialization for EM often leads to non-increasing likelihood when the complexity of the mixture model increases (numerical experiments not reported here).

The main conclusions are as follows:

- BIC does not work so well since it chooses the model $[\tau_g, \varepsilon]$ with ten clusters for maximum likelihood and with six clusters for RML, and these solutions do not seem attractive options for this data set. On the other hand, ICL and AIC select the model $[\tau_g, \varepsilon_g^j]$ with three clusters described in Table 6.1.

Table 6.1 *Maximum likelihood estimates of a_g^j and ε_g^j for model $[\tau_g, \varepsilon_g^j]$.*

	1	1	1	1	1	1	1		.06	.14	.00	.00	.06	.00	.00
$a =$	2	2	1	1	2	1	2	$\varepsilon =$	.49	.00	.00	.06	.25	.00	.37
	2	2	2	2	2	1	2		.00	.02	.14	.41	.00	.48	.00

- Fixing the model to be $[\tau_g, \varepsilon_g^j]$, all criteria select three clusters.
- The results obtained by model $[p, \varepsilon_g]$ with three clusters (Table 6.2), even if not selected by any selection criterion, are interesting because they provide two very homogeneous classes (slightly more than those provided by the most general model). Furthermore, the values of ε_g (0.06, 0.16, 0.08) confirm the results generally obtained on this data with two homogeneous classes and a more diffuse class. The results obtained with the simplest model $[p, \varepsilon]$ confirm such a latent structure (same a_g^j's and $\varepsilon = 0.089$).

Table 6.2 *Maximum likelihood estimates of a_g^j and ε_g for model $[p, \varepsilon_g]$.*

	1	1	1	1	1	1	1		.03
$a =$	2	2	1	1	2	1	2	$\varepsilon =$	.16
	2	2	2	2	2	2	2		.08

- All these results have been obtained with the maximum likelihood methodology. The results with the RML approach are quite similar, as can be seen by comparing Table 6.3 with Table 6.1. But for the most complex models $[\tau, \varepsilon_g^j]$ and $[\tau_g, \varepsilon_g^j]$, the regularization tends to produce some identical mixture components. For the model $[\tau, \varepsilon_g^j]$ (resp. $[\tau_g, \varepsilon_g^j]$), it was not possible to get more than five (resp. three) different classes.

Table 6.3 *RML estimates of a_g^j and ε_g^j for model $[\tau_g, \varepsilon_g^j]$.*

	1	1	1	1	1	1	1		.08	.18	.02	.02	.08	.02	.02
$a =$	2	2	1	1	2	1	2	$\varepsilon =$	.48	.05	.07	.09	.24	.05	.36
	2	2	2	2	2	1	2		.02	.04	.17	.41	.03	.47	.02

Finally, to illustrate the best estimation (model $[\tau_g, \varepsilon_g^j]$ with three clusters), Figure 6.6 represents the projection of the data and the partition associated to the estimated model on the plane defined by the first two components obtained by multiple correspondence analysis (MCA), a factor analysis technique for nominal categorical data which is an extension of simple correspondence analysis (Greenacre and Blasius, 2006).

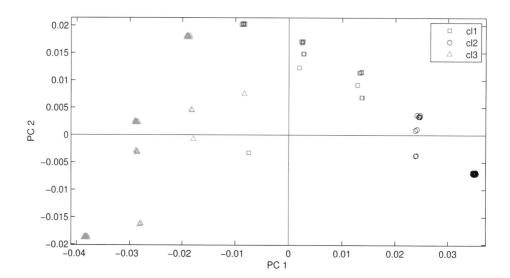

Figure 6.6 Carcinoma data: representation of the optimal three latent classes derived from the model $[\tau_g, \varepsilon_g^j]$ in the first plane of a multiple correspondence analysis.

6.2.6 Illustration on the Credit Data Set

This data set consists of 66 observations described by 11 categorical variables. Some of them are ordinal, but the order of the categories has not been considered here; we treat these variables as nominal. The aim of the study is to characterize the different banking behavior profiles. The variables in this data set have various numbers of levels and so we opt for the parsimonious model $[\tau_g, \varepsilon_{gj}]$. In this model, the scattering depends on the clusters and on the variables, but not on their levels, while the mixing proportions are free. Figure 6.7 displays the variations of BIC and ICL for a number of clusters varying from one to eight. BIC chooses the four-cluster solution while ICL prefers the three-cluster solution. Listing 6.1 presents some R code to run the LCM on the credit data set.

The three- and four-cluster solutions are displayed on the first plane of the multiple correspondence analysis in Figures 6.9 and 6.8 respectively. Figure 6.10 shows the $L/2$ most present levels and the $L/2$ most absent levels for each cluster

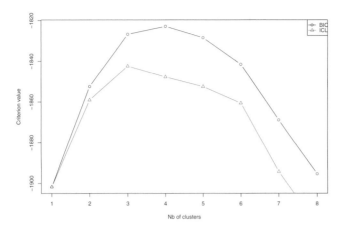

Figure 6.7 Credit data: BIC and ICL for the latent class model $[\tau_g, \varepsilon_{gj}]$.

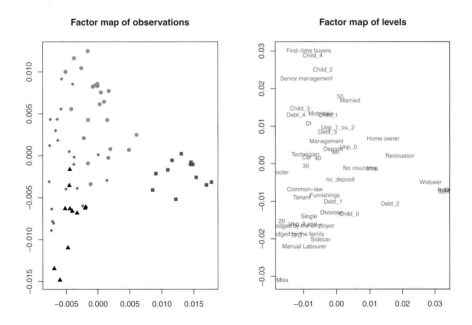

Figure 6.8 The four-cluster solution on the first MCA plane for the credit data set.

for the three-cluster solutions. It can be seen that Cluster 2 of the ICL clustering and Cluster 3 of the BIC clustering are identical. This is a cluster of home owners, retired seniors with no financial difficulties.

The two other clusters of the ICL clustering are easy to interpret. Cluster 1 is

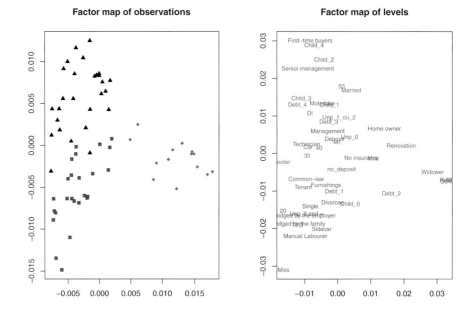

Figure 6.9 The three-cluster solution on the first MCA plane for the credit data set.

a cluster of young single people. They are tenants with a low social position who can have difficulty in making their loan payments. Cluster 3 is a cluster of men aged about 50 with a high social position. The BIC clustering splits Cluster 1 of the ICL clustering into two clusters, one cluster (Cluster 1) of young low-income families with children, and another (Cluster 4) of younger people with financial difficulties.

6.2.7 Bayesian Inference

Full Bayesian analysis is possible for the LCM. The conjugate prior distribution for a multinomial distribution $\mathcal{M}_r(q_1, \ldots, q_r)$ is the Dirichlet distribution $Dir(a, \ldots, a)$ with density

$$f(q_1, \ldots, q_r) = \frac{\Gamma(ra)}{\Gamma(a) \ldots \Gamma(a)} q_1^{a-1} \ldots q_r^{a-1},$$

where Γ denotes the gamma function. The mean and variance of each q_ℓ are $1/r$ and $(r-1)[/r^2(ra+1)]$ respectively. Thus, a Gibbs sampling implementation of Bayesian inference for the five latent class models can be derived in a straightforward way from the full conditional distributions which are now presented.

Listing 6.1: The analysis of the credit data set with the Latent Class Model

```
# Loading of the libraries and data
library(FactoMineR)
library(MBCbook)
data(credit)
X = credit
X$Age = as.factor(X$Age)


# Clustering with LCM through Rmixmod
res = mixmodCluster(X,nbCluster=1:8,dataType="qualitative"
                    ,model=mixmodMultinomialModel(listModels="
                        Binary_pk_Ekj"),
                    criterion = c("BIC","ICL"))


# Visualization of the best result according to BIC
par(mfrow=c(1,2))
lbl = res@results[[1]]@partition
acm = mca(X,abbrev = TRUE)
plot(predict(acm,X),col=lbl,pch=c(17,15,18,19)[lbl],
     xlab='',ylab='',main="Factor map of observations")
plot(acm,rows = FALSE,cex=0.75,main="Factor map of levels")


# Barplot of cluster-conditional level frequency
par(mfrow=c(1,1))
barplot(res)
```

Gibbs Sampling

In this paragraph, as in Section 6.2.3, conditional probabilities τ_{ig} are replaced by the labels z_{ig} in the definition of n_g, u_g^{jh}, v_g^j, v_g, v^j and v. For instance, $u_g^{jh} = \sum_{i=1}^n z_{ig} x_i^{jh}$ is now the number of occurrences of level h of variable j in cluster g.

For all the models, the prior distribution of the mixing weights $\tau_g, g = 1, \ldots, G$ is a $Dir(b \ldots, b)$ distribution. Then the full conditional distribution of $(\tau_g, g = 1, \ldots, G)$ is the Dirichlet distribution $Dir(b + n_1, \ldots, b + n_G)$.

For the standard LCM the prior distribution of $(\alpha_g^{j1}, \ldots, \alpha_g^{jm_j})$ is a $Dir(c, \ldots, c)$ for $g = 1, \ldots, G$ and $j = 1, \ldots, d$. Thus, the full conditional distribution for $(\alpha_g^{j1}, \ldots, \alpha_g^{jm_j}), j = 1, \ldots, d; g = 1, \ldots, G)$ is

$$(\alpha_g^{j1}, \ldots, \alpha_g^{jm_j}) \sim Dir(c + u_g^{j1}, \ldots, c + u_g^{jm_j}).$$

Results can be sensitive to the choice of the prior hyperparameters b and c. When b or c is equal to one, the Dirichlet prior distribution is the uniform distribution, which could be thought of as a natural choice. But, as discussed later and shown in Frühwirth-Schnatter (2011a), choosing b and c greater than 1 could be helpful to avoid numerical difficulties with the LCM.

For the parsimonious models presented in Section 6.2.2, the prior distribution will be defined by $p(\alpha_1, \ldots, \alpha_G) = p(a)p(\varepsilon)$ where $a = (a_1^1, \ldots, a_G^d)$ and $\varepsilon = (\varepsilon_1^1, \ldots, \varepsilon_G^d)$.

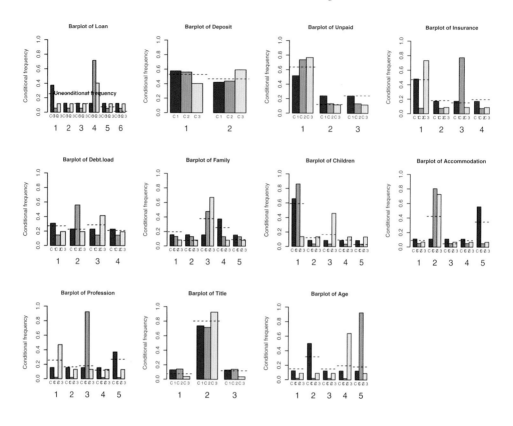

Figure 6.10 Credit data: for each cluster of the ICL clustering, the $L/2$ more present levels on the left side and the $L/2$ more absent levels on the right side.

A natural choice for the prior distribution of the modal levels a is

$$p(a) = \prod_{g=1}^{G} \prod_{j=1}^{d} p(a_g^j)$$

where the $p(a_g^j)$ are discrete uniform distributions on $\{1, \ldots, m_j\}$. Then the full conditional posterior probabilities $p(a_g^j = h \mid \ldots)$ are $\rho_g^{jh} = \frac{\gamma_g^{jh}}{\sum_{h=1}^{m_j} \gamma_g^{jh}}$ where

$\gamma_g^{jh} = \left(\frac{(m_j - 1)(1 - \varepsilon_g^j)}{\varepsilon_g^j} \right)^{u_{gj}^h}$, for $g = 1, \ldots, G$, $j = 1, \ldots, d$ and $h = 1, \ldots, m_j$.

Conditionally on the a_g^j and z, it can be shown that the distributions of the weighted means are binomial distributions:

- Model $[\varepsilon_g^j]$: $v_g^j \sim \mathcal{B}_i(n_g, 1 - \varepsilon_g^j)$ for $g = 1, \ldots, G$ and $j = 1, \ldots, d$;
- Model $[\varepsilon^j]$: $v^j \sim \mathcal{B}_i(n, 1 - \varepsilon^j)$ for $j = 1, \ldots, d$;
- Model $[\varepsilon_g]$: $v_g \sim \mathcal{B}_i(n_g d, 1 - \varepsilon_g)$ for $g = 1, \ldots, G$;
- Model $[\varepsilon]$: $v \sim \mathcal{B}_i(nd, 1 - \varepsilon)$.

Choosing a prior distribution is less simple for the scattering parameters and different views are possible. However, it seems that the most reasonable choice is to ensure the exchangeability of the levels. This encourages the method not to favor the modal level.

We denote by $\mathcal{B}t_{[a,b]}(p,q)$ the truncated Beta distribution of the parameters p and q defined on the interval $[a,b] \subseteq [0,1]$. Then the prior and full conditional posterior distributions of the scattering parameters for the different parsimonious models are

- Model $[\varepsilon_g^j]$. For $g = 1, \ldots, G$ and $j = 1, \ldots, d$
  - Prior: $\varepsilon_g^j \sim \mathcal{B}t_{[0,(m_j-1/m_j)]}((m_j-1)(c-1)+1, c)$.
  - Full conditional posterior: $\varepsilon_g^j \sim \mathcal{B}t_{[0,(m_j-1)/m_j]}(n_g - v_g^j + (m_j-1)(c-1)+1, v_g^j + c)$.

- Model $[\varepsilon^j]$. For $j = 1, \ldots, d$
  - Prior: $\varepsilon^j \sim \mathcal{B}t_{[0,(m_j-1/m_j)]}(g(m_j-1)(c-1)+1, G(c-1)+1)$.
  - Full conditional posterior: $\varepsilon^j \sim \mathcal{B}t_{[0,(m_j-1)/m_j]}(n - v^j + G(m_j-1)(c-1)+1, v^j + G(c-1)+1)$.

- Model $[\varepsilon_g]$. For $g = 1, \ldots, G$
  - Prior: $\varepsilon_g \sim \mathcal{B}t_{[0,(m-1/m)]}(d(m-1)(c-1)+1, d(c-1)+1)$.
  - Full conditional posterior: $\varepsilon_g \sim \mathcal{B}t_{[0,(m-1)/m]}(n_g d - v_g + d(m-1)(c-1)+1, v_g + d(c-1)+1)$.

- Model $[\varepsilon]$:
  - Prior: $\varepsilon \sim \mathcal{B}t_{[0,(m-1/m)]}(Gd(m-1)(c-1)+1, d(c-1)+1)$.
  - Full conditional posterior: $\varepsilon \sim \mathcal{B}t_{[0,(m-1)/m]}(nd - v + Gd(m-1)(c-1)+1, v + dG(c-1)+1)$.

Recall that for the last two models the number of levels is assumed to be the same for each variable, i.e. $m_j = m$ for $j = 1, \ldots, d$.

The choice of the hyperparameter c is important. Comparing the maximum likelihood and Bayesian formulae shows that c can be regarded as the number of prior "observations" that have been added to the observed sample in each variable level and each cluster by the Bayesian "expert". Thus, the choice $c = 1$ could be regarded as a (minimal) default choice. However, taking $c > 1$ could help to avoid numerical problems, as for the standard LCM with Dirichlet prior distributions.

The Label-switching Problem

Because the prior distribution is symmetric in the components of the mixture, the posterior distribution is invariant under a permutation of the component labels (McLachlan and Peel, 2000, Chapter 4). This lack of identifiability of a mixture is known as the *label-switching* problem. In order to deal with this problem, several authors have proposed efficient clustering-like procedures to possibly change the component labels of the simulated values for θ during a Gibbs sampling as Stephens (2000b), Celeux et al. (2000) or Papastamoulis and Iliopoulos (2010). However, those procedures have to choose among $G!$ permutations of the labels and become difficult to perform as soon as $G > 6$. Alternative procedures using a point-process representation avoid this difficulty but require data analyst skills (Frühwirth-Schnatter, 2011a). In any case, label-switching remains a difficulty when dealing with full Bayesian inference for mixture models. This could be a reason for preferring maximum likelihood inference to Bayesian inference via an MCMC algorithm for estimating the parameters of an LCM (see also Uebersax, 2010).

Regularized Maximum Likelihood through Bayesian Inference

Bayesian inference can be useful as an alternative to maximum likelihood to avoid numerical traps related to the curse of dimensionality. Bayesian inference introduces additional information which can be useful for solving ill-posed problems. It turns out that Bayesian maximum a posteriori (MAP) estimation by maximizing the posterior probability, $p(\theta \mid y)$, using the prior distributions defined previously, can help solve the numerical difficulties while avoiding the label-switching problem.

It can be carried out using the EM algorithm with little additional effort beyond what is needed for maximum likelihood estimation. The E-step remains unchanged, and updating the mixture parameter estimates in the M-step leads to

$$p_g^{(r+1)} = \frac{n_g^{(r)} + b - 1}{n + G(b-1)}, g = 1, \ldots, G,$$

for the mixing proportions, and to

$$(a_g^{jh})^{(r+1)} = \frac{(u_g^{jh})^{(r)} + c - 1}{n_g^{(r)} + m_j(c-1)},$$

with $g = 1, \ldots, G$, $j = 1, \ldots, d$ and $h = 1, \ldots, m_j$ for the standard LCM model ($[\varepsilon_g^{jh}]$).

For the parsimonious models, the update of the modal level is as before, namely

$$(a_g^j)^{(r+1)} = \arg\max_h (u_g^{jh})^{(r)},$$

as for the EM algorithm. The updates of the scattering parameters are as follows:

- Model $[\varepsilon_g^j]$. For $g = 1, \ldots, G$ and $j = 1, \ldots, d$

$$\varepsilon_g^j = \frac{n_g - v_g^j + (m_j - 1)(c-1)}{n_g + m_j(c-1)}.$$

- Model $[\varepsilon^j]$. For $j = 1, \ldots, d$

$$\varepsilon^j = \frac{n - v^j + G(m_j - 1)(c-1)}{n + Gm_j(c-1)}.$$

- Model $[\varepsilon_g]$. For $g = 1, \ldots, G$

$$\varepsilon_g = \frac{n_g d - v_g + d(m-1)(c-1)}{n_g d + dm(c-1)}.$$

- Model $[\varepsilon]$.

$$\varepsilon = \frac{nd - v + Gd(m-1)(c-1)}{nd + Gdm(c-1)}.$$

Note the similarity between this M-step and the M-step of the standard EM algorithm. This algorithm does not give the maximum likelihood estimators, but provides a regularized alternative to the maximum likelihood estimators. Moreover, it is not jeopardized by the label-switching problem since it is restricted to find the maximum a posteriori of the model parameters.

The role of the hyperparameters b and c is important: the larger they are, the more the estimates are shrunk to a common value. We refer to Frühwirth-Schnatter (2011a) for a thorough analysis of the role of these hyperparameters and we note that she advocates taking them greater than four. In any case, looking at the formulae of the EM-Bayes algorithm, it is clear that it is necessary to take b and c greater than one to get regularized parameter estimates. Taking b and c equal to one produces no regularization, and taking b or c smaller than one if anything exacerbates numerical problems. Typically this EM-Bayes algorithm, which avoids the label switching problem, can be preferred to EM when the sample size is not large with respect to the number of parameters of the LCM to be estimated.

6.3 Model-based Clustering for Ordinal and Mixed Type Data

We now consider the analysis of data sets where the categorical data are ordinal or combine categorical, ordinal and continuous variables.

6.3.1 Ordinal Data

A simple way to deal with ordinal data is to consider them as categorical data without taking the order into account. But ignoring the order can lead to an important loss of information and produce poor results. However, few specific latent class models for the clustering of ordinal data have been proposed. A fully developed model is proposed in the Latent Gold software (Vermunt and Magidson, 2005). It is related to the adjacent category logit model (Agresti, 2002), which was conceived to model an ordinal variable x with m levels when a covariate y is available. It is defined by the equation

$$\log \frac{x_h}{x_{h+1}} = \beta_h + \gamma y, \tag{6.11}$$

for $h = 1, \ldots, m-1$.

The extension of this model to a latent class model for ordinal variables proposed by Vermunt and Magidson (2005) under the local independence assumption consists of applying the adjacent category logit model where the latent class indicator is considered as a covariate. With the notation used in this chapter, it leads to the following conditional model:

$$\alpha_g^{jh} = \beta_h^j + \gamma_g, \tag{6.12}$$

for $j = 1, \ldots, d$, $h = 1, \ldots, m_j$ and $g = 1, \ldots, G$.

6.3.2 Mixed Data

When facing mixed data, the main problem is to deal with nominal and continuous data in the same exercise. In principle there is no difficulty in dealing with such data sets in the model-based clustering framework. Assuming that the continuous and the categorical variables are conditionally independent knowing the cluster memberships, the continuous data could be assumed to arise from a multivariate Gaussian distribution while the categorical variables are assumed to arise from an LCM. In these conditions, it is straightforward to derive the formulae for the EM algorithm, and the parameters of this mixture model can be estimated without particular numerical difficulty.

In this setting, a sensible strategy when concerned with a similar number of continuous and categorical variables is to use a mixed mixture model imposing a local independence assumption for all the variables in order to deal with both types of variables in a similar way. This leads to the assumption that the Gaussian mixture components for the continuous variables have diagonal covariance matrices while the categorical variables are assumed to arise from a mixture of locally independent multinomial variables. This is the option proposed in `Rmixmod`.

It is often the case that there are many more categorical variables than continuous variables. This typically arises in the analysis of social surveys, where there are many asked to which the answers are categorical, and a small number of continuous variables are also involved in the analysis, such as age and income. In such a case, a possibility could be to transform the few continuous variables into categorical data and then to use an LCM on the resulting data set. Transforming continuous data to categorical data could be thought of as a natural and relatively easy way to get a homogeneous resulting data set, and it is certainly an often used way of dealing with mixed data. However, a drawback of this strategy is that recoding continuous variables into categorical variables could lead to a serious loss of information in many situations (see for instance Celeux and Robert (1993) for a case study illustrating this issue).

6.3.3 The ClustMD Model

When there are many more continuous variables than categorical or ordinal variables, a promising model-based approach has been proposed in McParland and Gormley (2016), implemented in the clustMD R package. The clustMD model is a Gaussian mixture model where ordinal variables are the categorical manifestations of latent univariate Gaussian variables and nominal variables are the categorical manifestations of latent multivariate variables.

For an ordinal variable y with L levels, let γ denote a $(K + 1)$-vector of thresholds that partition the real line. The associated latent variable z which follows a Gaussian distribution $\mathcal{N}(\mu, \sigma^2)$ determines the observed y by the equation

$$P(y = k) = \Phi(\frac{\gamma_k - \mu}{\sigma}) - \Phi(\frac{\gamma_{k-1} - \mu}{\sigma}), \tag{6.13}$$

where Φ is the cumulative distribution function of a standard Gaussian distribution. A natural choice of the threshold γ_k is $\Phi^{-1}(\delta_k)$, where δ_k is the proportion of the observed values which are less than or equal to level k.

For a nominal variable with K levels, the underlying continuous vector z with $K - 1$ dimensions is assumed to follow a multivariate Gaussian distribution with mean μ and covariance matrix Σ. The observed nominal response y is a manifestation of the values of the elements of z relative to each other and to a threshold assumed to be 0, namely

$$y = \begin{cases} 1 & \text{if } \max_s z^s < 0 \\ k & \text{if } z^{k-1} = \max_s z^s \text{ and } z^{k-1} > 0 \text{ for } s = 2, \ldots, K. \end{cases}$$

In this framework McParland and Gormley (2016) considered different models corresponding to different assumptions on the Gaussian covariance matrices.

The parameters of the clustMD models are estimated by maximum likelihood with the EM algorithm. In the presence of nominal variables, the E-step of the EM algorithm requires approximating the conditional probabilities of the latent data and cluster labels by Monte Carlo simulations of the current multivariate Gaussian distributions. The BIC criterion to select a clusterMD model and in particular the number of clusters cannot be evaluated since the observed likelihood relies on the calculation of intractable integrals. Thus McParland and Gormley (2016) proposed approximating the observed likelihood by the product of the marginal distributions to get an approximation to the BIC criterion.

6.3.4 Illustration of ClustMD: Prostate Cancer Data

An illustration of the clustMD methodology is now given on a prostate cancer data set (Byar and Green, 1980), used in McParland and Gormley (2016). Twelve mixed measurements (eight are continuous, three are ordinal and one is nominal) are available for 475 patients who were diagnosed as having either stage 3 or 4 prostate cancer. In Figure 6.11, the mean values of the 12 variables for each stage are displayed. It can be seen that these mean values are similar except for the last two continuous variables.

We ran the R package clustMD and the package Rmixmod which allow us to treat mixed variables by assuming that the variables are conditionally independent given the cluster memberships. Recall that this means that the mixture covariance matrices for the continuous variables are assumed to be diagonal. ClustMD chose a three-cluster solution. Using BIC, Rmixmod chose the most possible complex model with three components and it chose the same model with two components with the ICL criterion. From Table 6.4, it appears that the two-cluster solution selected by Rmixmod with ICL is closely related to the two stages of cancer. Figures

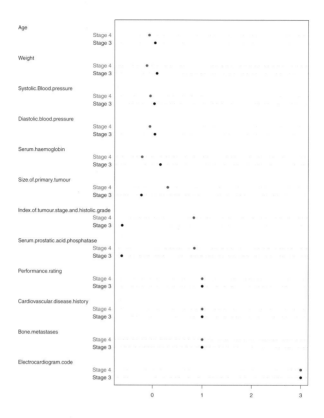

Figure 6.11 Prostate cancer data: the mean values of the two cancer stages 3 and 4 for the 12 variables; from top to bottom: the 8 continuous variables, the 3 ordinal variables and the nominal variable.

6.12 and 6.13 displays the mean values of the variables according to the 3-cluster `clustMD` and `Rmixmod` solutions.

Table 6.4 *Confusion tables of the clusterings (in row) and the two stages of cancer (in column).*

	3	4
1	207	14
2	45	13
3	21	175

(a) clustMD $G = 3$

	3	4
1	18	180
2	141	17
3	114	5

(b) LCM $G = 3$

	3	4
1	23	180
2	250	22

(c) LCM $G = 2$

Any of the above-mentioned strategies could be preferred if one wants a clustering that takes all variables into account. See Hennig and Liao (2013) for an interesting case study comparing different clustering approaches from this point of view. Otherwise, other strategies could be more beneficial in some situations.

Listing 6.2: Comparison of clustMD and LCM on the prostate data set

```
# Loading of libraries and data
library(clustMD)
library(MBCbook)
data(Byar)

# Transformation of skewed variables and
# ordering according to variable types
Byar$Size.of.primary.tumor <- sqrt(Byar$Size.of.primary.tumor)
Byar$Serum.prostatic.acid.phosphatase <- log(Byar$
    Serum.prostatic.acid.phosphatase)
Y <- as.matrix(Byar[,c(1,2,5,6,8,9,10,11,3,4,12,7)])
Y[, 9:12] <- Y[, 9:12] + 1
Y[, 1:8] <- scale(Y[, 1:8])
Yekg <- rep(NA, nrow(Y))
Yekg[Y[,12]==1] <- 1
Yekg[(Y[,12]==2)|(Y[,12]==3)|(Y[,12]==4)] <- 2
Yekg[(Y[,12]==5)|(Y[,12]==6)|(Y[,12]==7)] <- 3
Y[, 12] <- Yekg

# Clustering with clustMD
res <- clustMD(X=Y, G=3, CnsIndx=8, OrdIndx=11, Nnorms=20000,
    MaxIter=500, model="EVI", store.params=FALSE, scale=TRUE,
    startCL="kmeans")

colnames(res$means) = paste('Grp.',1:3)
rownames(res$means) = c(colnames(Y)[1:11],paste(colnames(Y)
    [12],1:2))
dotchart(t(res$means),col=1:3,pch=19)

# Clustering with Rmixmod
Y = as.data.frame(Y)
Y[,9] = as.factor(Y[,9]); Y[,10] = as.factor(Y[,10])
Y[,11] = as.factor(Y[,11]); Y[,12] = as.factor(Y[,12])
out = mixmodCluster(Y,1:4,criterion='BIC')

means = matrix(NA,ncol(Y),4)
means[1:8,] = t(out@bestResult@parameters@g_parameter@mean)
means[9:12,] = t(out@bestResult@parameters@m_parameter@center)
colnames(means) = paste('Grp.',1:4); rownames(means) = colnames(Y)
dotchart(t(means),col=1:3,pch=19)
```

The difficulty when dealing with mixed data is not numerical, it rather lies in a balanced consideration of the two types of variables. When there are many more continuous variables than categorical variables, it could be recommended to use a Gaussian mixture model to cluster the continuous data and then to use the categorical data as illustrative variables to interpret the resulting clusters. When there are many more categorical variables than continuous variables, it could be recommended to use an LCM to cluster the categorical data and then to use the continuous data as illustrative variables to interpret the resulting clusters through a classification procedure for instance.

When there are roughly equal numbers of continuous and categorical variables,

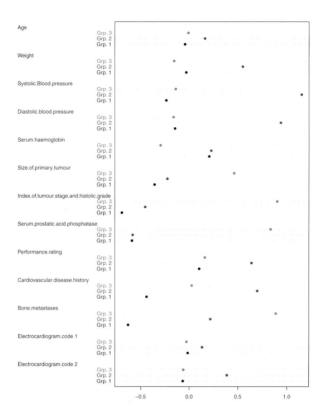

Figure 6.12 The mean values for the clustMD solutions; from top to bottom, the 8 continuous variables, the 3 ordinal variables and the nominal variable.

an alternative strategy would be to treat the continuous and the categorical variables in two separate clustering analyses (derived from an unrestricted Gaussian mixture model for the continuous variables and from an LCM for the categorical variables). The final clustering would then be obtained by intersecting the two resulting clusterings. This approach of separating the two cluster analyses is often relevant because the continuous variables and the categorical variables often do not give the same view on the objects to be analyzed.

6.4 Model-based Clustering of Count Data

In this section we consider count data where statistical units are described by counts on variables. High-throughput sequencing data sets to study genome expression are examples of such count data.

Contingency tables that display relationships between two categorical variables can also be analyzed using such methods, as we will see. Conversely, count data

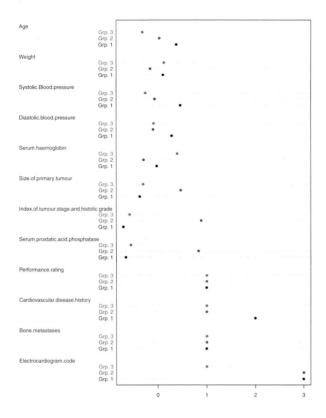

Figure 6.13 The mean values for the 3-cluster Rmixmod solution: from top to bottom, the 8 continuous variables, the 3 ordinal variables and the nominal variable.

can be viewed as a special case of a contingency table, in spite of possible semantic differences. Thus we adopt a unique notation for the two kinds of data.

The contingency table is denoted by y; it is a $n \times d$ data matrix defined by $y = \{(y_{ij}); i \in I, j \in J\}$, where I is a categorical variable with n categories and J a categorical variable with d categories. We write $y_{i.} = \sum_j y_{ij}$, $y_{.j} = \sum_i y_{ij}$ and $s = \sum_{ij} y_{ij}$. We shall also use the frequency table $\{(f_{ij} = y_{ij}/s); i \in I, j \in J\}$, the marginal frequencies $f_{i.} = \sum_j f_{ij}$ and $f_{.j} = \sum_i f_{ij}$, the row profiles $f_J^i = (f_{i1}/f_{i.}, \ldots, f_{is}/f_{i.})$ and the average row profile $f_J = (f_{.1}, \ldots, f_{.s})$.

6.4.1 Poisson Mixture Model

The Poisson distribution is a reference distribution for count data (Agresti, 2002). The probability distribution function of a Poisson distribution Y with parameter

λ is

$$\text{Pois}(y|\lambda) = \frac{e^{-\lambda}\lambda^y}{y!}, \tag{6.14}$$

where y is a non-negative integer. It satisfies $E(Y) = \text{var}(Y) = \lambda$. In the multivariate setting, the most used multivariate Poisson distribution assumes that the variables are independent. Alternative multivariate Poisson distributions exist (Karlis, 2003) but are difficult to analyze, especially for high-dimensional data.

For this reason, the variables are generally assumed to be independent conditionally on the class the observation belongs to. Also, $y_{i|g}$, the conditional distribution of the variable y_i given that observation i belongs to class g, is assumed to follow a Poisson distribution. We denote this by $y_{i|g} \sim \prod_{j=1}^{d} \text{Pois}(y_{ij} \mid \mu_{ijg})$, where $\text{Pois}(\cdot)$ denotes the Poisson density. This leads to the following Poisson mixture model (PMM):

$$f(y_i|\mu_i, \tau) = \sum_{g=1}^{G} \tau_g \prod_{j=1}^{d} \text{Pois}(y_{ij}|\mu_{ijg}),$$

where $\sum_{g=1}^{G} \tau_g = 1$ and $\tau_g \geq 0$ for $g = 1, \ldots, G$. The unconditional mean and variance of Y_{ij}, respectively, are

$$E(Y_{ij}) = \sum_{g=1}^{G} \tau_g \mu_{ijg}$$

and

$$\text{Var}(Y_{ij}) = \sum_{g=1}^{G} \tau_g \mu_{ijg} + \sum_{g=1}^{G} \tau_g \mu_{ijg}^2 - \left(\sum_{g=1}^{G} \tau_g \mu_{ijg}\right)^2.$$

Following Rau et al. (2015), we consider the following parameterization for the mean μ_{ijg}:

$$\mu_{ijg} = w_i \lambda_{jg} \tag{6.15}$$

where w_i is the intensity level for observation i and $\lambda_g = (\lambda_{1g}, \ldots, \lambda_{dg})$ corresponds to the clustering parameters that define the profiles of the observations in cluster g across all variables. The y_{ij} are assumed to be independent given the z_i (latent class model). Each y_{ij} will be distributed according to the Poisson distribution with mean $w_i \lambda_{jg}$. Thus the Poisson parameter is expressed as the product of two quantities, namely w_i, the effect of the row i, and λ_{jg}, the effect of the component g on the variable j. The probability distribution $f_g(y_i; \theta)$ is then a product of Poisson distributions

$$f_g(y_i; \theta) = \prod_j \frac{e^{-w_i \lambda_{jg}}(w_i \lambda_{jg})^{y_{ij}}}{y_{ij}!},$$

where $\theta = (\tau_1, \ldots, \tau_g, w_1, \ldots, w_n, \lambda_{11}, \ldots, \lambda_{dg})$.

If the parameters w_i and λ_{jg} in Equation (6.15) are left unconstrained, the model is not identifiable. As such, we consider the constraint $\sum_j \lambda_{jg} = 1$ for all

$g = 1, \ldots, G$. The interpretation of this constraint is as follows: the parameters λ_{jg} represent the percentage of total counts per observation that are attributed to each variable.

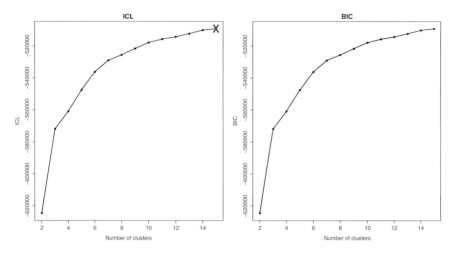

Figure 6.14 Behavior of BIC and ICL for the Poisson mixture model on the Vélib data set for a number of clusters varying from 2 to 15.

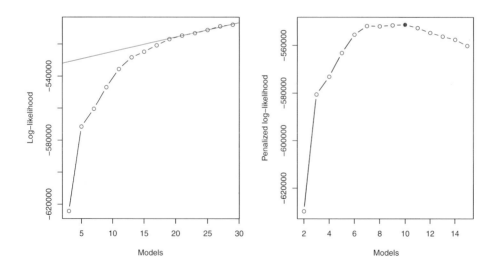

Figure 6.15 Behavior of the slope heuristic on the Vélib data set for a number of clusters varying from 2 to 15.

The parameters τ, w_i and λ_{jg} of the resulting mixture of Poisson distributions can be estimated with the EM algorithm. This algorithm has been implemented in the R package `HTSCluster`.

The M-step is straightforward and consists of updating the conditional probabilities $z_{ig}^{(r)}$ that y_i arises from component g, for $i = 1, \ldots, n$ and $g = 1, \ldots, G$. Then the M-step consists of updating the parameter estimation with the formulae

$$\tau_g^{(r)} = \frac{\sum_{i=1}^n z_{ig}^{(r)}}{n}$$

where $\mu_{ijg}^{(r-1)} = (w_i^{(r-1)} \lambda_{jg}^{(r-1)})$ for $g = 1, \ldots, G$, $i = 1, \ldots, n$, and $j = 1, \ldots, d$,

$$\hat{w}_i^{(r)} = y_{i.}$$

and

$$\hat{\lambda}_{jg}^{(r)} = \frac{\sum_i z_{ig}^{(r)} y_{ij}}{r_j \sum_i z_{ig}^{(r)} y_{i.}}.$$

The estimation of the w_i remains the same for all the iterations of EM. Thus, there is no need to estimate these parameters inside the EM algorithm. As a result, to estimate the mixture of Poisson distributions, it makes no difference to work conditionally on the marginal distributions $y_i, i = 1, \ldots, n$ of the rows y_i of the contingency table. Then the partition $\mathbf{z} = (\mathbf{z}_1, \ldots, \mathbf{z}_n)$ into g clusters can be viewed as n independent random draws from a multinomial distribution with one draw and with probabilities $(\tau_1, \ldots, \tau_g)$. Each $y_i = (y_{i1}, \ldots, y_{id})$ that belongs to the gth cluster of the partition z arises from a multinomial distribution $\mathcal{M}(y_{i.}, \lambda_{1g}, \ldots, \lambda_{dg})$.

Thus a mixture of multinomial distributions can be viewed as a mixture of Poisson distributions with fixed marginal parameters $w_i = y_{i.}$. The EM algorithm to estimate the mixing proportions $\tau_1, \ldots, \tau_G$ and the parameters $(\lambda_{1g}, \ldots, \lambda_{dg}), g = 1, \ldots, G$, is identical to the EM algorithm described above.

6.4.2 Illustration: Vélib Data Set

We apply the Poisson mixture model on the 1,230 stations on the count data grouped by hour for the five working days of a week. The data have been extracted from the Vélib system, as explained in Chapter 1. The discrete version (count data) of the Vélib data set is available in MBCbook package as the velibCount data set. Notice that this analysis is similar to that of Côme and Oukhellou (2014).

We ran the PoisMixClusWrapper function of the HTSCluster package (Rau et al., 2015) on the velibCount data with 2 to 15 clusters; see Listing 6.3. BIC and ICL have similar behavior to one another, see Figure 6.14;[1] they both select 15 clusters, the largest value considered. The slope heuristic (see Baudry et al., 2012), on the other hand, selects a much more parsimonious solution with 10 clusters; see Figure 6.15. We choose to interpret this 10-cluster solution which leads to meaningful and simple interpretations. The cluster profiles are displayed in Figure 6.17 and the positions of the clustered stations on the Paris map are displayed in Figure 6.18.

---

[1] There should be a red cross in both the BIC and the ICL plots, or in neither.

Listing 6.3: Clustering of the Vélib (`velibCount`) data with `HTSCluster`

```
# Loading libraries and data
library(HTSCluster)
library(MBCbook)
data(velibCount)
X = velibCount$data

# Visualization of some stations
Sys.setlocale("LC_TIME", "en_US.UTF-8")
dates = strftime(as.POSIXct(velibCount$dates,origin = "1970-01-01")
    ,format="%a-%I%P")
matplot(t(X[1:5,]),type='l',lwd=2,main='The Velib data set',xaxt='n
    ',xlab='',ylab='Number of available bikes')
axis(1,at=seq(5,181,6),labels=dates[seq(5,181,6)],las=2)

# Saturdays and Sundays are removed
days =  strftime(as.POSIXct(velibCount$dates,origin = "1970-01-01")
    ,format="%u")
X = X[,days %in% 1:5]

# An hour effect is defined
conds = strftime(as.POSIXct(velibCount$dates,origin = "1970-01-01")
    ,format="%H")
conds = conds[days %in% 1:5]

# Clustering with HTSCluster (may take several minutes!)
run <- PoisMixClusWrapper(as.matrix(X),gmin=2,gmax=15,conds=conds)

# The slope heuristic (DDSE) suggests G=10
summary(run)
G = 10

# Display of the cluster proportions
run$all.results[[G-1]]$pi

# Plot of the Estimated lambdas
matplot(run$all.results[[G-1]]$lambda,type='l',xaxt='n',lwd=2,lty
    =1,xlab='Hour',ylab='Estimated lambdas')
axis(1,at=1:24,labels=rownames(run$lambda),las=1)

# Map of the results with leaflet (no API key needed)
library(leaflet); library(RColorBrewer)
df = velibCount$position
colors = brewer.pal(12,'Paired')
leaflet(df) %>% addTiles() %>%
  addCircleMarkers(color = colors[run$all.results[[G-1]]$lab],
                   radius = 5, fillOpacity = 1, stroke = FALSE)
```

The mixing proportions are depicted by a barplot in Figure 6.16. The dark blue cluster is a cluster of stations where people are working. They are full during the day and empty during the night. By contrast, the light green cluster clearly represents a residential cluster. The pink cluster is a cluster of stations where people are living, but with considerable activity at night. The sky blue cluster

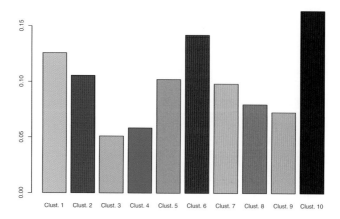

Figure 6.16 A barplot of the proportions of the ten clusters obtained with the slope heuristics on the Vélib data set.

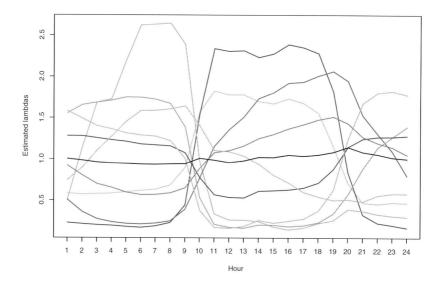

Figure 6.17 The profile by hours of the estimated α for the ten-cluster Poisson mixture solution.

is an intermediate cluster of stations where many people work but with a great deal of housing too. The red cluster is a cluster of mixed housing and moderate business activity. The dark green cluster is a cluster of tourist activities which fills slowly to reach a peak in the evening. The purple cluster has an almost constant number of bikes throughout the day: it is typically a residential area without

Figure 6.18 Representation of the 10-cluster solution on the Paris map.

much local activity. The light purple cluster is a cluster of stations filling up in the early morning and diminishing slowly during the rest of the day. The orange cluster has a peak at dinner time, probably stations near restaurant areas. And finally, the light orange cluster is made up of stations in residential hilly areas where people park the bikes but do not pick them up. These stations are filled by vans during the night.

6.5 Bibliographic Notes

The latent class model is much employed in psychology, educational and social science and behavior system analysis. It has been the subject of several books (Lazarsfeld and Henry, 1968; McCutcheon, 1987; Collins and Lanza, 2013). Uebersax (2010) developed an excellent website on latent class analysis with many bibliographic references organized by theme and presentation of software[2].

---

[2] www.john-uebersax.com/stat/index.htm

Software

There are many software programs proposing model-based clustering methods for qualitative data. They are of two kinds.

- The first category of software is concerned with Latent Structure Analysis methodology and is not restricted to Latent Class Methodology and cluster analysis. Among them, the Latent Gold software of Vermunt and Magidson (2005) is the most complete and elaborate one. Another important software package is LEM, a free Latent Class Analysis software developed by Vermunt.
- The second category of software consists of Model-Based Clustering software programs that analyze discrete data. These include the SNOB software (Dowe, 2008) and the Mixmod software (Biernacki et al., 2006) for which there is also an R interface Rmixmod. Mixmod allows one to analyze all the Latent Class Models presented in this chapter through maximum likelihood or classification maximum likelihood. Note that the majority of software for model-based clustering is for continuous data and does not include the analysis of categorical data.

Local Dependence

A limitation of the Latent Class Model is the local independence assumption which can appear to be unrealistic. Thus some authors have proposed mixture models on categorical data to relax this assumption. These include Bock (1986), who proposed a mixture of loglinear models, Hagenaars (1988), who proposed modeling direct effects between the indicators conditionally on group memberships, and Zhang (2004) who proposed hierarchical latent class models; see also Chen et al. (2012). More recently the mixture of latent trait models of Gollini and Murphy (2014), sketched in Section 6, relaxes the local independence assumption. Another approach is that of Marbac et al. (2015) where, conditionally on a class, the variables are grouped into independent blocks, each one following a specific distribution that takes into account the dependence between variables. More details and additional references can be found on the website of John Uebersax (www.john-uebersax.com/stat/condep.htm).

Dynamic Discrete Data

A common situation is where multivariate discrete data on the same variables are observed at multiple time points. Hasnat et al. (2017) has proposed a method based on parametric links between multinomial mixtures that allows one to interpret the evolution of clusters over time. Melnykov (2016) describes ClickClust, an R package for model-based clustering of categorical sequences.

7

Variable Selection

Variable selection is clearly a crucial task in supervised classification because irrelevant and noise variables increase the prediction error of classifiers designed from finite training data sets. This has been known for a long time but, more recently, it has been realized that the performance of model-based clustering (an *unsupervised* classification method) can also be degraded if irrelevant or noise variables are present. As a result, there has been considerable interest in variable selection for model-based clustering and more broadly for unsupervised classification. From a technical point of view the variable selection problem is simpler in a supervised context since the classification to be recovered is known for a training data set. We therefore concentrate on the model-based clustering approach in this chapter. We consider the continuous and categorical variable selection problems separately, because they have different characteristics.

The next two sections are focused on the continuous variable setting. The most used general approaches for variable selection have been model selection and regularization. Variable selection methods are described first and regularization methods are considered afterwards. Then, we consider the categorical data setting for which it is more difficult to embed the variable selection problem in a linear regression setting.

7.1 Continuous Variable Selection for Model-based Clustering

Let n observations $y = (y_1, \ldots, y_n)$ to cluster be described by d continuous variables ($y_i \in \mathbb{R}^d$). In the model-based clustering framework, the multivariate continuous data y_i are assumed to come from several subpopulations (clusters), each one modeled by a multivariate Gaussian density. The observations are assumed to arise from a finite Gaussian mixture with G components and a model m, which defines the form of the mixture component covariance matrices, namely

$$f(y_i|G, m, \alpha) = \sum_{k=1}^{G} \tau_g \phi(y_i|\mu_g, \Sigma_g),$$

where $\tau = (\tau_1, \ldots, \tau_G)$ is the mixing proportion vector ($\tau_g \in (0, 1)$ for all $g = 1, \ldots, G$ and $\sum_{g=1}^{G} \tau_g = 1$), $\phi(.|\mu_g, \Sigma_g)$ is the d-dimensional Gaussian density function with mean μ_g and variance matrix Σ_g and $\eta = (\tau, \mu_1, \ldots, \mu_G, \Sigma_1, \ldots, \Sigma_G)$ is the parameter vector.

This framework yields a range of possible models, each model m corresponding to different assumptions about the forms of the covariance matrices, related to their eigenvalue decomposition. These assumptions include whether the volume, shape and orientation of each mixture component vary between components or are constant across clusters, as seen in Chapter 2. Typically, the mixture parameters are estimated via maximum likelihood using the EM algorithm, and both the model structure and the number of components G are chosen using the BIC or other penalized likelihood criteria. Here we view each mixture component as corresponding to one cluster, and so the term cluster is used hereafter.

7.1.1 Clustering and Noisy Variables Approach

Model selection approaches to the problem were pioneered by Law et al. (2004) and Tadesse et al. (2005), who partitioned the set of candidate variables into two sets. One set S is the set of the relevant variables for clustering, and the other set W is the set of variables that are irrelevant for clustering. They assumed that the irrelevant variables in W are statistically independent of the relevant variables in S.

The approach of Law et al. (2004) requires, in addition, that the variables be conditionally independent of the clusters. They defined $\rho_j = P(j \in S)$, for $j = 1, \ldots, d$, a quantity that they called *feature saliency*. Under their assumptions, the marginal distribution of the data is

$$p(y_i \mid \theta, \rho) = \sum_{g=1}^{G} \tau_g \prod_{j=1}^{d} \left[\rho_j p(y_{ij} \mid \theta_{gj}) + (1 - \rho_j)\phi(y_{ij} \mid \mu_j, \sigma_j^2) \right], \quad (7.1)$$

where $\phi(\cdot \mid \mu, \sigma)$ is the probability density function of a Gaussian distribution with mean μ and variance σ^2. The parameters of this mixture model, including the $\rho_j, j = 1, \ldots, d$, can be estimated using the EM algorithm. The resulting $\rho_j, j = 1, \ldots, d$ are weights representing the importance of the variables in the clustering. Law et al. (2004) used the MML criterion of Wallace and Freeman (1987) to select a mixture model. They therefore made use of a sequential version of the EM algorithm proposed by Celeux et al. (2001) to avoid solutions that are too sensitive to their initial position.

Tadesse et al. (2005) proposed a fully Bayesian approach using a reversible jump MCMC algorithm assuming also that the irrelevant variables in W are statistically independent of the relevant variables in S.

7.1.2 Clustering, Redundant and Noisy Variables Approach

Raftery and Dean (2006) (hereafter RD) realized that irrelevant variables are often correlated with relevant ones, and developed a model selection method that takes account of this, and a greedy search algorithm to implement it. Their method assumes that each irrelevant variable depends on all the relevant variables according to a linear regression model.

Maugis et al. (2009a) pointed out that the RD method implies a very non-parsimonious model for all the variables jointly, explaining the method performance in some comparative studies (Steinley and Brusco, 2008; Witten and Tibshirani, 2010). They proposed modifying it by selecting the predictor variables in the linear regression part of the model. In their model, the irrelevant variables are allowed to depend on a subset $R \subseteq S$ of the relevant variables through a linear regression model. Not only could this model be more realistic, but it is also more parsimonious.

Maugis et al. (2009b) went further and allowed explicitly for an irrelevant variable to be independent of all the relevant variables. Although at first sight these assumptions may not seem like major changes to the method, they can actually make a big difference to the results and have led to greatly improved performance. The resulting method provides both more parsimonious and realistic models. Following Celeux et al. (2011), we refer to it here as the RD-MCM method.

The RD-MCM method, as described by Maugis et al. (2009b), involves three possible roles for the variables: the relevant clustering variables (S), the redundant variables (U) and the independent variables (W). Moreover, the redundant variables U are explained by a subset R of the relevant variables S, while the variables W are assumed to be independent of the relevant variables. Thus the data density is assumed to be decomposed into three parts as follows:

$$f(y_i | G, m, r, l, \mathbf{V}, \theta) = \sum_{g=1}^{G} \tau_g \phi(y_i^S | \mu_g, \Sigma_{g(m)}) \times \phi(y_i^U | a + y_i^R b, \Omega_{(r)}) \times \phi(y_i^W | \gamma, \Phi_{(\ell)})$$

where $\theta = (\eta, a, b, \Omega, \gamma, \Phi)$ is the full parameter vector and $\mathbf{V} = (S, R, U, W)$. We denote the form of the regression covariance matrix Ω by r; it can be spherical, diagonal or general. The form of the covariance matrix Φ of the independent variables W is denoted by ℓ and can be spherical or diagonal. A graphical synopsis of this model is given in Figure 7.1.

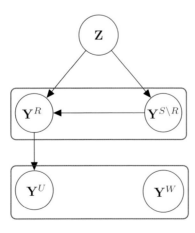

Figure 7.1 Graphical representation of the RD-MCM model for variable selection in model-based clustering.

The RD-MCM method recasts the variable selection problem for model-based clustering as a model selection problem. This model selection problem is solved using a model selection criterion, decomposed into the sum of the three values of the BIC criterion associated with the Gaussian mixture, the linear regression and the independent Gaussian density respectively.

More precisely the criterion to be optimized is

$$\text{CRIT}(G, m, r) = \text{BIC}_{\text{clust}}(y^S \mid G, m) + \text{BIC}_{\text{reg}}(y^U \mid r, y^R) + \text{BIC}_{\text{indep}}(y^W), \tag{7.2}$$

where G is the number of mixture components and m is the Gaussian mixture at hand. In (7.2), $\text{BIC}_{\text{clust}}$ represents the BIC criterion of the mixture model with the variables S, BIC_{reg} represents the BIC criterion for the regression model of the redundant variables U on the variables R, where R is a subset of S, and $\text{BIC}_{\text{indep}}$ represents the BIC value for the Gaussian model with the independent variables W.

We aim to find the model m, the number of clusters G and the set S, R, U and W optimizing (7.2). This is done by using two embedded backward or forward stepwise algorithms for variable selection, one for the clustering and one for the linear regression. A backward algorithm allows one to start with all variables in order to take variable dependencies into account. A forward procedure, starting with an empty clustering variable set or a small variable subset, could be preferred for numerical reasons if the number of variables is large. The method is implemented in the `clustvarsel` R package.

However, such forward and backward procedures can become painfully slow when the number of variables is large. In order to avoid the long CPU time required by these stepwise algorithms, an alternative variable selection procedure in two steps has been proposed by Celeux et al. (2018b). First, the variables are ranked by a lasso-like procedure derived from Zhou et al. (2009), and second, the variable roles are determined using the RD-MCM method on these ranked variables.

Then, for fixed G, the RD-MCM method is run on these ranked variables. The optimal partitioning of the variables can be obtained if the ranking of the variables from the lasso-like procedure is the following ideal ranking:

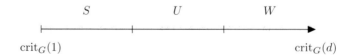

Figure 7.2 The ideal position of the sets S, U and W in the ranking of the variables.

Obviously, there is little chance that the variable ranking obeys this ideal situation. This was already the case with the orders provided by the forward and backward stepwise algorithms.

To increase the ability of the RD-MCM methodology to retrieve the sets S, R,

U and W in the forward or backward selection of variables, we do not stop the algorithms (clustering or regression) as soon as adding, or removing, a variable provides a smaller value of the criterion (7.2). Instead, we explore the possibility of increasing the criterion (7.2) by adding, or removing, ℓ variables, where ℓ is a fixed number. We have found taking $\ell = 5$ to give good results in practice. This variable selection procedure is implemented in the `SelvarMix` R package.

The RD-MCM method generalizes several previous model selection methods. The procedure of Law et al. (2004), where irrelevant variables are assumed to be independent of all the relevant variables, corresponds to $W = S^c$, $R = \emptyset$, $U = \emptyset$. The RD method (Raftery and Dean, 2006) assumes that the irrelevant variables are regressed on the whole relevant variable set ($W = \emptyset$, $U = S^c$ and $R = S$). The `clustvarsel` R package (Scrucca and Raftery, 2018) selects the variables using the software `Mclust` with ($W = \emptyset$, $U = S^c$ and $R \subset S$) according to the methodology of Maugis et al. (2009a).

7.1.3 Numerical Experiments

The methods of Law, RD and RD-MCM are compared through numerical experiments on the `wine27` data set analyzed in Section 2.7.1. Recall that this data set gives 27 physical and chemical measurements on 178 wine samples (Forina et al., 1986). There are three types of wine, Barolo, Grignolino and Barbera, and the type of wine is known for each of the samples. Obviously, since an unsupervised context is considered here, we will use only the 27 measurements to select the variables and cluster the samples, and assess to what extent the clusters reproduce the known types.

We ran the sequential EM algorithm of Law et al. (2004) with $G = 3$, implemented in the `MBCbook` package in the `varSelEM` function. Recall that in the Law model the variables are assumed conditionally independent knowing the clusters. The variable weights provided by this method are displayed in Figure 7.3. The confusion matrix between the clusters and the wine types is given in Table 7.1. The data in the planes of the two most, and the two least relevant variables are represented in Figure 7.4, where the three clusters are shown with three different colors.

Table 7.1 *Wine27 data: confusion matrix for 3-cluster model and wine types: the clustering variables are selected with the Law procedure.*

Cluster	Wine type		
	Barolo	Grignolino	Barbera
1	58	14	0
2	1	56	0
3	0	1	48

Then we applied the RD method using the clustvarsel R function of Scrucca

Listing 7.1: Clustering and variable selection with the `varSelEM` (Law et al., 2004) method on the `wine27` data set

```
# Loading libraries and data
library(MBCbook)
data(wine27)
X = scale(wine27[,1:27])
cls = wine27$Type

# Clustering and variable selection with VarSelEM
res = varSelEM(X,G=3)

# Clustering table
table(cls,res$cls)
```

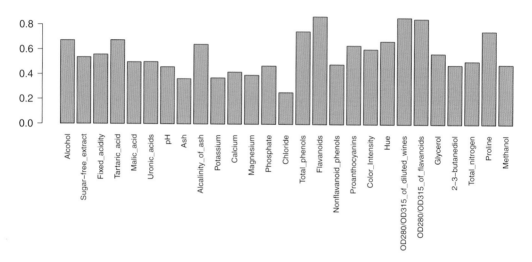

Figure 7.3 The variable weights for the three clusters with the Law method.

and Raftery (2018) with $G = 3$ clusters, by imposing $R = S$ (see Listing 7.2). This selects the model with different proportions and diagonal covariance matrices with equal volumes and different shapes. Seven relevant clustering variables were selected, so that $S = \{$Chloride, Flavanoids, Uronic acids, Color Intensity, Malic acid, Proline, Calcium$\}$. The corresponding confusion matrix between the clusters and the wine types is given in Table 7.2. Figure 7.5 displays the data for the two most relevant variables with the clusters (resp. the types of wine) shown on the left (resp. right) map.

We now run the RD-MCM method on the `wine27` data with the `SelvarMix` R package that ranks the variables via a lasso-like procedure (See Listing 7.3). The selected model has $G = 3$ clusters and assumes different proportions and diagonal covariance matrices with different volumes and shapes. The set of relevant variables is $S = \{$Tartaric acid, Malic acid, Uronic acids, Alkalinity of ash, Chloride, Flavanoids, Color Intensity, Hue, Proline$\}$, the subset $R \subset S$ is $R = \{$Tartaric

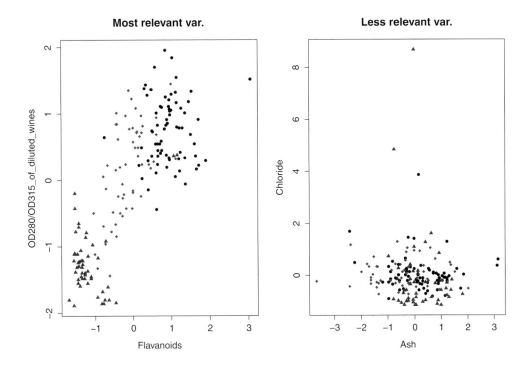

Figure 7.4 A cluster representation for the two most and two least relevant variables with the Law method.

Listing 7.2: Clustering and variable selection with the `clustvarsel` method (Raftery and Dean, 2006) on the `wine27` data

```
# Loading libraries and data
library(clustvarsel)
library(MBCbook)
data(wine27)
X = scale(wine27[,1:27])
cls = wine27$Type

# Variable selection with clustvarsel
out = clustvarsel(X,G=3)

# Clustering with Mclust on the subset of variables
res = Mclust(X[,out$subset],G=3)

# Clustering table
table(cls,res$class)
```

acid, Malic acid, Alkalinity of ash, Flavanoids, Color Intensity, Proline} and the set U contains all the variables not in S. The set W of independent variables is empty.

The corresponding confusion matrix between the clusters and the wine types is

Table 7.2 *Wine27 data: confusion matrix for 3-cluster model and wine types: the clustering variables are selected with the RD method.*

Cluster	Wine type		
	Barolo	Grignolino	Barbera
1	48	4	0
2	11	67	1
3	0	0	47

Listing 7.3: Running the RD-MCM method on the wine27 data

```
# Loading libraries and data
library(SelvarMix)
library(MBCbook)
data(wine27)
X = scale(wine27[,1:27])
cls = wine27$Type

# Clustering and variable selection with SelvarClustLasso
lambda <- seq(20,   100, by = 10)
rho <- seq(1, 2, length=2)
models <- mixmodGaussianModel(family = "diagonal")
out = SelvarClustLasso(X, nbcluster=3, lambda=lambda, rho=rho,
   hsize=3, criterion='BIC', models=models,rmodel=c("LI","LB","LC")
   , imodel=c("LI","LB"))

# The function SortvarClust provides the ranking of the variables
order <- SortvarClust(X, nbCluster, lambda, rho)

# Clustering table
table(cls,out$partition)
```

given in Table 7.3. Figure 7.6 displays the data for the two most relevant variables with the clusters (resp. the types of wine) shown on the left (resp. right) map.

Table 7.3 *Wine27 data: confusion matrix for 3-cluster model and wine types. The clustering variables are selected with the RD-MCM method.*

Cluster	Wine type		
	Barolo	Grignolino	Barbera
1	49	2	0
2	10	69	1
3	0	0	47

The RD and RD-MCM methods selected different numbers of clusters for the wine27 data set. The RD-MCM method selected $G = 3$ clusters while the RD method is less parsimonious and selected $G = 5$ clusters. The confusion matrix between the five clusters with the RD method and the three types of wine is

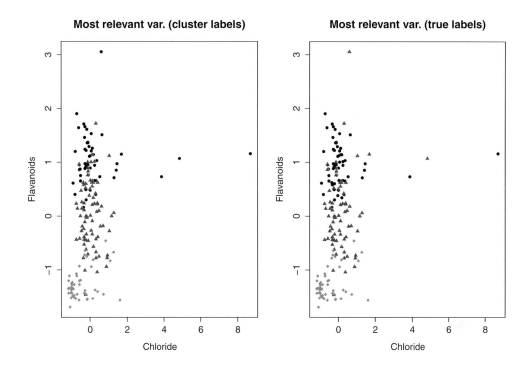

Figure 7.5 The cluster representation for the two most relevant variables on the left and for the three types of wine on the right with the RD method.

given in Table 7.4. The set of relevant clustering variables was $S = \{$Chloride, Flavanoids, Uronic acids, Color Intensity, Malic acid, Proline, Calcium$\}$.

In a further experiment, we added five noisy Gaussian independent variables to the wine27 data set, to assess the ability of the RD-MCM method to detect independent variables. The method performed well in this case. The sets S and R were unchanged, the set U was increased with the Gaussian variable 'Noise4' and the set of independent variables was $W = \{$Noise1, Noise2, Noise3, Noise5$\}$.

Table 7.4 *Wine27 data: confusion matrix for 5-cluster model and wine types. The clustering variables were selected by the RD method.*

Cluster	Wine type		
	Barolo	Grignolino	Barbera
1	43	0	0
2	2	49	1
3	6	6	0
4	8	15	0
5	0	16	47

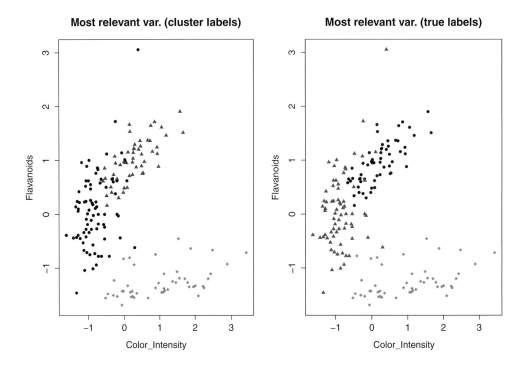

Figure 7.6 The cluster representation for the two most relevant variables on the left and for the three types of wine on the right with the RD-MCM method.

7.2 Continuous Variable Regularization for Model-based Clustering

A different approach, via regularization, was proposed by Pan and Shen (2007) for the specific case of spherical Gaussian mixtures. This methodology has been generalized to more complex models (Xie et al., 2008; Wang and Zhu, 2008; Zhou et al., 2009; Guo et al., 2010).

Pan and Shen (2007) proposed a penalized model-based clustering approach with a lasso-like ℓ_1 penalty. They used an appropriate ℓ_1 penalty function to adaptively shrink the mean parameters towards the average cluster means, resulting in automatic variable selection. The selection of the relevant variables and the estimation of the parameters were performed in the same exercise. They constructed a random model collection derived from a collection of sets of potentially relevant variables determined by ℓ_1-penalization as follows. The regularization procedure is sketched for the most general case with unknown and free component covariance matrices described in Zhou et al. (2009):

1 Each variable is centered by subtracting its mean.

2 A collection of models is obtained by minimizing the penalized log-likelihood

$$\alpha \mapsto \sum_{i=1}^{n} \log \left[\sum_{g=1}^{G} \tau_g \phi(\bar{y}_i \mid \mu_g, \Sigma_g) \right] - \lambda \sum_{g=1}^{G} \|\mu_g\|_1 - \rho \sum_{g=1}^{G} \left\| \Sigma_g^{-1} \right\|_1, \qquad (7.3)$$

as a function of θ, where

$$\|\mu_g\|_1 = \sum_{j=1}^{d} |\mu_{gj}|, \qquad \left\| \Sigma_g^{-1} \right\|_1 = \sum_{\substack{j',j=1 \\ j' \neq j}}^{d} \left| (\Sigma_g^{-1})_{vj} \right|,$$

$\bar{y}_i = (y_{ij} - \bar{y}_j)_{1 \le j \le d}$ with $\bar{y}_j = \frac{1}{n} \sum_{i=1}^{n} y_{ij}$ and where λ and ρ are two non negative regularization parameters defined on two grids of values $\mathcal{G}_\lambda$ and $\mathcal{G}_\rho$ respectively. The estimated mixture parameters for fixed tuning parameters λ and ρ, $\widehat{\theta}(\lambda, \rho)$, are computed with the EM algorithm of Zhou et al. (2009).

3 Pan and Shen and the authors who generalized their method used BIC as model selection criterion. First, models are grouped according to their dimension v in order to obtain a model collection $(m_v)_v$. Following a result of Zou et al. (2007), they calculate a model dimension by taking into account the sparsity of the model: the mean parameters set to zero in the EM algorithm are not considered in the dimension calculation. For each dimension v, the model maximizing the likelihood among the models of dimension v is shortlisted. Finally, the selected model is the model optimizing the BIC criterion among the shortlisted models.

7.2.1 Combining Regularization and Variable Selection

The variable selection model presented in Section 7.1 leads to relevant interpretation of the clustering, but the required forward or backward stepwise algorithms become painfully slow as soon as the number of variables reaches the order of a few tens. Placing the ℓ_1 penalty of (7.3) on the Gaussian mixture mean vectors and precision matrices leads to a ranking of the variables, as described above, which allows one to avoid using stepwise variable selection algorithms, with their associated combinatorial problems. It is hoped that using this lasso-like ranking of the variables instead of stepwise algorithms would not degrade the identification of the sets S, R, U and W (Celeux et al., 2018b).

This lasso-like criterion does not take into account the typology of the variables induced by the RD-MCM model. It distinguishes only two possible roles for the variables: a variable is declared relevant to or independent of the clustering. A variable is declared independent of the clustering if for all $j = 1, \ldots, p$, and $g = 1, \ldots, G$, $\widehat{\mu}_g(\lambda, \rho) = 0$. The variance matrices are not considered in this definition. In fact their role is secondary in clustering, and taking them into account could lead to numerical difficulties.

Varying the regularization parameters (λ, ρ) in $\mathcal{G}_\lambda \times \mathcal{G}_\rho$, a score is defined for each variable $j \in \{1, \ldots, d\}$ and for fixed G by

$$\mathcal{O}_G(j) = \sum_{(\lambda, \rho) \in \mathcal{G}_\lambda \times \mathcal{G}_\rho} \mathcal{B}_{(G, \rho, \lambda)}(j), \qquad (7.4)$$

where

$$\mathcal{B}_{(G,\rho,\lambda)}(j) = \begin{cases} 0 & \text{if } \widehat{\mu}_{1j}(\lambda,\rho) = \cdots = \widehat{\mu}_{Gj}(\lambda,\rho) = 0 \\ 1 & \text{else.} \end{cases} \tag{7.5}$$

The larger $\mathcal{O}_G(j)$ is, the more relevant variable j is expected to be for the clustering. The variables are thus ranked by their decreasing values on $\mathcal{O}_G(j)$. This variable ranking is denoted $\mathcal{I}_G = (j_1, \ldots, j_d)$ with $\mathcal{O}_G(j_1) > \mathcal{O}_G(j_2) > \ldots > \mathcal{O}_G(j_p)$.
     Some comments are in order.

- It is possible to use a lasso procedure instead of the stepwise variable selection algorithm in the linear regression step. However, this replacement is not expected to be highly beneficial since stepwise variable selection in linear regression is not too expensive and the number of variables in the set S is not expected to be high.
- There is no guarantee that the variable order designed with the lasso-like procedure agrees with the ideal ranking of the variables. In particular when the variables are highly correlated, lasso-like procedures could be expected to produce confusions between the sets S and U. This is the reason why we wait a few steps ($c > 1$ steps) before deciding the variable roles in the procedure. We give the procedure a chance to catch more variables in S and in W.

7.3 Continuous Variable Selection for Model-based Classification

All the variable selection strategies presented in the previous section can be straightforwardly extended to the supervised classification context. In fact, the variable selection algorithms are simpler since the groups to be characterized are known. The approaches of Raftery and Dean (2006) and Maugis et al. (2009b) have been extended to the classification context by Murphy et al. (2010) and Maugis et al. (2011) respectively. The only difference between the clustering and the classification contexts is that the variable selection procedures described in Section 7.1 are not embedded in an EM algorithm for classification. Variable selection is thus much faster in the supervised case. It could be also more useful.
     Variable selection could help to improve the classification performance of nonlinear Gaussian classification models. Those models involve many parameters when the number of variables is large with respect to the training sample size. But the proposed variable selection procedures allow us to overcome the dimensionality problem, leading to powerful classifiers with a nice interpretation of variable roles. Proper variable selection could make nonlinear generative classification methods such as quadratic discriminant analysis much more efficient in high dimensional settings, and competitive with reference classifiers such as LDA, logistic regression, k-nearest neighbor classifier or support vector classifiers in many situations.
     In order to illustrate variable selection in supervised classification, we again consider the wine27 data set. This time the problem is to classify the three wine types using the 27 physical and chemical measurements. We repeated the following procedure 50 times: (i) we drew at random a training data set of size 150 to

build the QDA classifiers with and without variable selection, (ii) we assessed the misclassification error rates on the classifiers on the remaining test sample of size $178 - 150 = 28$.

For almost all the 50 random samples, the same five variables (Flavanoids, Color intensity, Hue, OD280/OD315 of diluted wines, Proline) were selected out of the 27 candidate variables. The misclassification error rate without variable selection of QDA was 0.068 with a standard error of 0.060. The misclassification error rate with the five selected variables was substantially lower, at 0.043 with a standard error of 0.035. Note that with only 5 of the 27 variables, a lower misclassification error rate is obtained with the standard QDA classifier than with all 27 variables.

7.4 Categorical Variable Selection Methods for Model-based Clustering

In the model-based clustering context with continuous data the covariance parameter of a multivariate normal distribution provides a natural tool to model the conditional dependence between continuous variables. When the variables are categorical, the reference model-based clustering model is the latent class model which makes use of a conditional independence assumption of the variables knowing the clusters and, as discussed by White et al. (2016), there is no simple way to measure the dependence between variables. In other words, when a variable is proposed to be included or excluded by the sampling scheme, it is possible to answer the question "does the proposed variable contain information about the clustering?", but the question "does the proposed variable contain *additional* information about the clustering?" is hard to answer. As a result, it is hard to take into account the links between variables that are not explained by the latent clusters. Thus, the categorical variable selection for the latent class model involves only two roles: a variable either belongs to the set S of relevant clustering variables, or it belongs to the set W of independent variables as represented in Figure 7.7.

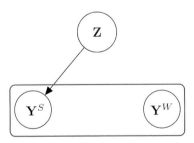

Figure 7.7 Graphical representation of the simplest variable selection model.

The observations to be classified are described by d categorical variables, with variable j having m_j response levels. The data are $(y_1, \ldots, y_n)$, where y_i is encoded by the vector $(y_i^j; j = 1, \ldots, d)$ of response level h for each variable j. The data point y_i can be written equivalently as the binary vector $(y_i^{jh}; j = 1, \ldots, d; h = 1, \ldots, m_j)$ where

$$
\begin{cases}
y_i^{jh} = 1 & \text{if } y_i^j = h \\
y_i^{jh} = 0 & \text{otherwise.}
\end{cases}
$$

7.4.1 Stepwise Procedures

The variable selection model for latent class analysis assumes that the data are distributed according to the following probability distribution function:

$$
f(y_i; \theta) = \left[\sum_g \tau_g \prod_{j \in S} \prod_{h \in j} (\alpha_g^{jh})^{y_i^{jh}} \right] \prod_{j \in W} \prod_{h \in j} (\beta^{jh})^{y_i^{jh}}, \tag{7.6}
$$

where, for $j \in S$, α_g^{jh} denotes the probability that variable j is at level h if object i is in cluster g, and, for $j \in W$, β^{jh} denotes the probability that variable j is at level h on the whole data set. Also, $\theta = (\tau_1, \ldots, \tau_G; \alpha_g^{jh}; j \in S; h = 1, \ldots, m_j; \beta^{jh}, j \in W, h = 1, \ldots, m_j)$ denotes the whole vector parameter. The variables in W are assumed to be independent while the variables in S are assumed to be conditionally independent given the latent clusters.

In this framework, Toussile and Gassiat (2009) determined the set S and W as in Maugis et al. (2009a) using the BIC criterion through a backward-stepwise procedure. Dean and Raftery (2010) used the same model for variable selection in the latent class model with the two sets S (clustering variables) and W (independent variables). Instead of the backward-stepwise algorithm, they used a forward-stepwise algorithm with the BIC criterion, and they included a preselection of the variables which are expected to be more useful for clustering using a headlong search algorithm. This headlong search algorithm makes it possible to start the forward selection procedure with enough variables to avoid unstable selection. It also gives good results when it is used as the final method, and not simply as a starting point, and it can be much more computationally efficient.

7.4.2 A Bayesian Procedure

The same model was considered in a Bayesian setting by White et al. (2016). They used non-informative Dirichlet prior distributions $\mathcal{D}(\alpha, \ldots, \alpha, \lambda)$ for all the parameters of the model (7.6). They chose $\alpha = 0.5$ to discourage overfitting. Now, denoting $\nu = (\nu_C, \nu_I)$ where ν_C (resp. ν_I) contains these indexes of the clustering (resp. independent) variables, and defining ρ as the prior probability that any variable is a clustering variable leads to the following joint prior probability for ν:

$$
p(\nu \mid \rho) = \prod_{j \in \nu_C} \rho \prod_{j \in \nu_I} (1 - \rho). \tag{7.7}
$$

To reduce the sensitivity of the results to the hyperparameter ρ one can use a hyperprior distribution, defined by assuming that ρ itself has a prior distribution, $\beta(a_0, b_0)$. Following Nobile and Fearnside (2007), they used a Poisson prior distribution with mean 1, $\mathcal{P}(1)$ for G. Since the Dirichlet prior distribution is a proper conjugate prior, it is possible to get an analytic expression for the posterior distribution $p(G, z, \nu)$. It is therefore possible to use the collapsed sampler of Nobile and Fearnside (2007) rather than a reversible jump MCMC algorithm.

In summary, each sweep of the collapsed sampler involves the following steps:

- Update the class membership of observations using a Gibbs sampler step.
- Propose creating a new component with probability p_G, otherwise propose to remove a component.
- Choose one variable at random. If it is not included in the clustering variables propose to do so. If it is currently included as a clustering variable, propose to exclude it.

Details can be found in White et al. (2016). From a converged set of iterations from this collapsed sampler, it is possible to estimate summaries of the integrated parameters with empirical expectations:

$$E(\theta \mid y) \approx \frac{1}{T} \sum_{l=1}^{T} E(\theta \mid y, z^{(t)}),$$

where $(z^{(t)}), t = 1, \ldots, T$ is the sequence of the thinned collapsed sampler simulated labels.

White et al. (2016) noticed that this collapsed sampler is more computationally efficient than exhaustively computing information criteria. They also found that it clearly outperforms reversible jump MCMC algorithms (e.g. Tadesse et al., 2005) in variable selection, since it mixes better.

All the methods for variable selection in the LCM that we have described so far consider a variable to be added to or removed from the already selected set of clustering variables assuming that the former is independent of the latter. By this assumption, informative linked variables are selected even if they contain similar group information. Thus, these methods are capable of discarding non-informative variables but not the redundant variables. Selecting only a subset of them can lead to a similar clustering with a more parsimonious variable selection.

Fop et al. (2017) developed a variable selection method for LCM which overcomes the limitation of the above independence assumption. This model is represented in Figure 7.8. They take possible redundancy into account by modeling the conditional distribution of the proposed variable Y^P given the clustering variables Y^S with a multinomial logistic regression function:

$$p(Y^P = \ell \mid Y^R \in Y^S) = \frac{\exp(Y^R \beta_\ell)}{\sum_{\ell=1}^{m_P} \exp(Y^R \beta_\ell)}, \tag{7.8}$$

where β_ℓ is the vector of regression parameter for level ℓ and $\ell = 1, \ldots, m_P$ is the level of a proposed variable.

The clustering variables are selected with a swap-stepwise selection algorithm where in the swapping step a non-clustering variable is swapped with a clustering

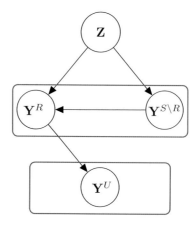

Figure 7.8 Graphical representation of the Fop et al. (2017) variable selection model.

variable. The number of classes G is re-estimated at each step of the algorithm. The set of clustering variables can change at each step (by adding, removing or swapping) and so G needs to be re-estimated. The procedure stops when no change has been made to the set Y^S after consecutive exclusion, inclusion and swapping steps.

7.4.3 An Illustration

We illustrate the methods using the UCI Congressional Voting data set[1] recording the votes (yes, no, abstained or absent) of 435 members of the 98th US Congress on 16 different key issues. This data set involves three-level categorical data. With the `LCAvarsel` R package, it is possible to run the conditional independence variable selection of Dean and Raftery (2010) and the variable selection using the model 7.8 with and without the swapping step. Listing 7.4 provides the code to run such a task. The three methods select $G = 4$ clusters. With the Dean and Raftery (2010) method, all the variables are declared useful for the clustering except the variable v_2. With the Fop et al. (2017) method without swapping, all the variables are declared useful for the clustering except the variables v_2 and v_{12}. The selection is more stringent with the swapping step: the clustering variables are v_1, v_4, v_6, v_8, v_{12}, v_{13}, v_{14} and v_{15}.

It appears from Table 7.5, which gives the confusion matrix of the three clusterings with the Democratic party and the Republican party, that the clusterings obtained under the conditional independence assumption or with the logistic method without swapping are nearly identical, while the logistic method with swapping provides a solution which is not so different (essentially a few Democrats go from Cluster 2 to Cluster 4). This is a common situation. Often variable

---

[1] http://archive.ics.uci.edu/ml/datasets/Congressional+Voting+Records

Listing 7.4: Variable selection with `LCAvarsel`

```
# Loading libraries and data
library(LCAvarsel)
library(MBCbook)
data(UScongress)

# Pretreatment of the data
grp <- UScongress[,1]
X <- UScongress[,-1]
colnames(X) <- paste0( "V", 1:ncol(X))

# LCA without variable selection
lca <- fitLCA(X, G = 1:6)
plot(lca$BIC, type = "b")
compareCluster(grp, lca$class)

# \texttt{LCAvarsel} with independence assumption
selInd <- LCAvarsel(X,G=1:6,swap=FALSE,independence=TRUE,parallel=
    TRUE)
selInd

# \texttt{LCAvarsel} without indep. assumption, no swap steps
selNoSwap <- LCAvarsel(X,G=1:6,swap=FALSE,parallel=TRUE)
selNoSwap

# \texttt{LCAvarsel} without indep. assumption, with swap steps
selSwap <- LCAvarsel(x,G=1:6,swap=TRUE,parallel=TRUE)
selSwap
```

selection in model-based clustering does not greatly change the clusters, but the interpretation of the clustering is improved by making the variable roles clearer.

Table 7.5 *Confusion tables of the clusterings (in the columns) and the two political parties (in the rows).*

	1	2	3	4	1	2	3	4	1	2	3	4
Dem.	19	190	7	51	22	188	4	53	17	162	6	82
Rep.	131	4	9	24	131	4	5	28	136	1	5	26

| (a) Cond. ind. $G = 4$ | (b) no swapping $G = 4$ | (c) swapping $G = 4$ |

7.5 Bibliographic Notes

There are several other recent proposals for variable selection in model-based clustering. Fraiman et al. (2008) proposed a method for variable selection *after* the clustering has been carried out that also assumes the number of clusters known. Thus it is not fully comparable with the methods considered here, which carry out clustering and variable selection simultaneously. Nia and Davison (2012) proposed a fully Bayesian approach using a spike and slab prior, while Kim et al. (2012) proposed a Bayesian approach using Bayes factors to compare the different models.

Wallace et al. (2018) proposed a method for variable selection in non-Gaussian model-based clustering, where the components have skew-normal distributions, motivated by applications in sleep research.

Lee and Li (2012) proposed an approach to variable selection for model-based clustering using ridgelines. Silvestre et al. (2015) proposed a method adapted from Law et al. (2004) based on feature saliency for selecting categorical variables in clustering.

In place of forward and backward algorithms, genetic algorithms have been proposed for subset selection for model-based clustering of continuous data (Scrucca, 2016a). A genetic algorithm has also been considered in Galimberti et al. (2017) where these authors considered a variable selection method that allows one to describe two different clustering structures according to a scheme analogous to Maugis et al. (2009a).

In addition, Poon et al. (2013) proposed a method for facet determination in model-based clustering. This is related to but not the same as variable selection.

Finally, Fop and Murphy (2018) provide a review of variable selection for model-based clustering, including worked examples and a discussion of software.

8

High-dimensional Data

In the previous chapters we have seen that strengths of model-based clustering and classification include their probabilistic foundation, flexibility and good performance in many practical situations. Up to now, however, we have focused on examples with small to medium numbers of dimensions (up to 57 variables in our previous examples).

However, the nature of data has been changing in recent decades, and it is common nowadays, in many scientific domains, to measure hundreds or thousands of variables on each observation; this is the world of high-dimensional data.

Unfortunately, until recently, model-based techniques were not well prepared to deal with these kinds of data and thus clustering or classifying high-dimensional data was a challenging problem for model-based methods. Indeed, model-based methods sometimes have disappointing behavior in high-dimensional spaces. Model-based methods suffer from the well-known *curse of dimensionality* (Bellman, 1957) which is mainly due to the fact that the models used are over-parameterized in high-dimensional spaces. Ironically, model-based methods can suffer in high-dimensional spaces from the parametric nature which makes them attractive and flexible in low-dimensional spaces.

8.1 From Multivariate to High-dimensional Data

Let us begin with one of our previous examples, which is of high enough dimension (while not truly high-dimensional) to illustrate some of the issues. Figure 8.1 shows the first seven variables of the Wine data set (`wine27` data set in the `MBCbook` package) which is made up of 178 observations (Italian wines) described by 27 variables (physicochemical measures). The wines come from three different regions of Italy which are indicated by the colors on the figure. Figure 8.2 shows an alternative representation of those 27-dimensional data, with and without the group labels. Although it was often fairly easy to identify the groups by eye in the lower-dimensional examples in previous chapters, here the larger number of variables increases the difficulty of doing this. Indeed, it appears clearly in the latter figures that the clustering of those data (i.e. going from the left to the right panel of Figure 8.2) is a difficult problem.

For such a data set with a moderate number of variables, the methodologies in previous chapters should still work well. Listing 8.1 provides R code to obtain the figures and to cluster the data with `mclust`. The comparison of the `mclust` results

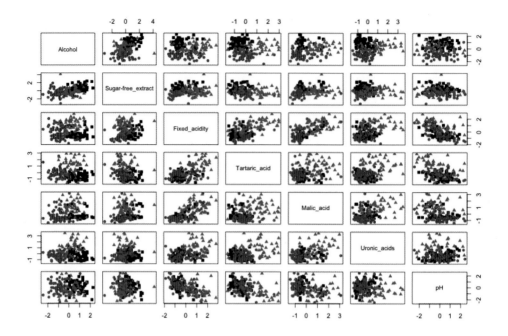

Figure 8.1 Matrix of scatter plots for the first seven variables of the Wine data set. Colors correspond to the (known) three types of wine.

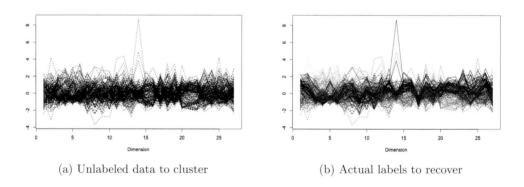

(a) Unlabeled data to cluster　　　　　　(b) Actual labels to recover

Figure 8.2 Multivariate representation of the Wine data (a) without and (b) with the actual labels.

with the true labels shows that `mclust` globally succeeds in recovering the wine types. Listing 8.1 also gives the R code to cluster the same data with a clustering method designed for high-dimensional data (HDDC, available in the `HDclassif` package, see Section 8.4.4). The clustering results turn out to be closer to the actual labels. This suggests that several dozens of variables may be the upper limit of good performance for classical model-based methods. In this chapter we

Listing 8.1: Visualization and clustering of the `wine27` data

```
# Loading libraries and data
library(MBCbook)
data(wine27)

# Extract data and scale for visualization purpose
X = scale(wine27[,1:27])
cls = as.numeric(wine27$Type)

# Visualization of data
pairs(X[,1:7],col=cls,pch=(15:17)[cls])
matplot(t(X),col=1,type='l',lty=2,xlab='Dimension',ylab='')
matplot(t(X),col=cls,type='l',lty=cls,xlab='Dimension',ylab='')

# Clustering with Mclust (not HD-ready)
res.mclust = Mclust(X,3)
table(cls,res.mclust$classification)
     cls   1   2   3
       1  59   0   0
       2   1  70   0
       3   0   4  44

# Clustering with HDDC (HD-ready)
library(HDclassif)
res.hddc = hddc(X,3)
table(cls,res.hddc$class)
     cls   1   2   3
       1  58   1   0
       2   2  69   0
       3   0   0  48
```

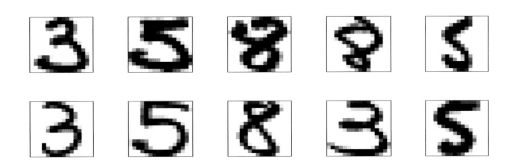

Figure 8.3 A sample of handwritten digits from the US postal services (usps358 data set).

show that it is possible to go beyond that point with methodologies specifically designed for model-based clustering in high-dimensional data.

As an example of more truly high-dimensional data, Figure 8.3 shows a sample of handwritten digits from the US postal services (usps358 data set in the MBCbook package). The data set is a subset of the famous USPS data from UCI. The

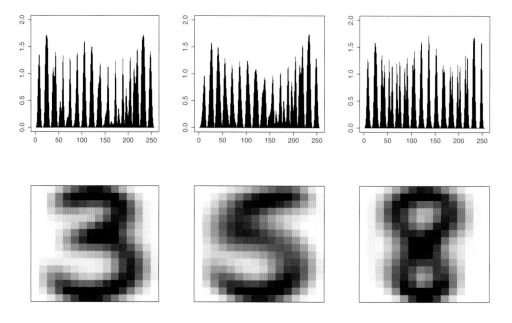

Figure 8.4 Image (bottom) and multivariate representation (top) of the means of digits 3, 5 and 8 of the `usps358` data set.

original set consisted of 7,291 images divided into 10 classes corresponding to the digits from 0 to 9. The `usps358` set contains only the 1,756 images of the digits 3, 5 and 8 which are the most difficult digits to discriminate. Each digit is a 16×16 gray level image and is represented as a 256-dimensional vector in the the `usps358` data set. Figure 8.4 shows the images and multivariate representations of the means of digits 3, 5 and 8. The R code to produce this figure is presented in Listing 8.2.

Although the classification task may seem not so difficult when looking at the images, the high-dimensional nature of the data makes the task more difficult for automatic methods. Indeed, the multivariate representation of the digit means appear to be quite similar. In addition, if one wants to classify these data with, for instance, quadratic discriminant analysis (QDA, equivalent to EDDA/MclustDA with model VVV), the number of parameters to estimate would be $(G-1) + Gd + Gd(d-1)/2 = 98,690$ here (according to Table 2.2). From this, it is clear that the 2,039 observations will not be enough to correctly estimate model parameters and then build an efficient QDA classifier.

Furthermore, in several applications such as chemometrics, mass spectrometry or genomics, the number of available observations is small compared to the number of variables. Such situations increase the difficulty of the problem and often make estimation infeasible from the numerical point of view. Figure 8.5 shows the 202

Listing 8.2: Images and multivariate representations of the usps358 means

```
# Loading libraries and data
library(MBCbook)
data(usps358)
X = usps358[,-1]; cls = usps358$cls

# Computing group means
m1 = colMeans(X[cls==1,]); m2 = colMeans(X[cls==2,]); m3 = colMeans
    (X[cls==3,])

# Plotting means as vectors
par(mfrow=c(2,3))
matplot(m1,type='h',col=1,lty=1,ylim=c(0,2),ylab='')
matplot(m2,type='h',col=1,lty=1,ylim=c(0,2),ylab='')
matplot(m3,type='h',col=1,lty=1,ylim=c(0,2),ylab='')

# Plotting means as images
image(t(matrix(t(m1),ncol=16,byrow=TRUE)[16:1,]),col=gray(255:0/
    255),axes=F);box()
image(t(matrix(t(m2),ncol=16,byrow=TRUE)[16:1,]),col=gray(255:0/
    255),axes=F); box()
image(t(matrix(t(m3),ncol=16,byrow=TRUE)[16:1,]),col=gray(255:0/
    255),axes=F); box()
```

near infra-red (NIR) spectra presented in Devos et al. (2009) (NIR data set in the MBCbook package) described by 2,800 wavelengths. The data were obtained from the analysis of three types of textiles and the colors on Figure 8.5 indicate the textile type associated with each curve. As one can see, the classification problem seems difficult without considering the problem of model over-parameterization.

8.2 The Curse of Dimensionality

In this section, we focus on some of the causes of the curse of dimensionality in model-based clustering and classification.

When reading research articles or books related to high-dimensional data, it is common to find the term "curse of dimensionality" to refer to problems caused by the analysis of high-dimensional data. This term was first used by R. Bellman in the preface of his book (Bellman, 1957) promoting dynamic programming. To illustrate the difficulty of working in high-dimensional spaces, Bellman recalled that if one considers a regular grid with step 0.1 on the unit cube in a 10-dimensional space, the grid is made up of 10^{10} points. Consequently, the search for the optimum of a given function in this unit cube requires 10^{10} evaluations of this function which was an intractable problem in the 1960s (it remains a difficult problem nowadays). Although the term "curse of dimensionality" used by Bellman is rather pessimistic, the paragraph of the preface in which the term first appeared is in fact more optimistic:

All this [the problems linked to high dimension] *may be subsumed under the heading "the curse*

Listing 8.3: Clustering of the usps358 data set

```
# Loading libraries and data
library(MBCbook)
data(usps358)
X = usps358[,-1]; cls = usps358$cls

# Clustering with Mclust (non HD-ready)
res.mclust = Mclust(X,3,initialization=list(subset=sample(nrow(X)
    ,300)))
table(cls,res.mclust$classification)
   cls    1    2    3
     1  315    0  343
     2  298    2  256
     3  519    0   23

# Clustering with HDDC (HD-ready)
library(HDclassif)
res.hddc = hddc(X,3)
table(cls,res.hddc$class)
   cls    1    2    3
     1   19  596   43
     2    2    5  549
     3  492   24   26
```

of dimensionality". Since this is a curse, [...], there is no need to feel discouraged about the possibility of obtaining significant results despite it.

We will see that Bellman's thought was indeed correct since, at least for classification, high-dimensional spaces have nice properties that allow us to obtain excellent performance.

High-dimensional spaces are hard to handle because simple ideas which are true and well established in low-dimensional spaces (two or three dimensions for example) turn out to be incorrect in high-dimensional spaces (Scott, 1992; Donoho, 2000; Verleysen, 2003; Verleysen and François, 2005). A simple and classical example is the volume of the unit hypersphere which can be easily computed with respect to the dimension of the space as follows:

$$V(d) = \frac{\pi^{d/2}}{\Gamma(d/2+1)},$$

where Γ is the usual Gamma function. Figure 8.6 shows the surprising behavior of the unit hypersphere volume according to the dimension of the space. The R code to get this figure is presented in Listing 8.4. As expected, the volume of the sphere increases when moving from dimension 1 to 2, 2 to 3, 3 to 4, and 4 to 5. However, after dimension 5, the volume stops increasing and, surprisingly, decreases quickly toward 0. For example, the volume of the unit hypersphere in 30 dimensions is 2×10^{-5}. This is one example of how high-dimensional spaces have very different features from those of low-dimensional spaces.

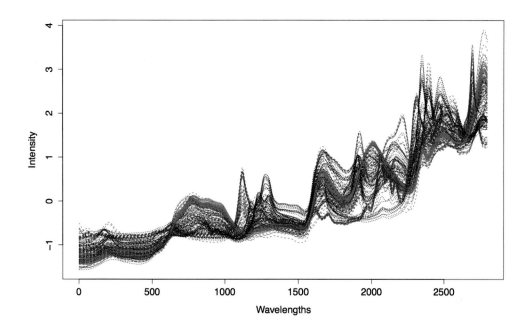

Figure 8.5 202 observed spectra of the three class NIR data set.

Listing 8.4: Volume of the unit hypersphere according to dimension

```
# Volume of the unit hypersphere
d = 1:100
V = pi^(d/2) / gamma(d/2+1)
plot(d,V,type='b',xlab='Dimension',ylab='Volume')
```

8.2.1 The Curse of Dimensionality in Model-based Clustering and Classification

The curse of dimensionality takes a particular form in the context of model-based clustering. Indeed, model-based clustering methods require the estimation of a number of parameters which depend directly on the dimension of the observed space. If we consider the Gaussian mixture model (GMM) with unconstrained component covariance matrices, the total number of parameters in the model is equal to

$$\nu = (G - 1) + Gd + Gd(d - 1)/2,$$

where $(G-1)$, Gd and $Gd(d-1)/2$ are the numbers of parameters in the proportions, the means and the covariance matrices respectively. When fitting this model, it turns out that the number of parameters to be estimated is a quadratic function of d in this case and a large number of observations will be necessary to correctly estimate those model parameters.

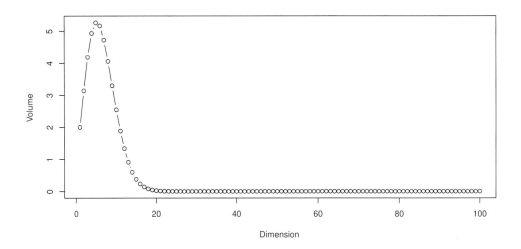

Figure 8.6 Volume of the unit hypersphere according to the dimension of the space.

Furthermore, a computational problem occurs in the EM algorithm when computing the conditional probabilities $t_{ig} = E[Z_{ig}|y_i, \theta]$ which depends, in the GMM context, on the quantity $H_g(x) = -2\log(\tau_g\phi(x; \mu_g, \Sigma_g))$. The computation of H_g, which can be written as

$$H_g(x) = (x - \mu_g)^T \Sigma_g^{-1}(x - \mu_g) + \log(\det \Sigma_g) - 2\log(\tau_g) + d\log(2\pi), \qquad (8.1)$$

requires the inversion of the covariance matrices Σ_g, for $g = 1, ..., G$. Consequently, if the number of observations n is small compared to ν, the estimated covariance matrices $\hat{\Sigma}_g$ are ill-conditioned and their inversion leads to unstable classification functions. In the worst case where $n < d$, the estimated covariance matrices $\hat{\Sigma}_g$ are singular and GMM model-based classification methods cannot be used at all. Unfortunately, this situation occurs more and more frequently in biology (DNA sequences, genotype analysis) and in computer vision (facial recognition) for example.

The curse of dimensionality has also been demonstrated by Pavlenko (2003) and Pavlenko and Von Rosen (2001) in the Gaussian case from the estimation point of view. Let us consider the estimation of the normalized trace $\xi(\Sigma) = \text{tr}(\Sigma^{-1})/d$ of the inverse covariance matrix Σ of a multivariate Gaussian distribution $\mathcal{N}(\mu, \Sigma)$. The estimation of ξ from a sample of n observations $\{x_1, ..., x_n\}$ leads to

$$\hat{\xi}(\Sigma) = \xi(\hat{\Sigma}) = \frac{1}{d}\text{tr}(\hat{\Sigma}^{-1}),$$

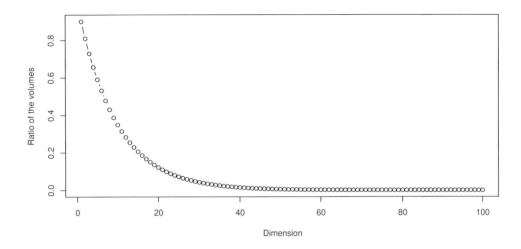

Figure 8.7 Ratio of the hypersphere of radius 0.9 to that of the unit hypersphere according to the dimension of the space.

and its expectation is

$$E[\hat{\xi}(\Sigma)] = \left(1 - \frac{d}{n-1}\right)^{-1} \xi(\Sigma).$$

As a result, if the ratio $d/n \to 0$ when $n \to +\infty$, then $E[\hat{\xi}(\Sigma)] \to \xi(\Sigma)$. However, if the dimension d is comparable with n, then $E[\hat{\xi}(\Sigma)] \to c\xi(\Sigma)$ when $n \to +\infty$, where $c = \lim_{n\to+\infty} d/n$. We refer to Pavlenko (2003) and Pavlenko and Von Rosen (2001) for further details on the effect of the dimensionality on classification in the asymptotic context, i.e. as d and $n \to +\infty$.

8.2.2 *The Blessing of Dimensionality in Model-based Clustering and Classification*

In the context of classification, high-dimensional spaces do have useful characteristics which ease the classification of data in those spaces. In particular, Scott and Thompson (1983) showed that high-dimensional spaces are mostly empty. A simple experiment illustrates this "empty space" phenomenon. We consider the shell between the hypersphere of radius 0.9 and the unit hypersphere in p-dimensional space. In order to study the behavior of the volume of this shell with respect to the dimension of the space, we consider the ratio of the volumes of the two hyperspheres. Figure 8.7 shows how this ratio changes as the dimension d of the space increases. The ratio decreases quickly toward 0 and suggests that the d-dimensional shell between the two hyperspheres tends to have in fact an intrinsic dimension equal to $d-1$.

A similar experiment, suggested by Huber (1985), consists of considering a d-dimensional random vector X with uniform probability distribution on the hypersphere of radius 1. The probability that X belongs to the shell between the hypersphere of radius 0.9 and the unit hypersphere is therefore

$$\mathbb{P}\left(X \in S_{0.9}(d)\right) = 1 - 0.9^d.$$

In particular, the probability that X belongs to the shell between the hypersphere of radius 0.9 and the unit hypersphere in a 20-dimensional space is roughly equal to 0.88. Therefore, most of the realizations of the random vector X live near a $d-1$ dimensional subspace and the remainder of the space is mostly empty. This suggests that clustering methods should model the groups in low-dimensional subspaces instead of modeling them in the full observation space. Furthermore, it seems reasonable to expect that different groups live in different subspaces and this may be a useful property for discriminating the groups. Subspace clustering methods, presented in Section 8.4, exploit this characteristic of high-dimensional spaces.

A simple way to verify this property of high-dimensional spaces is to look at the behavior of the optimal Bayes classifier (i.e. the MAP classifier with the actual conditional distribution, which does not suffer from estimation problems) for different values for the dimension of the space. Listing 8.5 implements a simple experiment which draws some three-class data from a mixture of Gaussians and computes the classification error rate of the Bayes' classifier and compares it with that of EDDA (with model $p_k L_k I$). Let us highlight that the Bayes' classifier will obtain the lowest possible error rate since it uses the true density to classify the data. The data were drawn from a mixture of three Gaussians with means

$$\begin{cases} \mu_1 = & (0, 0, ..., 0) \\ \mu_2 = & (0, -2, ..., 0) \\ \mu_3 = & (0, 2, ..., 0) \end{cases}$$

and covariance matrices $\Sigma_1 = I_d$, $\Sigma_2 = 0.8 \times I_d$ and $\Sigma_3 = 0.9 \times I_d$; the results are shown in Figure 8.8.

It appears on Figure 8.8 that the Bayes' classifier has a lower classification error in high-dimensional space than in spaces of lower dimensionality. Indeed, the classification error goes from 0.2 to less than 0.15 when the dimensionality increases from 10 to 200. In comparison, the EDDA classification error rate strictly increases when d increases. This is due to the difficult estimation of parameters in high-dimensional spaces, even though we are assuming the correct model. From this experiment, we can conclude that high-dimensional spaces are more a blessing for classification than a curse, provided that we succeed in overcoming the parameter estimation problem.

Listing 8.5: Behavior of Bayes' classifier according to dimension

```r
# Loading libraries
library(mvtnorm)
library(Rmixmod)

# Bayes' classifier function for Gaussian distributions
BayesClass <- function(X,tau,m,S){
  G = length(m); d = ncol(X)
  P = matrix(NA,nrow(X),G)
  for (g in 1:G) P[,g] = tau[g] * dmvnorm(X,m[[g]],S[[g]])
  P = P / rowSums(P) %*% matrix(1,1,G)
  list(P = P, cls = max.col(P))
}

# Simulation
n = 120; nbrep = 25
dims = seq(10,210,20)
err = err2 = matrix(NA,length(dims),nbrep)
for (i in 1:length(dims)){
  for (j in 1:nbrep){
    # Data simulation
    d = dims[i]
    m1 = c(0,0,rep(0,d-2)); m2 = c(0,-2,rep(0,d-2)); m3 = c(2,0,rep
        (0,d-2));
    S1 = diag(d); S2 = 0.8 * diag(d); S3 = 0.9 * diag(d)
    X = as.data.frame(rbind(mvrnorm(n/3,m1,S1),mvrnorm(n/3,m2,S2),
        mvrnorm(n/3,m3,S3)))
    X2 = as.data.frame(rbind(mvrnorm(10*n/3,m1,S1),mvrnorm(10*n/3,
        m2,S2),mvrnorm(10*n/3,m3,S3)))
    cls = rep(1:3,rep(n/3,3)); cls2 = rep(1:3,rep(10*n/3,3))
    # Classification with the Bayes' classifier
    res1 = BayesClass(X2,rep(1/3,3),list(m1,m2,m3),
                      list(S1,S2,S3))$cls
    # Classification with EDDA
    mod = mixmodLearn(X,cls,models=mixmodGaussianModel(listModels =
        'Gaussian_pk_Lk_I'))
    res2 = mixmodPredict(X2,mod["bestResult"])@partition
    # Computing error rate
    err[i,j] = sum(res1 != cls2) / length(cls2)
    err2[i,j] = sum(res2 != cls2) / length(cls2)
  }
}

# Plotting error rate
boxplot(t(err),ylim=c(0.1,0.33),names=dims,xlab='Dimension',ylab='
    Classification error rate',col=3)
boxplot(t(err2),names=dims,xlab='Dimension',ylab='Classification
    error rate',col=4,add=TRUE)
legend("bottomleft",legend = c('Bayes classifier','EDDA'),col=c
    (3,4),lty=1,pch=19)
```

8.3 Earlier Approaches for Dealing with High-dimensional Data

The earliest approaches for clustering or classification of high-dimensional data can be split into three families: dimension reduction methods, regularization methods and parsimonious methods.

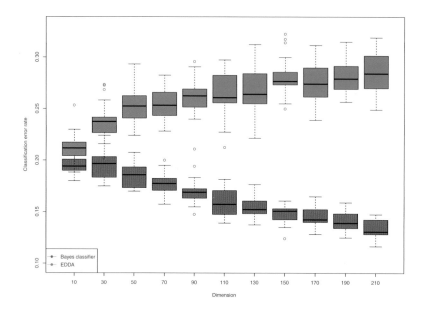

Figure 8.8 Classification error rate for the Bayes' classifier (blue) and EDDA (green) according to the space dimensionality for data drawn from a mixture of three Gaussians.

8.3.1 Unsupervised Dimension Reduction

An intuitive way to deal with high-dimensional data is to try to reduce the data dimensionality in such a way that the curse of the dimensionality vanishes. Approaches based on dimension reduction assume that the number d of measured variables is too large and, implicitly, that the data at hand live in a space of lower dimension, let us say $q < d$. Such approaches first look for a subspace which contains most of the data information and then project the high-dimensional data into this subspace. Once the data have been projected into a low-dimensional space, it is then possible to classify or cluster the data within this subspace. For instance, it may be possible to cluster or classify the data by applying `mclust` or `Rmixmod` to the projected observations to obtain a partition of the original data.

The most popular linear method used for dimension reduction is certainly principal component analysis (PCA). It was introduced by Pearson (1901) who defined PCA as a linear projection that minimizes the average projection cost. Later, Hotelling (1933) proposed another definition for PCA as reducing the dimension of the data by keeping as much as possible the variation of the data set. In other words, this method aims to find an orthogonal projection of the data set into a low-dimensional linear subspace, such that the variance of the projected data is maximized. This leads to the classical result where the principal axes $\{u_1, ..., u_q\}$ are the eigenvectors associated with the largest eigenvalues of the

Listing 8.6: PCA on the `NIR` data

```
# Loading libraries and data
library(MBCbook)
data(NIR)
Y = NIR[,-1]
cls = NIR$cls

# PCA on NIR data (through SVD since n < d here)
U = svd(Y)$v
X = as.matrix(Y) %*% U[,1:2]
plot(X,col=cls,pch=(15:17)[cls],cex=1.25,xlab='PC 1',ylab='PC 2')
```

empirical covariance matrix (S) of the data, where

$$S = \frac{1}{n}\sum_{i=1}^{n}(y_i - \overline{y})(y_i - \overline{y})^T.$$

Several decades later, Tipping and Bishop (1997) described a probabilistic view of PCA by assuming that the observations are independent realizations of a random variable $Y \in \mathbb{R}^d$ which is linked to a low-dimensional latent variable $X \in \mathbb{R}^q$ through the linear relation

$$Y = \Lambda X + \varepsilon,$$

where Λ is a $(d \times q)$ matrix. It is further assumed that $X \sim \mathcal{N}(\mu, I_q)$ and $\varepsilon \sim \mathcal{N}(0, \sigma^2 I_d)$, such that the marginal distribution of Y is

$$Y \sim \mathcal{N}(\Lambda\mu, \Lambda\Lambda^T + \sigma^2 I_d).$$

With this point of view, probabilistic PCA (PPCA) reduces to the estimation of the model parameters μ, Λ and σ^2. Estimation by maximum likelihood reduces in particular to estimate Λ by the eigenvectors associated with the largest eigenvalues of the empirical covariance matrix S of the data. PCA is strongly related to another latent variable model, called factor analysis (FA) (Spearman, 1904), and its generalized, structural equation modeling (Jöreskog, 1978). FA aims to both reduce the dimensionality of the space and keep the observed covariance structure of the data. Many nonlinear data reduction techniques have also been proposed such as Kernel PCA (Schölkopf et al., 1998), nonlinear PCA (Hastie and Stuetzle, 1989) and neural network based techniques (Kohonen, 1995).

Figure 8.9 shows the projection of the chemometrical `NIR` data (202 observations in a 2,800-dimensional space) onto the first two principal component axes; Listing 8.6 provides the R code to obtain the figure. One can see that the projection onto the first PCA axes mostly preserves the information in the data (see Figure 8.5). Indeed, the PCA projection makes it easy to discriminate two of the three classes (red and black) as in the high-dimensional representation of Figure 8.5.

Let us also point out that for this specific example, where $n < d$, it is not possible to use the classical eigen-decomposition of the empirical covariance matrix

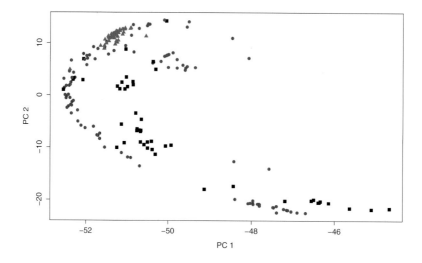

Figure 8.9 Visualization of the `NIR` data onto the first two principal component axes. Colors correspond to the known three types of textiles.

for computing the PCA axes. The empirical covariance matrix is indeed not of full rank and it is more efficient to use instead the singular value decomposition (SVD) of the data. It can be shown that the first δ right singular vectors of Y are equal to the first δ eigenvectors of $Y^T Y$.

8.3.2 The Dangers of Unsupervised Dimension Reduction

Despite the popularity of the dimension reduction approach, we would like to caution the reader that reducing the dimension without taking into consideration the classification or clustering goal may be dangerous. Such a dimension reduction may yield a loss of information which could have been useful for discriminating the classes or groups. In particular, when PCA is used for reducing the data dimensionality, only the components associated with the largest eigenvalues are kept. The risks of this were illustrated by Chang (1983) who showed that the first components do not necessarily contain more discriminative information than the later ones. In addition, reducing the dimension of the data may not be a good idea since, as discussed earlier, it may be easier to discriminate groups of points in high-dimensional spaces than in lower dimensional spaces, assuming that one can build a good classifier in high-dimensional spaces.

A simple experiment to reveal the potential danger of using an unsupervised dimension reduction method before classification or clustering is proposed by Listing 8.7. The experiment considers also the `NIR` data set. First, the data are projected on the first four PCA axes and, as shown by Figure 8.10, the subspace spanned by the first two PCA axes is not always the best subspace to discriminate

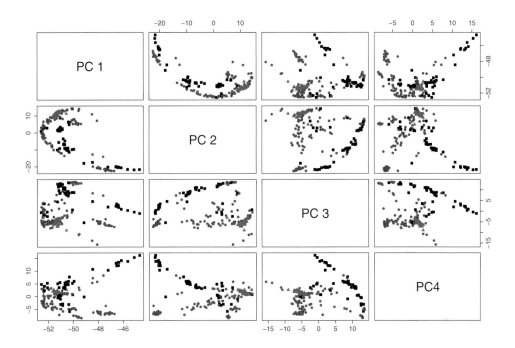

Figure 8.10 Matrix of scatter plots for the first four PCA axes of the NIR data. Colors correspond to the known three types of textiles.

the classes. Here, the subspace of PC3 and PC4 appears to be a better choice than the subspace of PC1 and PC2.

Then, it is also easy to verify this observation by comparing the performance of a supervised classifier, QDA for instance, within both subspaces. The last part of Listing 8.7 makes this comparison using a cross-validation scheme. The results are presented in Figure 8.11. As expected, the classification error of QDA in the subspace PC3-4 is substantially lower than QDA in the subspace PC1-2 for these data. This practical example has therefore shown that unsupervised dimension reduction may not be well adapted to the classification and clustering tasks and should be used with caution in these contexts.

8.3.3 Supervised Dimension Reduction for Classification

Although unsupervised dimension reduction should be avoided as a preprocessing step for clustering or classification, supervised dimension reduction may be useful in the classification context. Indeed, supervised techniques for dimension reduction take into account the classification goal by incorporating the objective through the known labels of the learning data.

The earliest methodology for supervised dimension reduction was proposed by Fisher (1936). Fisher posed the problem of the discrimination of three species of iris described by four measurements. His main goal was to find the linear subspace

Listing 8.7: The dangers of unsupervised dimension reduction in classification

```
# Loading libraries and data
library(MBCbook)
data(NIR)
Y = NIR[,-1]
cls = NIR$cls

# PCA on NIR data (through SVD since n < d here)
U = svd(Y)$v
X = as.matrix(Y) %*% U
pairs(X[,1:4],col=cls,pch=(15:17)[cls],cex=1.25,
        labels=c('PC 1','PC 2','PC 3','PC4'))

# Classification on different couples of PCA axes
library(MASS)
n = nrow(Y); nb = 50
Err = matrix(NA,2,nb)
for (i in 1:nb){
  ind = sample(nrow(Y),round(n/2))
  U = svd(Y[ind,])$v
  X = as.matrix(Y) %*% U
  Err[1,i] = sum(predict(qda(X[-ind,1:2],cls[-ind]),X[ind,1:2])$cl
    != cls[ind]) / length(ind)
  Err[2,i] = sum(predict(qda(X[-ind,3:4],cls[-ind]),X[ind,3:4])$cl
    != cls[ind]) / length(ind)
}

boxplot(t(Err),col=2:3,names=c('QDA on PC axes 1/2','QDA on PC axes
    3/4'),ylim=c(0,0.7),ylab='Error rate')
```

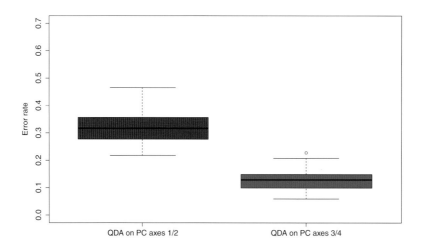

Figure 8.11 Cross-validated classification error rates of QDA on PCA axes 1-2 (left) and on PCA axes 3-4 (right) for the NIR data.

that best separates the classes according to a criterion (see Duda et al., 2000, for more details). For this, it is assumed that the dimension d of the original space is greater than the number G of classes. Fisher's linear discriminant analysis (FDA or LDA) looks for a linear transformation matrix U which projects the observations $\{y_1, ..., y_n\}$ in a discriminative low-dimensional subspace of dimension q. To this end, the $d \times q$ transformation matrix U maximizes a criterion which is large when the between-class covariance matrix (S_B) is large and when the within-covariance matrix (S_W) is small. Since the rank of S_B is at most equal to $G - 1$, the dimension q of the discriminative subspace is therefore at most equal to $G - 1$.

Four criteria satisfy such a constraint (see Fukunaga, 1990, for a review). The criterion which is traditionally used is:

$$J(U) = \mathrm{tr}((U^T S_W U)^{-1} U^T S_B U), \qquad (8.2)$$

where:

$$S_W = \frac{1}{n} \sum_{g=1}^{G} \sum_{y_i \in C_g} (y_i - m_g)(y_i - m_g)^T,$$

$$S_B = \frac{1}{n} \sum_{g=1}^{G} n_g (m_g - \bar{y})(m_g - \bar{y})^T,$$

are respectively the within and the between covariance matrices, n_g is the number of observations in the gth class, $m_g = \frac{1}{n_g} \sum_{i \in C_g} y_i$ is the empirical mean of the observed column vector y_i in the class g and $\bar{y} = \frac{1}{n} \sum_{i=1}^{n} y_i$ is the mean column vector of the observations.

The maximization of the criterion (8.2) is equivalent to the generalized eigenvalue problem (Krzanowski, 2003) $(S_W^{-1} S_B - \lambda I_d) U = 0$, and the classical solution to this problem is given by the eigenvectors associated with the q largest eigenvalues of the matrix $S_W^{-1} S_B$. Once the discriminative axes have been determined, the MAP rule is applied to classify the data in this subspace.

Optimization of the Fisher criterion supposes that the matrix S_W is non-singular. However, S_W is often singular, particularly in the cases of very high-dimensional data or small sample sizes. Different solutions have been proposed to deal with this problem in the supervised classification framework (Friedman, 1989; Fukunaga, 1990; Hastie et al., 1995; Jin et al., 2001; Howland and Park, 2004).

Listing 8.8 proposes a simple experiment for comparing dimension reduction with an unsupervised technique (PCA, left) and with a supervised technique (FDA, right) on the usps358 data. Figure 8.12 shows both projections. Although PCA works quite well here, FDA clearly provides a better projection of the data, considering the classification point of view.

An alternative to FDA was proposed by Barker and Rayens (2003) for supervised dimension reduction of high-dimensional data, with corollated inputs. The method, called PLS-DA, is based on partial least squares (PLS) regression. The idea behind PLS is to find q latent variables which both maximize the projected variance of

Listing 8.8: PCA vs. Fisher discriminant analysis on the `usps358` data

```
# Loading libraries and data
library(MBCbook)
data(usps358)
Y = usps358[,-1]
cls = usps358$cls

# PCA on usps358 data
U = svd(Y)$v
X.pca = as.matrix(Y) %*% U[,1:2]
plot(X.pca,type='n',xlab='PC 1',ylab='PC 2')
text(X.pca,col=cls,labels=c(2,6,9)[cls])

# FDA on usps358 data
V = lda(Y,cls)$scaling
X.fda = as.matrix(Y) %*% V
plot(X.fda,type='n',xlab='FDA axis 1',ylab='FDA axis 2')
text(X.fda,col=cls,labels=c(2,6,9)[cls])
```

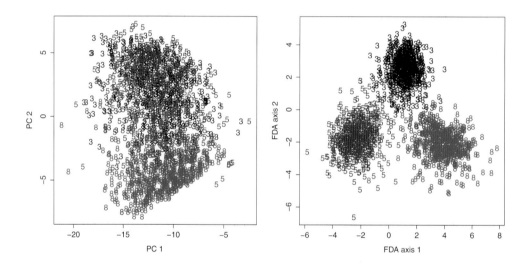

Figure 8.12 Dimension reduction with an unsupervised technique (PCA, left) and with a supervised technique (FDA, right) on the `usps358` data.

explanatory variable $Y \in \mathbb{R}^d$ (as in PCA) and maximize the correlation between the target variable $Z \in \mathbb{R}^\ell$ and Y. The underlying model of PLS assumes that there exist two latent variables $X \in \mathbb{R}^q$ and $T \in \mathbb{R}^r$ such that:

$$Y = U^T X + \varepsilon,$$
$$Z = V^T W + e,$$

where U and V are loading matrices of sizes $q \times d$ and $r \times \ell$ respectively, and ε and e are Gaussian noises. PLS therefore looks for the latent variables X and W

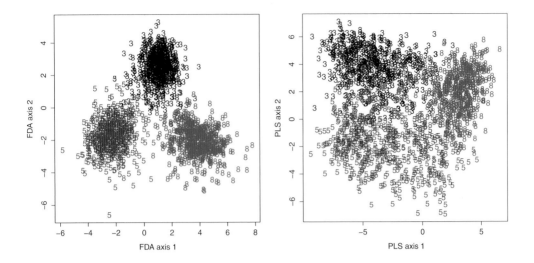

Figure 8.13 Dimension reduction with two supervised techniques: FDA (left) and PLS-DA (right) on the usps358 data.

maximizing the covariance of YU and ZV under orthogonality constraints on U and V. The PLS2 algorithm solves this problem by sequentially computing each column of U and V within the complementary spaces of the already computed components (Wold et al., 2001).

In the classification context, the target variable Z is categorical, $Z \in \{1, ..., G\}$, and must be recoded as an $n \times G$ dummy block matrix where each of the response categories (the classes) is coded by an indicator variable:

$$Z = \begin{pmatrix} 1 \\ 3 \\ 2 \\ 1 \\ 3 \end{pmatrix} \rightarrow Z' = \begin{pmatrix} 1 & 0 & 0 \\ 0 & 0 & 1 \\ 0 & 1 & 0 \\ 1 & 0 & 0 \\ 0 & 0 & 1 \end{pmatrix}.$$

Then PLS-DA consists simply of learning a PLS regression between the explanatory variable $Y \in \mathbb{R}^d$ and the recoded target variable $Z' \in \{0, 1\}^G$.

Listing 8.9 provides R code for computing the PLS-DA latent subspace. The results are pictured in Figure 8.13 in addition to the previous visualization obtained with FDA. The two visualizations turn out to be similar, with a slight advantage for FDA here. Recall that PLS and PLS-DA are designed for very high-dimensional data with small sample sizes and, for these reasons, PLS and PLS-DA are popular in fields such as chemometrics.

Listing 8.9: PLS discriminant analysis on the `usps358` data

```
# Loading libraries and data
library(pls)
library(MBCbook)
data(usps358)
Y = usps358[,-1]
cls = usps358$cls

# Dummy recoding of the labels
n = nrow(Y); G = max(cls)
Z = matrix(0,n,G)
for (g in 1:G) Z[cls==g,g] = 1

# PLS-DA on usps358 data
pl = plsr(Z ~ as.matrix(Y),ncomp=2,method='oscorespls')
W = unclass(pl$loadings)
X.pls = as.matrix(Y) %*% W

# Plotting the resulting projection
plot(X.pls,type='n',xlab='PLS axis 1',ylab='PLS axis 2')
text(X.pls,col=cls,labels=c(2,6,9)[cls])
```

8.3.4 Regularization

It is also possible to see the curse of dimensionality in classification as a numerical problem in the inversion of the covariance matrices Σ_g, $g = 1, ..., G$. From this point of view, a way to tackle the curse of dimensionality is to numerically regularize the estimates of the covariance matrices Σ_g before their inversion. As we will see, most of the regularization techniques have been proposed in the supervised classification framework, but they can be easily used for clustering as well.

A simple way to regularize the estimate of Σ_g is to consider a ridge regularization which adds a positive quantity ξ_g to the diagonal of the matrix:

$$\tilde{\Sigma}_g = \hat{\Sigma}_g + \xi_g I_d.$$

Notice that this regularization is often implicitly used in statistical software for performing linear discriminant analysis (LDA). For instance, the `lda` R function spheres the data before analysis. Tuning the regularization parameter ξ_g can be done by cross-validation in the supervised case and by model selection in clustering.

Friedman (1989) also proposed for his popular regularized discriminant analysis (RDA, see Chapter 4.3.2) a regularization which makes a trade-off between a quadratic classifier and a linear one (typically LDA). Another regularization has been proposed by Hastie et al. (1995) for LDA which also penalizes the correlations between the predictors. This regularization is used in particular in the supervised classification method penalized discriminant analysis (PDA). See also Mkhadri et al. (1997) for a comprehensive overview of regularization techniques in discriminant analysis. The solution based on regularization does not have the same drawbacks as dimension reduction and can be used more readily. However,

Table 8.1 *Number of free parameters to estimate for parsimonious Gaussian mixture models with G components and d variables.*

Model	Name	No. of parameters	$G = 4$ $d = 100$
$[\lambda_g D_g A_g D_g^T]$	VVV	$(G-1) + Gd + Gd(d+1)/2$	20603
$[\lambda D_g A_g D_g^T]$	EVV	$(G-1) + Gd + Gd(d+1)/2 - (G-1)$	20600
$[\lambda_g D_g A D_g^T]$	VEV	$(G-1) + Gd + Gd(d+1)/2 - (G-1)(d-1)$	20306
$[\lambda D_g A D_g^T]$	EEV	$(G-1) + Gd + Gd(d+1)/2 - (G-1)d$	20303
$[\lambda_g D A_g D^T]$	VVE	$(G-1) + Gd + d(d+1)/2 + (G-1)d$	5753
$[\lambda D A_g D^T]$	EVE	$(G-1) + Gd + d(d+1)/2 + (G-1)(d-1)$	5750
$[\lambda_g D A D^T]$	VEE	$(G-1) + Gd + d(d+1)/2 + (G-1)$	5456
$[\lambda D A D^T]$	EEE	$(G-1) + Gd + d(d+1)/2$	5453
$[\lambda_g B_g]$	VVI	$(G-1) + Gd + Gd$	803
$[\lambda B_g]$	EVI	$(G-1) + Gd + Gd - (G-1)$	800
$[\lambda_g B]$	VEI	$(G-1) + Gd + d + (G-1)$	506
$[\lambda B]$	EEI	$(G-1) + Gd + d$	503
$[\lambda_g \mathbf{I}_d]$	VII	$(G-1) + Gd + G$	407
$[\lambda \mathbf{I}_d]$	EII	$(G-1) + Gd + 1$	404

all regularization techniques require the tuning of a parameter which is difficult to tune in the unsupervised context, whereas this can be done easily in the supervised context using cross-validation.

8.3.5 Constrained Models

A third way to look at the the curse of dimensionality in classification is to consider it as a problem of over-parameterized modeling. Indeed, as we have seen earlier in this section, the Gaussian mixture model turns out to be highly parameterized which naturally yields inference problems in high-dimensional spaces. Consequently, the use of constrained models is another solution to avoid the curse of dimensionality in model-based classification.

A traditional way to reduce the number of free parameters of Gaussian models is to add constraints on the model through their parameters. Let us recall that the unconstrained Gaussian model is a highly parameterized model and requires the estimation of 20 603 parameters when the number of components is $G = 4$ and the number of variables is $d = 100$. A first possible constraint for reducing the number of parameters to estimate is to constrain the covariance matrices to be the same across all mixture components, i.e. $\Sigma_g = \Sigma$, $\forall g = 1, ..., G$. This model yields linear discriminant analysis (LDA, Fisher, 1936) in the supervised classification case. It is also possible to make use of the parsimonious models of Banfield and Raftery (1993) and Celeux and Govaert (1995), which are discussed in detail in Chapters 1 and 2.

Table 8.1 lists those models and provides the number of free parameters to estimate for each model. The number of parameters to estimate can be decomposed into the number of parameters to estimate for the proportions $(G - 1)$, for the means (Gd) and for the covariance matrices (last terms). The table also gives the number of parameters to estimate when $G = 4$ and $d = 100$.

This approach proposes a trade-off between the perfect modeling and what one can correctly estimate in practice in model-based classification (Bensmail and Celeux, 1996). However those models are still not appropriate in very high dimensions because they are not parsimonious enough. In the following sections, we will see that recent solutions for classifying or clustering high-dimensional data are based on alternative methods of constrained modeling and can perform better.

8.4 Subspace Methods for Clustering and Classification

We now present model-based subspace methods for the supervised and unsupervised classification of high-dimensional data. In what follows, we will present only the models. Unless otherwise noted, inference for these models is done either via the EM algorithm in the unsupervised case (see Chapter 2) or by direct likelihood maximization in the supervised context (see Chapter 4).

Unlike the previous solutions, subspace methods exploit the "empty space" phenomenon to ease the discrimination between groups of points. To do so, they model the data in low-dimensional subspaces and introduce some restrictions while keeping all dimensions. Subspace methods are mostly related to the factor analysis model (Spearman, 1904; Rubin and Thayer, 1982), which assumes that the observation space is linked to a latent space through a linear relationship.

8.4.1 Mixture of Factor Analyzers (MFA)

The mixture of factor analyzers (MFA, Ghahramani and Hinton, 1997; McLachlan et al., 2003) was the earliest subspace classification method to both classify the data and locally reduce the dimension of each cluster. The MFA model differs from the FA model in that it allows different local factor models, in different regions of the input space, whereas the standard FA assumes a common factor model. Figure 8.14 illustrates the modeling of MFA for two groups with intrinsic dimension 2. The MFA model was introduced by Ghahramani and Hinton (1997), and then extended by McLachlan et al. (2003). We focus here only on the version of Baek et al. (2009) which can be considered as the most general.

The MFA model is an extension of the factor analysis model to a mixture of G factor analyzers. Let $\{y_1, \ldots, y_n\}$ be independent observed realizations of a random vector $Y \in \mathbb{R}^d$. We assume that the distribution of Y is determined by an unobserved random vector $X \in \mathbb{R}^q$, named the factor and belonging to a lower dimensional space of dimension $q < d$. The unobserved labels $\{z_1, \ldots, z_n\}$ indicating the group memberships are assumed to be independent unobserved realizations of a random vector $Z \in \{1, \ldots, G\}$, where $z_i = g$ indicates that y_i is generated by the gth factor analyzer. The relationship between these two spaces

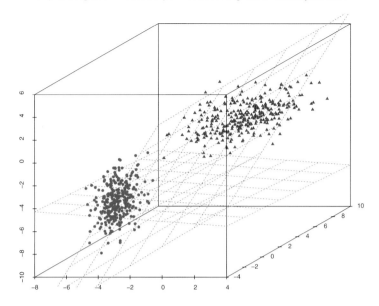

Figure 8.14 Illustration of mixture of factor analyzers (MFA) modeling for two groups with intrinsic dimension 2 in R^3.

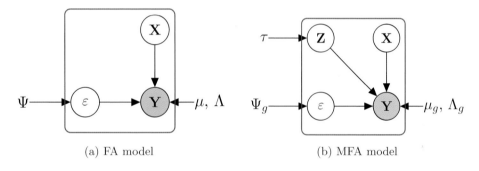

(a) FA model   (b) MFA model

Figure 8.15 Graphical summary of (a) factor analysis (FA) model and (b) mixture of factor analyzers (MFA).

is assumed to be linear, conditionally on Z:

$$Y|Z = g = \Lambda_g X + \mu_g + \varepsilon, \qquad (8.3)$$

where Λ_g is a $d \times q$ matrix and $\mu_g \in \mathbb{R}^d$ is the mean vector of the gth factor analyzer. Conditionally to Z, $\varepsilon \in \mathbb{R}^d$ is assumed to be a zero-centered Gaussian noise term with a diagonal covariance matrix Ψ_g:

$$\varepsilon|Z = g \sim \mathcal{N}(0, \Psi_g). \qquad (8.4)$$

The conditional independence structure of the model is illustrated in Figure 8.15.

As in the FA model, the factor $X \in \mathbb{R}^q$ is assumed to be distributed according to a Gaussian density function such as $X \sim \mathcal{N}(0, \mathbf{I}_q)$. This implies that the

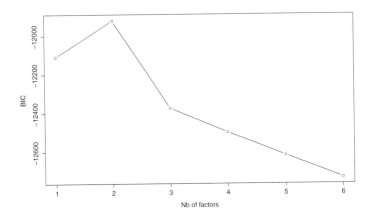

Figure 8.16 Choice of the number q of factors for the MFA model on the `wine` data.

conditional distribution of Y is also Gaussian:

$$Y|X, Z = g \sim \mathcal{N}(\Lambda_g X + \mu_g, \Psi_g). \tag{8.5}$$

The marginal density of Y is thus a Gaussian mixture model such that

$$f(y) = \sum_{g=1}^{G} \tau_g \phi(y; \mu_g, \Lambda_g \Lambda_g^T + \Psi_g),$$

where τ_g is the mixture proportion for the gth component. The number of independently estimated parameters in the MFA model is then $\nu_{MFA} = (G-1) + Gd + Gq(d - (q-1)/2) + d$ (McLachlan et al., 2003). If $d = 100$, $G = 4$ and $q = 3$, then 1,691 parameters have to be estimated for this MFA model.

The alternative version of the MFA model proposed by Ghahramani and Hinton (1997) restricts the previous model such that ε has a diagonal covariance matrix Ψ common to all factors. For inference, Ghahramani and Hinton (1997) proposed an exact EM algorithm for their model whereas McLachlan et al. (2003) used an AECM algorithm.

We now apply the MFA approach to the `wine27` data. Listing 8.10 provides some R code using the `pgmm` package which provides the MFA model under the name CCU (see Section 8.4.3). A key issue for the method is choosing the number of factors q. A classical way to do this is to consider MFA models with different numbers of factors as different models and choose between them with model selection criteria. BIC is used here for picking q within the range $1, \ldots, 6$. Figure 8.16 shows the BIC curve which has its maximum when $q = 2$. The clustering result turns out to be comparable with the one obtained earlier with `mclust` but with a much shorter computing time (a couple of seconds for the whole procedure).

Listing 8.10: Mixture of factor analyzers on the `wine27` data

```
# Loading libraries and data
library(pgmm)
library(MBCbook)
data(wine27)
Y = wine27[,1:27]
cls = wine27$Type

# Selecting the number of factors by model selection
res.mfa = pgmmEM(Y,rG=3,rq=1:6,modelSubset='CCU')
plot(1:6,res.mfa$bic$CCU,type='b',xlab='Nb of factors',ylab='BIC')

# Clustering results
table(cls,res.mfa$map)
           cls          1   2   3
           Barbera      1  47   0
           Barolo       0   0  59
           Grignolino  71   0   0
```

8.4.2 Extensions of the MFA Model

Baek et al. (2009) proposed an alternative formulation that aims both to lower the complexity of the MFA model and to ease the visualization of the clustered data. To that end, they reparameterized the mixture model with restrictions on the means and on the covariance structure. The means are restricted by specifying $\mu_g = A\rho_g$, where A is a $d \times q$ orthonormal matrix (that is, $A^T A = \mathbf{I}_d$) and ρ_g is a q-dimensional vector. The covariance matrices are assumed to be structured as $S_g = A\Omega_g A^T + \Psi$ where Ω_g is a $q \times q$ positive definite symmetric matrix and Ψ a diagonal $d \times d$ matrix. This model is referred to as the mixture of factor analyzers with common factor loadings (MCFA) by its authors, as the matrix A is common to the factors.

According to the MCFA assumptions, there are only Gq mean parameters to estimate instead of Gd in the MFA model. Since the matrix A is constrained to have orthonormal columns and to be common to all classes, then only $dq - q(q+1)/2$ loadings are required to estimate it. According to the restriction on the matrices Ω_g, the number of parameters to estimate for these G matrices is $Gq(q+1)/2$. The complexity of the MCFA model is therefore $\nu_{MCFA} = (G-1) + Gq + d + (dq - q(q+1)/2) + Gq(q+1)/2$. When $G = 4$ and $d = 100$, this complexity is equal to 433, which is much more parsimonious than the previous MFA models. Besides, this MCFA approach is a special case of the MFA model, but has the advantage of allowing the data to be displayed in a common low-dimensional plot.

The MCFA approach is a generalization of the work of Yoshida et al. (2004, 2006) since they constrained the covariance of the noise term to be spherical ($\Psi = \lambda \mathbf{I}_d$) and the component-covariance matrices of the factors to be diagonal ($\Omega_g = \Delta_g = \text{diag}(\sigma_{g1}^2, \ldots, \sigma_{gd}^2)$). This approach, called mixtures of common uncorrelated factor spherical-error analyzers (MCUFSA), is therefore more parsimonious than MCFA

Table 8.2 *Nomenclature of the MFA models developed by Ghahramani and Hinton (G-MFA), McLachlan et al. (M-MFA), and several MCFA models with their corresponding covariance structure.*

Model name	Cov. structure	No. of parameters	$G = 4$ $q = 3$ $d = 100$
M-MFA	$S_g = \Lambda_g \Lambda_g^T + \Psi_g$	$(G-1) + Gd + Gq[d - (q-1)/2] + Gd$	1991
G-MFA	$S_g = \Lambda_g \Lambda_g^T + \Psi$	$(G-1) + Gd + Gq[d - (q-1)/2] + d$	1691
MCFA	$S_g = A\Omega_g A^T + \Psi$	$(G-1) + Gq + d + q[d - (q+1)/2] + Gq(q+1)/2$	433
HFMA	$S_g = V\Omega_g V^T + \Psi$	$(G-1) + (G-1)q + d + q[d - (q-1)/2] + (G-1)q(q+1)/2$	427
MCUFSA	$S_g = A\Delta_g A^T + \lambda \mathbf{I}_d$	$(G-1) + Gq + 1 + q[d - (q+1)/2] + Gq$	322

A is defined such as $A^T A = \mathbf{I}_d$, V such as $V\Psi^{-1}V^T$ is diagonal with decreasing order and Δ_g is a diagonal matrix.

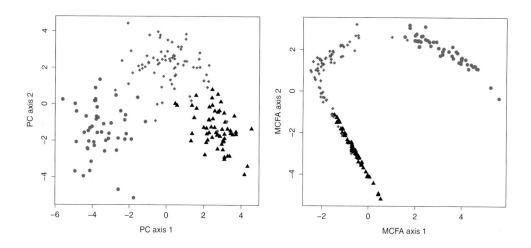

Figure 8.17 Projection of the `wine27` data on the two first components of PCA (left) and on the MCFA subspace (right). Colors indicate the actual group labels of data.

according to the additional assumptions made on the parameters of the MFA model.

Montanari and Viroli (2010) presented the heteroscedastic mixture factor model (HMFA) which is similar to the model described in MCFA. Their model differs from the MCFA approach only in the definition of the common loadings matrix A, which does not need to have orthonormal columns. However, to obtain a unique solution for the matrix A, Montanari and Viroli added restrictions on this matrix such as $A^T \Psi^{-1} A$ is diagonal with elements in decreasing order. The links and differences between all these MCFA models are summarized in Table 8.2, which shows both the covariance structure and the model complexity of each approach.

Inference for both the MCFA and MCUFSA models is done using the AECM algorithm whereas a standard EM algorithm is used for the HMFA model.

We now apply the MCFA model to the `wine27` data. Listing 8.11 gives the R code for that using the `EMMIXmfa` package. We fix the number q of common factors to $q = 2$, such that we can project the data onto a plane. Indeed, one of the main advantages of this model is the possibility of visualizing the clustered data in a common and low-dimensional subspace. Figure 8.17 shows the projection of the data on the estimated subspace with, as colors, the actual groups. As a comparison, the left panel of the figure shows the PCA projection for these data.

Listing 8.11 also provides the classification table for the MCFA results. One can notice that, compared to the MFA results, constraining the FA subspace to be common to the groups implies a moderate but significant deterioration of the clustering results. This may be viewed as the price for getting a useful visualization of the clustered data. From Figure 8.17, one may also remark that

Listing 8.11: Mixture of common factor analyzers on the `wine27` data

```
# Loading libraries and data
library(EMMIXmfa)
library(MBCbook)
data(wine27)
Y = wine27[,1:27]
cls = as.numeric(wine27$Type)

# Learning the common FA subspace and compare with PCA
res.mcfa <- mcfa(Y,g=3,q=2)
par(mfrow=c(1,2))
plot(predict(princomp(Y)),col=cls+1,pch=c(17,18,19)[cls],xlab='PC
    axis 1',ylab='PC axis 2')
plot(res.mcfa$Uscores,col=cls+1,pch=c(17,18,19)[cls],xlab='MCFA
    axis 1',ylab='MCFA axis 2')

# Clustering results
table(cls,res.mcfa$clust)
   cls  1  2  3
     1 47  0  1
     2  0 51  8
     3  0  4 67
```

at least two (black and blue) of the three groups have intrinsic dimension lower than 2.

8.4.3 Parsimonious Gaussian Mixture Models (PGMM)

A general framework for the MFA model was also proposed by McNicholas and Murphy (2008) which includes the previous work of Ghahramani and Hinton (1997) and of McLachlan et al. (2003) as special cases. By considering the previous framework, defined by Equations (8.3) and (8.5), McNicholas and Murphy (2010b) proposed a family of models known as the expanded parsimonious Gaussian mixture model (EPGMM) family. They defined 12 EPGMM models by constraining the terms of the covariance matrix to be either equal or not, considering an isotropic variance for the noise term, or reparameterizing the factor analysis covariance structure. The nomenclature of both PGMM and EPGMM is illustrated in Table 8.3 in which the covariance structure of each model is detailed as well.

The terminology of the PGMM family is as follows: the first letter refers to the loading matrix, which is constrained to be common between groups (C..) or not (U..), the second term indicates whether the noise variance is common between factors (.C.) or not (.U.), and the last term refers to the covariance structure, which can be either isotropic (..C) or not (..U). Thus, the CCC model refers to a model with common factors ($\Lambda_g = \Lambda, \forall g \in \{1, \ldots, G\}$) and a common and isotropic noise variance ($\Psi_g = \psi \mathbf{I}_p$).

In the terminology of the EPGMM family, the structure of the noise covariance matrix is as follows: Δ_g can be common (.C..) or not (.U..), $\omega_g = \omega \, \forall g \in \{1, \ldots, G\}$

Table 8.3 Nomenclature of the members of the PGMM and EPGMM families and the corresponding covariance structure.

Model name	Cov. structure	No. of parameters	$G = 4, d = 3$ $p = 100$
UUUU - UUU	$S_g = \Lambda_g \Lambda_g^T + \Psi_g$	$(G-1) + Gd + Gq[d-(q-1)/2] + Gd$	1991
UUCU -	$S_g = \Lambda_g \Lambda_g^T + \omega_g \Delta_g$	$(G-1) + Gd + Gq[d-(q-1)/2] + [1+G(d-1)]$	1988
UCUU -	$S_g = \Lambda_g \Lambda_g^T + \omega_g \Delta$	$(G-1) + Gd + Gq[d-(q-1)/2] + [G+(d-1)]$	1694
UCCU - UCU	$S_g = \Lambda_g \Lambda_g^T + \Psi$	$(G-1) + Gd + Gq[d-(q-1)/2] + d$	1691
UCUC - UUC	$S_g = \Lambda_g \Lambda_g^T + \psi_g \mathbf{I}_p$	$(G-1) + Gd + Gq[d-(q-1)/2] + G$	1595
UCCC - UCC	$S_g = \Lambda_g \Lambda_g^T + \psi \mathbf{I}_p$	$(G-1) + Gd + Gq[d-(q-1)/2] + 1$	1592
CUUU - CUU	$S_g = \Lambda \Lambda^T + \Psi_g$	$(G-1) + Gd + q[d-(q-1)/2] + Gd$	1100
CUCU -	$S_g = \Lambda \Lambda^T + \omega_g \Delta_g$	$(G-1) + Gd + q[d-(q-1)/2] + [1+G(d-1)]$	1097
CCUU -	$S_g = \Lambda \Lambda^T + \omega_g \Delta$	$(G-1) + Gd + q[d-(q-1)/2] + [G+(d-1)]$	803
CCCU - CCU	$S_g = \Lambda \Lambda^T + \Psi$	$(G-1) + Gd + q[d-(q-1)/2] + d$	800
CCUC - CUC	$S_g = \Lambda \Lambda^T + \psi_g \mathbf{I}_d$	$(G-1) + Gd + q[d-(q-1)/2] + G$	704
CCCC - CCC	$S_g = \Lambda \Lambda^T + \psi \mathbf{I}_d$	$(G-1) + Gd + q[d-(q-1)/2] + 1$	701

where $\omega_g \in \mathbb{R}^+$ and $|\Delta_g| = 1$.

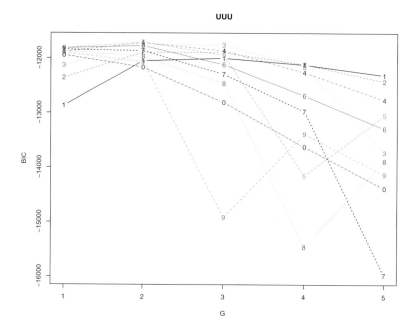

Figure 8.18 BIC curves for tested models ($G = 1, ..., 5$ and $q = 1, ..., 10$) of PGMM.

(..C.) or not (..U.) and finally $\Delta_g = \mathbf{I}_d$ (...C) or not (...U). The table also gives the maximum number of free parameters to be estimated according to G, d and q for the 12 models. Once again, this number of free parameters to be estimated can be decomposed in the number of parameters to estimate for the proportions $(G - 1)$, for the means (Gd) and for the covariance matrices (last terms).

According to this family of 12 models, the previous approaches (Ghahramani and Hinton, 1997; Tipping and Bishop, 1999; McLachlan et al., 2003; McNicholas and Murphy, 2008; Baek et al., 2009) then become submodels of the EPGMM approach. For example, by constraining only the noise variance to be isotropic on each class ($\Psi_g = \sigma_g^2 \mathbf{I}_d$), which corresponds to the UUC and UUUC models, it produces the mixture of probabilistic PCA (Mixt-PPCA) of Tipping and Bishop (1999). Similarly, by considering the covariance structure of the UCU and UCCU models such that $\Psi_g = \Psi$ and Λ_g, then we obtain the mixture of factor analyzers model developed by Ghahramani and Hinton (1997). The UUUU model is also equivalent to the MFA model of McLachlan et al. (2003). Finally, by parameterizing the covariance structure as $\Psi_g = \omega_g \Delta_g$, where Δ_g is a diagonal matrix and $|\Delta_g| = 1$, McNicholas and Murphy (2010b) proposed four additional models to their previous work (McNicholas and Murphy, 2008).

McNicholas and Murphy (2010b) proposed using the alternating expectation-conditional maximization (AECM) algorithm (Meng and Van Dyk, 1997) to speed up convergence for estimating the EPGMM model in the unsupervised

Listing 8.12: Mixture of probabilistic PCA on the `wine27` data

```
# Loading libraries and data
library(pgmm)
library(MBCbook)
data(wine27)
Y = wine27[,1:27]
cls = wine27$Type

# Clustering with PGMM
res.pgmm = pgmmEM(Y,rG = 1:5, rq = 1:10,modelSubset = 'UUC')

Based on k-means starting values, the best model (BIC) for the
    range of factors and components used is a UUC model with q = 1
    and G = 3. The BIC for this model is -12383.08.

# Plot results
plot(res.pgmm)

# Classification table
table(cls,res.pgmm$map)
   cls   1   2   3
     1   2   0  57
     2  65   2   4
     3   1  47   0
```

case. Note that the AECM could be used to infer most of the MFA-based models. The EPGMM models can also be used in the supervised context to perform discriminant analysis with high-dimensional data.

As an illustration of the use of the PGMM family, we applied the **pgmm** library to the `wine27` data with the UUC model (corresponding to the popular MPPCA model).

BIC selects a number G of components equal to 3 and a number q of factors equal to 1. Figure 8.18 shows the BIC curves for the different models that we asked to explore. The selected model is therefore very parsimonious, but, as shown by the above classification table, the clustering has a good overlap with known classes for these data. PGMM misclassifies only nine wines.

8.4.4 Mixture of High-dimensional GMMs (HD-GMM)

Bouveyron et al. (2007a,b) proposed a family of 28 parsimonious and flexible Gaussian models to deal with high-dimensional data. Conversely to the previous approaches, this family of GMMs was proposed in both supervised and unsupervised classification contexts. In order to ease the designation of this family, we propose to refer to these Gaussian models for high-dimensional data by the acronym HD-GMM. Bouveyron et al. (2007a) proposed constraining the GMM model through the eigen-decomposition of the covariance matrix Σ_g of the gth group:

$$\Sigma_g = Q_g \Lambda_g Q_g^T,$$

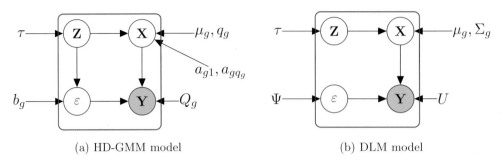

(a) HD-GMM model   (b) DLM model

Figure 8.19 Graphical representation of the HD-GMM and DLM models.

where Q_g is a $d \times d$ orthogonal matrix that contains the eigenvectors of Σ_g and Λ_g is a $d \times d$ diagonal matrix containing the associated eigenvalues (sorted in decreasing order).

The key idea of the work is to reparameterize the matrix Λ_g, such that Σ_g has only $q_g + 1$ different eigenvalues:

$$\Lambda_g = \mathrm{diag}\left(a_{g1}, \ldots, a_{gq_g}, b_g, \ldots, b_g\right),$$

where the q_g first values $a_{g1}, \ldots, a_{gq_g}$ parameterize the variance in the group-specific subspace and the $d - q_g$ last terms, the b_g model the variance of the noise and $q_g < d$. With this parameterization, these parsimonious models assume that, conditionally on the groups, the noise variance matrix of each cluster g is isotropic and is contained in a subspace that is orthogonal to the subspace of the gth group. Following the classical parsimony strategy, the authors proposed a family of parsimonious models from a very general model, the model $[a_{gj}b_gQ_gq_g]$, to very simple models.

Table 8.4 lists the 15 HD-GMM models with closed-form estimators and reports their complexity for comparison purposes. Once again, the first quantity $(G - 1) + Gd$ stands for the number of parameters for the mixture proportions and the means of G clusters. Then, there are $\sum_{g=1}^{G} q_g[d(q_g + 1)/2]$ loadings to estimate for the G orientation matrices Q_g and finally the last terms represent the number of free parameters for the covariance matrices in the latent and in the noise subspaces of G clusters and their intrinsic dimension.

Such an approach can be viewed in two different ways: on the one hand, these models make it possible to regularize the models in high dimensions. In particular, by constraining q_g such that $q_g = d - 1$ for $g = 1, \ldots, G$, the proposed approach can be viewed as a generalization of the works of Celeux and Govaert (1995) and Fraley and Raftery (2002). Indeed, the model $[a_{gj}b_gQ_g (d - 1)]$ is equivalent to the Full-GMM model or the $[\lambda_g D_g A_g D_g]$ model in Celeux and Govaert (1995). In the same manner, the model $[a_{gj}b_gQ (d - 1)]$ is equivalent to the Diag-GMM and the $[a_j bQ (d - 1)]$ is also the Com-Diag-GMM. On the other hand, this approach can also be viewed as an extension of the mixture of principal component analyzer (MPPCA) model Tipping and Bishop (1999) since it relaxes

Table 8.4 *Nomenclature for the members of the HD-GMM family and the number of parameters to estimate. For the numerical example, the intrinsic dimension of the clusters has been fixed to* $q_g = q = 3, \forall g = 1, \ldots, G.$

Model name	No. of parameters	$\begin{array}{l} d = 100 \\ G = 4, q = 3 \end{array}$
$[a_{gj}b_gQ_gq_g]$	$(G-1) + Gd + \sum_{g=1}^{G} q_g[d - (q_g+1)/2] + \sum_{g=1}^{G} q_g + 2G$	1599
$[a_{gj}bQ_gq_g]$	$(G-1) + Gd + \sum_{g=1}^{G} q_g[d - (q_g+1)/2] + \sum_{g=1}^{G} q_g + 1 + G$	1596
$[a_gb_gQ_gq_g]$	$(G-1) + Gd + \sum_{g=1}^{G} q_g[d - (q_g+1)/2] + 3G$	1591
$[ab_gQ_gq_g]$	$(G-1) + Gd + \sum_{g=1}^{G} q_g[d - (q_g+1)/2] + 1 + 2G$	1588
$[abQ_gq_g]$	$(G-1) + Gd + \sum_{g=1}^{G} q_g[d - (q_g+1)/2] + 2 + G$	1585
$[a_{gj}b_gQ_gq]$	$(G-1) + Gd + Gq[d - (q+1)/2] + Gq + G + 1$	1596
$[a_jb_gQ_gq]$	$(G-1) + Gd + Gq[d - (q+1)/2] + q + G + 1$	1587
$[a_{gj}bQ_gq]$	$(G-1) + Gd + Gq[d - (q+1)/2] + Gq + 2$	1593
$[a_jbQ_gq]$	$(G-1) + Gd + Gq[d - (q+1)/2] + q + 2$	1584
$[a_gb_gQ_gq]$	$(G-1) + Gd + Gq[d - (q+1)/2] + 2G + 1$	1588
$[ab_gQ_gq]$	$(G-1) + Gd + Gq[d - (q+1)/2] + G + 2$	1585
$[a_gbQ_gq]$	$(G-1) + Gd + Gq[d - (q+1)/2] + G + 2$	1585
$[abQ_gq]$	$(G-1) + Gd + Gq[d - (q+1)/2] + 3$	1582
$[a_jbQq]$	$(G-1) + Gd + q[d - (q+1)/2] + q + 2$	702
$[abQq]$	$(G-1) + Gd + q[d - (q+1)/2] + 3$	700

the equality assumption on q_g made in Tipping and Bishop (1999) and the model $[a_{gj}b_gQ_gq]$ is therefore equivalent to the MPPCA model.

In the case of the 15 HD-GMM models listed in Table 8.4, inference can be done easily using the EM algorithm since update formulae for mixture parameters are in closed form. We refer to Bouveyron et al. (2007a) regarding inference for the other 13 models which require an iterative M-step. Estimation of the intrinsic dimensions q_g, $g = 1, \ldots, G$, relies on the scree test of Cattell (1966), which looks for a break in the eigenvalue scree of the empirical covariance matrix of each group. Finally, Bouveyron et al. (2011) have demonstrated the surprising result that the maximum likelihood estimator of the intrinsic dimensions q_g is asymptotically consistent in the case of the model $[a_gb_gQ_gq_g]$.

As illustrations of the HD-GMM family, we focus now on the clustering of the usps358 data using the HDclassif package. The package provides the HDDC and HDDA methods which classify the data using the HD-GMM models, respectively, in the unsupervised and supervised cases. Listing 8.13 gives the R code to apply the hddc function to the data.

Conversely to the previous approaches, HDDC is parameterized by a threshold which varies the number of intrinsic dimensions per group. The choice of the dimensions q_g is indeed made using the Cattell scree-test (Cattell, 1966), which looks for a break in the eigenvalue scree of the group covariance matrices. The threshold is here fixed to 0.1, which results in selecting 14, 11 and 9 intrinsic dimensions for the three respective groups. It is worth noticing that a unique

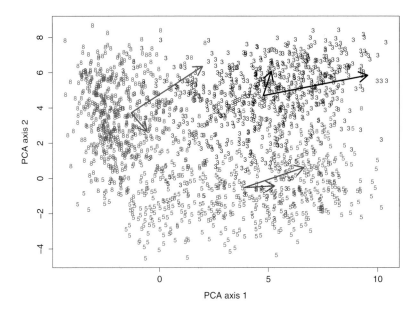

PCA axis 1

Figure 8.20 Projection into the two first principal components of the estimated group-specific subspaces with HDDC on the USPS358 data.

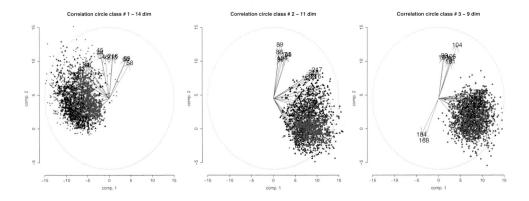

Figure 8.21 Projection of the USPS358 data into the estimated group-specific subspaces with HDDC and associated correlation circles.

choice of threshold leads to a selection of different number of dimensions for the three groups.

Figures 8.20 and 8.21 aim at illustrating the modeling of the groups in different subspaces. Figure 8.20 shows the projection into the two first principal components of the estimated group-specific subspaces with HDDC (only the two first axes of each subspace are shown). As expected, the subspaces are oriented according to the variance projection. Figure 8.21 presents group-specific biplots which allow

Listing 8.13: Clustering with HDDC of the usps358 data

```
# Loading libraries and data
library(MBCbook)
data(usps358)
Y = usps358[,-1]
cls = usps358$cls

# Clustering with HDDC (notice that HDclassif uses the index 'k'
    for the groups in model names)
library(HDclassif)
res = hddc(Y,3,model='AkjBkQkDk',threshold=0.1)

# Number of selected dimensions per group
res$d

# Clustering results
table(cls,res$class)
   cls    1    2    3
     1   43  605   10
     2  553    2    1
     3   10   44  488
```

to visualize both the projected data into the subspaces of the groups and the correlations between original variables and the axes carrying the group subspaces.

8.4.5 The Discriminative Latent Mixture (DLM) Models

Bouveyron and Brunet (2012b) proposed a family of mixture models which fit the data into a common discriminative subspace. This mixture model, called the discriminative latent mixture (DLM) model, differs from the FA-based models in the fact that the latent subspace is common to all groups and is assumed to be the most discriminative subspace of dimension q. This latter feature of the DLM model makes it significantly different from the other FA-based models. Roughly speaking, the FA-based models choose the latent subspace(s) maximizing the projected variance whereas the DLM model chooses the latent subspace which maximizes the separation between the groups.

Let us nevertheless start with an FA-like modeling. Let $Y \in \mathbb{R}^d$ be the observed random vector and let $Z \in \{1, \ldots, G\}$ be once again the unobserved random variable to predict. The DLM model then assumes that Y is linked to a latent random vector $X \in \mathbb{E}$ through a linear relationship of the form

$$Y = UX + \varepsilon,$$

where $\mathbb{E} \subset \mathbb{R}^q$ is assumed to be the most discriminative subspace of dimension $q \leq G-1$ such that $\mathbf{0} \in \mathbb{E}$, $G < d$, U is a $d \times d$ orthonormal matrix common to the G groups and satisfying $U^T U = \mathbf{I}_q$, and $\varepsilon \sim \mathcal{N}(0, \Psi)$ models the non-discriminative information.

Besides, within the latent space and conditionally to $Z = g$, X is assumed to be distributed as

$$X|Z = g \sim \mathcal{N}(\mu_g, \Sigma_g),$$

where $\mu_g \in \mathbb{R}^q$ and $\Sigma_g \in \mathbb{R}^{q \times q}$ are respectively the mean vector and the covariance matrix of the gth group. Given these distribution assumptions, the marginal distribution of Y is once again a mixture of Gaussians, i.e. $f(y) = \sum_{g=1}^{G} \tau_g \phi(y; m_g, S_g)$, where $m_g = U\mu_g$ and $S_g = U\Sigma_g U^T + \Psi$. Let $W = [U, V]$ be the $d \times d$ matrix such that $W^T W = W W^T = \mathbf{I}_d$ and V is an orthogonal complement of U. Finally, the noise covariance matrix Ψ is assumed to satisfy the conditions $V\Psi V^T = \beta \mathbf{I}_{d-q}$ and $U\Psi U^T = \mathbf{0}_q$, such that $\Delta_g = W^T S_g W$ is block-diagonal:

$$\Delta_g = \text{diag}\left(\Sigma_g, \beta I_{d-q}\right).$$

These last conditions imply that the discriminative and the non-discriminative subspaces are orthogonal, which suggests in practice that all the relevant clustering information remains in the latent subspace.

This model is referred to as $\text{DLM}_{[\Sigma_g \beta]}$ by Bouveyron and Brunet (2012b). Following the classical strategy, several other models can be obtained from the $\text{DLM}_{[\Sigma_g \beta]}$ model by relaxing or adding constraints on model parameters. Firstly, it is possible to consider a more general case than the $\text{DLM}_{[\Sigma_g \beta]}$ by relaxing the constraint on the variance term of the non-discriminative information. Assuming that $\varepsilon|Z = g \sim \mathcal{N}(0, \Psi_g)$ yields the $\text{DLM}_{[\Sigma_g \beta_g]}$ model, which can be useful in some practical cases.

From this extended model, ten parsimonious models can be obtained by constraining the parameters Σ_g and β_g to be common between and within the groups. The list of the 12 different DLM models is given by Table 8.5, which also provides the number of free parameters to estimate. As we can see, the DLM family yields very parsimonious models while being able to fit a wide range of situations. The complexity of the $\text{DLM}_{[\Sigma_g \beta_g]}$ model mainly depends on the number of clusters G since the dimension of the discriminative subspace is such that $q \leq G - 1$. A model similar to the $\text{DLM}_{[\alpha \beta]}$ model was considered by Sanguinetti (2008).

Unlike most of the MFA-based models, inference for the DLM models cannot be directly done using the EM algorithm because of the specific features of its latent subspace. To overcome this problem, Bouveyron and Brunet (2012b) proposed a Fisher-EM algorithm, for estimating both the discriminative subspace and the parameters of the mixture model. This algorithm is based on the EM algorithm to which an additional step is added between the E- and the M-step. This additional step, named the F-step, aims to compute the projection matrix U whose columns span the discriminative latent space. At iteration s, this step estimates the orientation matrix $U^{(s)}$ of the discriminative latent space given the posterior probabilities by maximizing Fisher's criterion (Fisher, 1936; Fukunaga, 1990) under orthonormality constraints:

$$\hat{U}^{(s)} = \max_U \quad \text{trace}\left((U^T S U)^{-1} U^T S_B^{(q)} U\right),$$

$$\text{w.r.t.} \quad U^T U = \mathbf{I}_d, \tag{8.6}$$

Table 8.5 *Number of free parameters to estimate when $q = G - 1$ for the DLM models and some classical models (see text for details).*

Model	No. of parameters	$d = 100$ and $G = 4$
$\text{DLM}_{[\Sigma_g \beta_g]}$	$(G-1) + G(G-1) + (G-1)(d-G/2) + G^2(G-1)/2 + G$	337
$\text{DLM}_{[\Sigma_g \beta]}$	$(G-1) + G(G-1) + (G-1)(d-G/2) + G^2(G-1)/2 + 1$	334
$\text{DLM}_{[\Sigma \beta_g]}$	$(G-1) + G(G-1) + (G-1)(d-G/2) + G(G-1)/2 + G$	319
$\text{DLM}_{[\Sigma \beta]}$	$(G-1) + G(G-1) + (G-1)(d-G/2) + G(G-1)/2 + 1$	316
$\text{DLM}_{[\alpha_{gj} \beta_g]}$	$(G-1) + G(G-1) + (G-1)(d-G/2) + G^2$	325
$\text{DLM}_{[\alpha_{gj} \beta]}$	$(G-1) + G(G-1) + (G-1)(d-G/2) + G(G-1) + 1$	322
$\text{DLM}_{[\alpha_g \beta_g]}$	$(G-1) + G(G-1) + (G-1)(d-G/2) + 2G$	317
$\text{DLM}_{[\alpha_g \beta]}$	$(G-1) + G(G-1) + (G-1)(d-G/2) + G + 1$	314
$\text{DLM}_{[\alpha_j \beta_g]}$	$(G-1) + G(G-1) + (G-1)(d-G/2) + (G-1) + G$	316
$\text{DLM}_{[\alpha_j \beta]}$	$(G-1) + G(G-1) + (G-1)(d-G/2) + (G-1) + 1$	313
$\text{DLM}_{[\alpha \beta_g]}$	$(G-1) + G(G-1) + (G-1)(d-G/2) + G + 1$	314
$\text{DLM}_{[\alpha \beta]}$	$(G-1) + G(G-1) + (G-1)(d-G/2) + 2$	311

where S stands for the empirical covariance matrix and $S_B^{(s)}$, defined as follows:

$$S_B^{(s)} = \frac{1}{n} \sum_{g=1}^{G} n_g^{(s)} (m_g^{(s)} - \bar{y})(m_g^{(s)} - \bar{y})^T, \qquad (8.7)$$

denotes the soft-between-covariance matrix with

$$n_g^{(s)} = \sum_{i=1}^{n} t_{ig}^{(s)}, m_g^{(s)} = 1/n_g^{(s)} \sum_{i=1}^{n} t_{ig}^{(s)} y_i \text{ and } \bar{y} = 1/n \sum_{i=1}^{n} y_i.$$

This optimization problem was solved by Bouveyron and Brunet (2012b) using the concept of the orthonormal discriminant vector developed by Foley and Sammon (1975) through a Gram–Schmidt procedure.

Such a process enables the fitting of a discriminative and low-dimensional subspace given the current soft partition of the data, while providing orthonormal discriminative axes. In addition, according to the rank of the matrix $S_B^{(s)}$, the dimension of the discriminative space q is strictly bounded by the number of clusters G and can be set to $G - 1$ in practice.

Two additional procedures were proposed by Bouveyron and Brunet (2011) for estimating the latent subspace orientation. We shall note that the Fisher-EM algorithm alternatively optimizes two objective functions. The convergence properties of the Fisher-EM algorithm were studied by Bouveyron and Brunet (2012c) from both a theoretical and a practical point of view.

We now use the Fisher-EM algorithm to cluster the wine27 data, using the FisherEM package for R. Listing 8.14 shows how to apply the package to these data. The fem function is called here with the model $\text{DLM}_{[\alpha_{gj} \beta_g]}$ and $G = 3$. The function can alternatively choose the most appropriate model and number of groups through model selection. Listing 8.14 also shows the estimated loading matrix U, which relates the observation space to the discriminative subspace. One

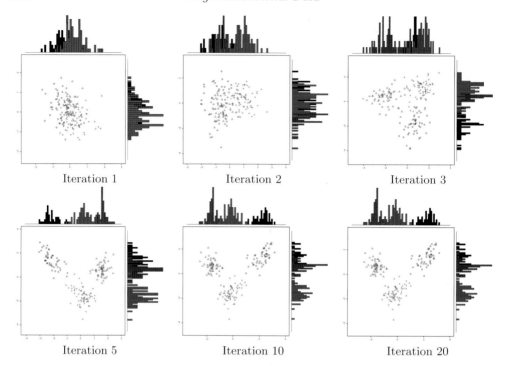

Figure 8.22 Estimated discriminative subspace at some iterations of the Fisher-EM algorithm on the `wine27` data set.

can already notice that some of the original variables are useful for discriminating the three groups since their loading coefficients are close to zero (and not displayed by the `print` function).

In order to better understand the building of the discriminative subspace, Figure 8.22 shows the evolution of the estimated subspace along the iterations of the algorithm. In this case, the algorithm was initialized by a random partition, which explains the messy first plot. The algorithm finds the discriminative subspace in a few iterations and converges within 20 iterations. The clustering result is very close to the known partition for these data. Only 2 of the 178 observations are misclassified. As one can see, Fisher-EM provides a much more informative subspace than other methods while producing relevant clustering results.

8.4.6 Variable Selection by Penalization of the Loadings

In the context of subspace clustering, it may also be of interest to perform variable selection while projecting data into a low-dimensional subspace. This can be done through penalization by directly applying the lasso penalty to the loading matrix of an MFA-based model. This was done by Bouveyron and Brunet (2012a) for the Fisher-EM algorithm, but it could also be applied to other MFA-based models. In the context of Fisher-EM, direct penalization of the loading matrix U makes sense

Listing 8.14: Clustering of the `wine27` data with Fisher-EM

```
# Loading libraries and data
library(FisherEM)
library(MBCbook)
data(wine27)
Y = scale(wine27[,1:27])
cls = wine27$Type

# Clustering with Fisher-EM
res = fem(Y,K = 3,model='AkjBk')

# Loading matrix
print(res$U)
Loadings:
                                 U1       U2
Alcohol                        -0.164   0.286
Sugar-free_extract
Fixed_acidity                           0.114
Tartaric_acid
Malic_acid                      0.101   0.193
Uronic_acids                    0.135
pH
Ash
Alcalinity_of_ash               0.161
Potassium                               0.146
Calcium                                -0.175
Magnesium
Phosphate
Chloride
Total_phenols
Flavanoids                     -0.660  -0.295
Nonflavanoid_phenols
Proanthocyanins                        -0.109
Color_Intensity                 0.389   0.234
Hue                            -0.142  -0.272
OD280/OD315_of_diluted_wines   -0.287   0.254
OD280/OD315_of_flavanoids              -0.394
Glycerol                       -0.113   0.130
2-3-butanediol                  0.143
Total_nitrogen                         -0.135
Proline                        -0.381   0.540
Methanol

# Clustering results
table(cls,res$cls)

cls   1   2   3
  1  59   0   0
  2   1  69   1
  3   0   0  48
```

since it is not estimated by likelihood maximization. The matrix U is estimated in the F-step of the Fisher-EM algorithm by maximizing the Fisher criterion given the current partition of the data.

Bouveyron and Brunet (2012a) proposed two ways of penalizing U. The first solution is a two-stage approach which first estimates U, at each iteration with the F-step, and then looks for its best sparse approximation as follows:

$$\min_{U} \left\| X^{(s)T} - Y^T U \right\|_F^2 + \lambda \sum_{j=1}^{q} \|u_j\|_1 ,$$

where u_j is the jth column vector of U, $X^{(s)} = \hat{U}^{(s)T} Y$ and $\|\cdot\|_F$ refers to the Frobenius norm. The solution of this penalized regression problem can be computed through the LARS algorithm.

The second solution directly recasts the maximization of the Fisher criterion as a regression problem and provides a sparse loading matrix by solving the lasso problem associated to this regression problem. Let us define, conditionally to the posterior probabilities $t_{ig}^{(s)}$, the matrices

$$H_W^{(s)} = \frac{1}{\sqrt{n}} \left[Y - \sum_{g=1}^{G} t_{1g}^{(s)} m_g^{(s)}, \ldots, Y - \sum_{g=1}^{G} t_{ng}^{(s)} m_g^{(s)} \right] \in \mathbb{R}^{d \times n}$$

and

$$H_B^{(s)} = \frac{1}{\sqrt{n}} \left[\sqrt{n_1^{(s)}} (m_1^{(s)} - \bar{y}), \ldots, \sqrt{n_G^{(s)}} (m_G^{(s)} - \bar{y}) \right] \in \mathbb{R}^{d \times G},$$

where $n_g^{(s)} = \sum_{i=1}^{n} t_{ig}^{(s)}$ and $m_g^{(s)} = \frac{1}{n} \sum_{i=1}^{n} t_{ig}^{(s)} y_i$. According to these definitions, the matrices $H_W^{(s)}$ and $H_B^{(s)}$ satisfy $H_W^{(s)} H_W^{(s)T} = S_W^{(s)}$ and $H_B^{(s)} H_B^{(s)T} = S_B^{(s)}$ where $S_W^{(s)}$ and $S_B^{(s)}$ are respectively the soft within and between covariance matrices computed at iteration s. Then, the best sparse approximation of the solution of the Fisher criterion is the solution $\hat{B}^{(s)}$ of the following penalized regression problem:

$$\min_{A,B} \sum_{g=1}^{G} \left\| R_W^{(s)T} H_{B,g}^{(s)} - AB^T H_{B,g}^{(s)} \right\|_F^2 + \rho \sum_{j=1}^{q} \beta_j^T S_W^{(s)} \beta_j + \lambda \sum_{j=1}^{q} |\beta_j|_1 ,$$

w.r.t. $A^T A = \mathbf{I}_q$,

where $A = [\alpha_1, \ldots, \alpha_q] \in \mathbb{R}^{d \times q}$, $B = [\beta_1, \ldots, \beta_q] \in \mathbb{R}^{d \times q}$, $R_W^{(s)} \in \mathbb{R}^{d \times d}$ is an upper triangular matrix resulting from the Cholesky decomposition of $S_W^{(s)}$, i.e. $S_W^{(s)} = R_W^{(s)T} R_W^{(s)}$, $H_{B,g}^{(s)}$ is the gth column of $H_B^{(s)}$ and $\rho > 0$. However, solving this lasso problem is not direct in this case and requires the use of an iterative algorithm.

Regarding the implementation details, Bouveyron and Brunet (2012a) proposed initializing the sparseFEM algorithm with the result of the Fisher-EM algorithm and determining the value of λ by model selection through the BIC criterion, for which the model complexity depends on the number of non-zero values.

Listing 8.15 shows how to run the sparse Fisher-EM algorithm on the `wine27` data set, with a sparsity of 90% (parameter `ll` set to 0.1). When comparing with the result of Listing 8.14, it is worth noticing that sparseFEM provides a clustering extremely close to the one of Fisher-EM (only one observation classified in a different group) while using only 3 variables among the 27 original ones. In addition, sparseFEM identified the variables `Alcohol`, `Flavanoids` and `OD280/OD315 of diluted wines` as the most relevant ones to discriminate the three wine types, which tells us about the differences between the three types.

8.5 Bibliographic Notes

The subspace clustering methods which have been presented in this section belong to a large family of Gaussian mixture models and several links exist between these approaches. Indeed, in the last paragraph, we saw that some constrained HD-GMM models are equivalent to traditional parsimonious models such as the Com-Diag-GMM or Diag-GMM, and consequently to the models proposed by Raftery and Fraley or by Celeux and Govaert. Besides, it also appears that the model $[a_{gj}b_gQ_gq]$ is equivalent to the Mixt-PPCA model.

In the same manner, several models which belong to the EPGMM family of McNicholas and Murphy (2010b) are also included in the HD-GMM family (and vice versa). In particular, the UCUC model (McNicholas and Murphy, 2010b) corresponds to the HD-GMM model $[a_{gj}b_gQ_gq]$. Moreover, the EPGMM family proposed by McNicholas and Murphy includes individual works on MFA such as the works of Ghahramani and Hinton (1997), Tipping and Bishop (1999), McLachlan et al. (2003), McNicholas and Murphy (2010b) and Baek et al. (2009). Let us recall that the hypotheses underlying HD-GMM models are stronger than those for the MFA models. Indeed, the subspace of each class is spanned by orthogonal vectors, whereas it is not a necessary condition in MFA, even if such a situation can occur sometimes as in the case of the model UCUC (CUC).

However, it is worth noting that, despite their effectiveness for clustering high-dimensional data, these probabilistic methods based on subspace clustering have several limitations. Since most models assume that each cluster lives in a specific subspace, the visualization of the clustering results is not easy. Only the MCFA, MCUFSA, HMFA and DLM, for which the latent subspace is assumed to be common, can provide low-dimensional plots of the clustered data. Among the related works, we may cite the robust versions (Andrews and McNicholas, 2011a; McLachlan et al., 2007) of the MFA models, which rely on t-distributions, and the GMMDR technique (Scrucca, 2010), which looks for the subspace containing the most relevant information for clustering.

Finally, Bouveyron and Brunet-Saumard (2014) and McParland and Murphy (2018) provide reviews of various approaches for model-based clustering and mixture modeling of high-dimensional data.

Listing 8.15: Clustering of the `wine27` data with the sparse Fisher-EM algorithm

```
# Loading libraries and data
library(FisherEM)
library(MBCbook)
data(wine27)
Y = scale(wine27[,1:27])
cls = wine27$Type

# Clustering with Fisher-EM
res = sfem(Y,K = 3,model='AkjBk',ll = 0.1)

# Loading matrix
class(res$U) = 'loadings'
print(res$U)

Loadings:
                                 U1      U2
Alcohol                                  -1.000
Sugar.free_extract
Fixed_acidity
Tartaric_acid
Malic_acid
Uronic_acids
pH
Ash
Alcalinity_of_ash
Potassium
Calcium
Magnesium
Phosphate
Chloride
Total_phenols
Flavanoids                       -0.487
Nonflavanoid_phenols
Proanthocyanins
Color_Intensity
Hue
OD280.OD315_of_diluted_wines -0.874
OD280.OD315_of_flavanoids
Glycerol
X2.3.butanediol
Total_nitrogen
Proline
Methanol


# Clustering results
table(cls,res$cls)
    cls              1  2  3
      Barbera        0  0 48
      Barolo        58  1  0
      Grignolino     1 69  1
```

9

Non-Gaussian Model-based Clustering

In this chapter, we review a number of model-based clustering approaches for multivariate continuous data that are based on finite mixture models other than the Gaussian mixture model. The models described are used in situations where the clusters exhibit heavy tails and/or non-elliptical cluster shapes which cannot be accounted for directly using a Gaussian mixture model. Thus, the use of non-Gaussian component distributions can be viewed as an alternative approach to the cluster merging procedures described in Section 3.3.

The chapter commences by introducing a number of models that have been proposed for non-Gaussian model-based clustering and these are illustrated in the eruption time/duration data for the Old Faithful geyser.

Subsequently, the models are compared on an application from flow cytometry, where clustering algorithms can be used to assist with the task of *gating*. In particular, we use the methods of this chapter to cluster the control patient data of the GvHD (Graft-versus-Host Disease) flow cytometry data (Brinkman et al., 2007); these data are available in the `mclust` package. The GvHD control sample consists of 6,809 observations. Each observation consists of four biomarker variables, namely, CD4, CD8b, CD3 and CD8. The objective of the analysis is to identify CD3+, CD4+ and CD8b+ cell sub-populations present in the sample; the data are shown in Figure 9.1 where clear clusters are visible, but the clusters appear to be non-elliptical and there are many outlying values.

9.1 Multivariate t-Distribution

The multivariate t-distribution can be formulated as a scale mixture of normal distributions. Suppose that

$$Y_i|\mu, \Sigma, V_i \sim N\left(\mu, \frac{\Sigma}{V_i}\right) \text{ and } V_i|\nu \sim \text{gamma}\left(\frac{\nu}{2}, \frac{\nu}{2}\right),$$

then the marginal distribution of Y_i is a multivariate t-distribution which has density function

$$f(y_i|\mu, \Sigma, \nu) = \frac{\Gamma\left(\frac{\nu+d}{2}\right)|\Sigma|^{-\frac{1}{2}}}{(\pi\nu)^{\frac{d}{2}}\Gamma\left(\frac{\nu}{2}\right)\left\{1 + (y_i - \mu)^T\Sigma^{-1}(y_i - \mu)\right\}^{\frac{\nu+d}{2}}},$$

where the parameter μ is the mean, Σ is the scale matrix and ν is the number of degrees of freedom. The multivariate t-distribution has the following mean and

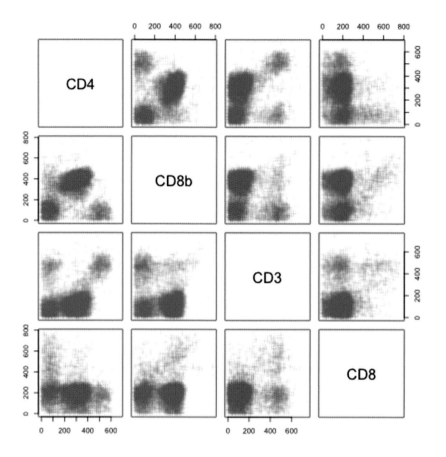

Figure 9.1 A scatter plot of matrix of the GvHD control data.

variance

$$\mathbb{E}(Y) = \mu \text{ and } \mathbb{V}\mathrm{ar}(Y) = \left(\frac{\nu}{\nu - 2}\right)\Sigma, \qquad (9.1)$$

provided that $\nu > 2$. Further, the multivariate t-distribution converges to an $N(\mu, \Sigma)$ as $\nu \to \infty$. Thus, the multivariate t-distribution provides an extension of the normal distribution where the distribution has heavier tails but where the contours of the density remain elliptical.

The mixture of multivariate t-distributions was proposed for model-based clustering in McLachlan and Peel (1998) and Peel and McLachlan (2000) to yield a model-based clustering method that can accommodate heavy-tailed clusters and which is robust to outliers. Peel and McLachlan (2000) develop a method for fitting the mixture of multivariate t-distributions using an ECM algorithm (Liu, 1997).

The mixture of multivariate t-distributions is of the form

$$f(y_i) = \sum_{g=1}^{G} \tau_g \frac{\Gamma\left(\frac{\nu_g+d}{2}\right)|\Sigma_g|^{-\frac{1}{2}}}{(\pi\nu_g)^{\frac{d}{2}}\Gamma\left(\frac{\nu_g}{2}\right)\left\{1+(y_i-\mu_g)^T\Sigma_g^{-1}(y_i-\mu_g)\right\}^{\frac{\nu_g+d}{2}}},$$

where $(\tau_g, \mu_g, \Sigma_g, \nu_g)$ are the component specific parameters; the model can be represented by a graphical model (Figure 9.2).

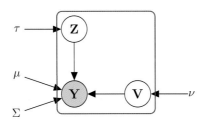

Figure 9.2 Graphical model of the mixture of multivariate t-distributions model.

Many parsimonious model families based on the mixture of multivariate t-distributions have been proposed. Peel and McLachlan (2000) consider constraining the degrees of freedom (ν_g) of the component densities to be equal and constraining the scale matrices (Σ_g) to be equal; such constraints were further explored in Andrews et al. (2011).

McLachlan et al. (2007) develop a version of the model with a factor analytic structure for the component scale matrix to facilitate using these models in high-dimensional settings; given that the component covariance is proportional to the scale matrix, as shown in Equation (9.1), this means that the factor analytic structure also holds for the component covariance. Andrews and McNicholas (2011a,b), Steane et al. (2012) and Lin et al. (2014) extend the mixtures of multivariate t-distributions with factor analytic scale matrix by constraining model parameters to be equal or unequal across clusters to yield the PtMM family of parsimonious models, analogous to the PGMM family of models (Section 8.4.3).

Andrews and McNicholas (2012) applied the eigen-decomposition of the component scale matrix $(\Sigma_g = \lambda_g D_g A_g D_g^T)$ as used in `mclust` and `Rmixmod` to yield a heavy-tailed alternative that is robust to outliers. However, the interpretation of the λ_g as volume is no longer valid because the cluster volume is determined by a combination of the component degrees of freedom ν_g and the λ_g value, as seen in the component variance (9.1).

Model-based clustering with multivariate t-distributions has been incorporated in the `EMMIX` software (McLachlan et al., 1999), which fits mixtures of multivariate t-distributions, `EMMIXskew` package (Wang et al., 2013), which fits mixtures of multivariate t-distributions with constraints on the covariance, and the `teigen` package (Andrews et al., 2015, 2018), which is an `R` package that fits mixtures of multivariate t-distributions with eigen-decomposed scale matrices.

In an analogous way to `mclust` (Fraley and Raftery, 2002), the `teigen` package has a nomenclature for reporting model constraints. The nomenclature reports whether the model has constrained λ_g, D_g, A_g or ν_g and uses the letter C for constrained, U for unconstrained and I for identity. Thus, the CCCC model is a mixture of multivariate t-distributions with λ_g, D_g, A_g and ν_g to be constrained to be equal across groups and the CIIU model is a mixture of multivariate t-distributions with λ_g constrained across groups, identity D_g matrix, identity A_g matrix and the degrees of freedom ν_g being unconstrained across groups.

The mixture of multivariate t-distributions was fitted to the Old Faithful data using the `teigen` package (Listing 9.1) and the model fit is shown in Figure 9.4; the optimal model was selected using BIC. A plot of the BIC values for the 28 possible models for $G = 1, 2, \ldots, 10$ is shown in Figure 9.3. The model with the highest BIC has $G = 3$ and is the CCCC model in the `teigen` nomenclature which has constrained scale matrix volume, constrained scale matrix orientation, constrained scale matrix shape and constrained degrees of freedom; the parameters of the fitted model are given in (9.2).

Listing 9.1: Clustering the Old Faithful data using the `teigen` package

```
# Load the Old Faithful data
data(faithful)

# Load the teigen package
library(teigen)

# Fit all of the models in the teigen family for G=1 to 10
# Use the default starting value strategy (k-means)
# The data are not scaled (scaling is the default in teigen)
tfaith <- teigen(faithful, models="all", Gs=1:10, scale=FALSE,
    parallel.cores=TRUE)

# Plot the BIC and ICL values for each model
matplot(t(tfaith$allbic), ylab="BIC", type="p")
matplot(t(tfaith$allbic), type="l", add=TRUE)

# View the output
tfaith

# View the model parameters
tfaith$parameters

# Plot the resulting density estimate and uncertainty plot (
    dimensions 1 and 2)
plot(tfaith, what="contour")
plot(tfaith, what="uncertainty")
```

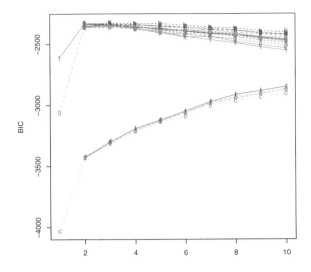

Figure 9.3 A plot of the BIC values for the 28 models fitted by `teigen` for $G = 1, 2, \ldots, 10$.

$$\tau = (0.45, 0.19, 0.36),$$

$$\mu_1 = \begin{pmatrix} 4.48 \\ 80.87 \end{pmatrix},$$

$$\mu_2 = \begin{pmatrix} 3.84 \\ 77.87 \end{pmatrix},$$

$$\mu_3 = \begin{pmatrix} 2.03 \\ 54.44 \end{pmatrix},$$

$$\Sigma_g = \begin{pmatrix} 0.074 & 0.462 \\ 0.462 & 32.764 \end{pmatrix}, \text{ for } g = 1, 2, 3,$$

$$\nu = (51.7, 51.7, 51.7).$$

(9.2)

Clearly, the model with the highest BIC does very well at finding the clusters of observations with the clusters consisting of short waiting/eruption time, medium waiting/eruption time and long waiting/eruption time. Looking closely at the density plot in Figure 9.4, the model is using one extra component to cluster the longer eruption times; thus dividing these into medium and long times. The points at the boundary of these two components have the highest clustering uncertainty, as evidenced in Figure 9.4. Although the clustering is good, the model uses two components to model the non-elliptical structure of the long waiting/eruption times and the fitted density for the short waiting time combined with short

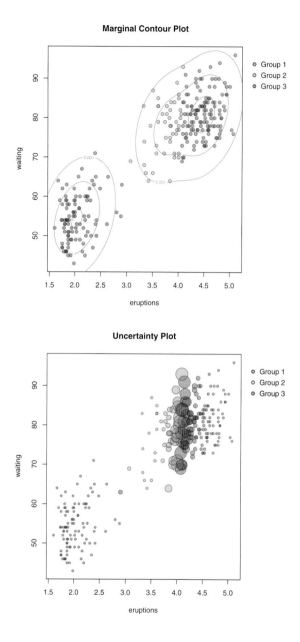

Figure 9.4 Plot of the `teigen` fitted density for the Old Faithful data, with the points colored by cluster membership. The uncertainty values are shown in the second panel (the larger the dot, the higher the uncertainty).

eruption time cluster does not appear to be a good fit because there are regions with high probability density that have no observations. It is also worth noting that the estimated degrees of freedom for each component is 51.7, which is quite large and thus the selected model is very similar to a normal mixture model with $G = 3$ components and the `mclust` EEE covariance structure.

If ICL is used for model selection, then the model with the highest ICL value is the UUUC model with $G = 2$; the UUUC model in the `teigen` nomenclature means that the model has unconstrained scale matrix volume, unconstrained scale matrix orientation, unconstrained matrix shape and constrained degrees of freedom.

A plot of the fitted density and clustering uncertainty is shown in Figure 9.5 and the parameters for this model are given in Equation (9.3); the code for extracting these model parameters and plotting the model fit is given in Listing 9.2.

Listing 9.2: Extracting the output for the highest ICL `teigen` model for the Old Faithful data

```
# View the highest ICL model parameters
tfaith$iclresults$parameters

# Plot the model with the highest ICL value
# We create a new object called tfaithICL and overwrite
# the highest BIC model with the corresponding highest
# ICL model.
# This is then used for plotting
tfaithICL<-tfaith
tfaithICL$parameters<-tfaithICL$iclresults$parameters
tfaithICL$fuzzy<-tfaithICL$iclresults$fuzzy
tfaithICL$G<-tfaithICL$iclresults$G
tfaithICL$bestmodel<-tfaithICL$iclresults$bestmodel
tfaithICL$classification<-tfaithICL$iclresults$classification

# Plot the highest ICL model
plot(tfaithICL,what="contour")
plot(tfaithICL,what="uncertainty")
```

$$
\begin{aligned}
\tau_1 &= 0.36, & \tau_2 &= 0.64, \\
\mu_1 &= \begin{pmatrix} 2.03 \\ 54.42 \end{pmatrix}, & \mu_2 &= \begin{pmatrix} 4.29 \\ 79.99 \end{pmatrix}, \\
\Sigma_1 &= \begin{pmatrix} 0.066 & 0.418 \\ 0.418 & 32.812 \end{pmatrix}, & \Sigma_2 &= \begin{pmatrix} 0.163 & 0.884 \\ 0.884 & 34.584 \end{pmatrix}, \\
\nu_1 &= 50.0 & \nu_2 &= 50.0.
\end{aligned}
\tag{9.3}
$$

The highest ICL model clearly separates the eruptions into a cluster of short waiting time and short eruption time observations and a cluster of long waiting time and long eruption time observations; this model has essentially merged two of the clusters from the highest BIC model fit. Further, it is worth noticing that the cluster parameters for the short waiting time and eruption time cluster are very similar to the equivalent cluster in the highest BIC model fit. This demonstrates the robustness of the cluster parameters for this cluster. Further, as with the highest BIC model, the fitted model has high degrees of freedom for

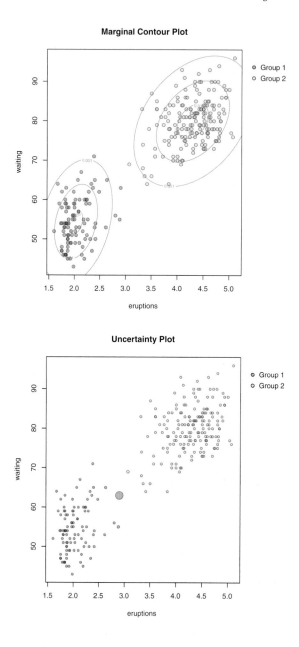

Figure 9.5 Plot of the `teigen` fitted density for the highest ICL `teigen` model for the Old Faithful data, with the points colored by cluster membership. The uncertainty values are shown in the second panel.

each component density ($\nu_g = 50.0$), so the fit is very similar to the $G = 2$ normal mixture model with a VVV covariance structure.

However, a close inspection of the fitted density for short waiting and eruption times shows that the fitted density is not capturing the skew in the data in this region; the fitted density is large in a region with very few data points. Thus, although the model is clustering the data in a robust manner, the fitted density estimate is not accurate in either the highest BIC or highest ICL models.

9.2 Skew-normal Distribution

A multivariate skew-normal distribution can accommodate skewed data clusters, but more than one type of multivariate skew-normal distribution has been developed. Azzalini and Dalla Valle (1996), Azzalini and Capitanio (1999) and Azzalini (2014) developed an extensive body of work for one version of the skew normal distribution but Sahu et al. (2003) developed an alternative version of multivariate skew-normal distribution. The advantages and disadvantages of the different versions of the skew normal have been debated in Azzalini et al. (2016) and McLachlan and Lee (2016).

The original multivariate skew-normal distribution (Azzalini and Dalla Valle, 1996) has density function of the form

$$f(y_i) = 2\phi_d(y_i; \mu, \Sigma)\Phi_1\left(\alpha^T \omega^{-1}(y_i - \mu)\right),$$

where ϕ_d is the d-dimensional normal pdf, Φ_1 is the 1-dimensional normal cdf, α is the skewing parameter, μ is the location parameter, Σ is the scale matrix, ω is a diagonal matrix formed by the square roots of the diagonal elements of Σ. This formulation of the skew-normal distribution is often termed the *restricted* skew-normal distribution.

The mean and covariance of the (restricted) skew-normal are

$$\mathbb{E}(Y) = \mu + \sqrt{\frac{2}{\pi}}\omega\delta \text{ and } \mathbb{Var}(Y) = \Sigma - \frac{2}{\pi}\omega\delta\delta^T\omega,$$

where

$$\delta = \frac{1}{\{1 + \alpha^T(\omega^{-1}\Sigma\omega^{-1})\alpha\}^{1/2}}(\omega^{-1}\Sigma\omega^{-1})\alpha.$$

Thus, the mean and covariance of the distribution are controlled by a combination of the parameters. The mass of the density function is skewed in the direction of the skewing parameter vector α. Examples of skew-normal distributions with different parameter settings are shown in Figure 9.6.

The skew normal distribution was extended by Arellano-Valle and Genton (2005) to yield the canonical Fundamental Unrestricted Skew Normal (cFUSN); in fact, Arellano-Valle and Genton (2005) developed an even more general family that we do not consider here. The cFUSN model has the following generative description. Let

$$\begin{pmatrix} Y \\ Y_0 \end{pmatrix} \sim N\left[\begin{pmatrix} \mu \\ 0 \end{pmatrix}, \begin{pmatrix} \Sigma & \Delta^T \\ \Delta & \Gamma \end{pmatrix}\right]$$

where Y is a vector of length d and Y_0 is a vector of length q. Suppose that Y is

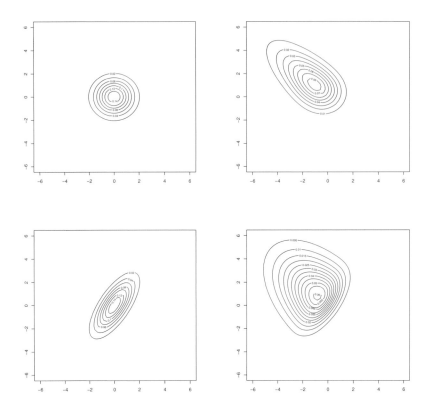

Figure 9.6 Examples of skew normal distributions. The left and right plots
in each row have the same location and scale parameters. The left-hand
examples have skew parameter $\alpha^T = (0,0)$ and the right-hand examples
have skew parameter $\alpha^T = (-2,2)$. The non-zero skewing parameter moves
the tail of the distribution towards lower values on the x-axis and high
values on the y-axis.

only observed when $Y_0 + \kappa$ is greater than 0, where κ is a vector of length q and
where the inequality holds elementwise if $q > 1$. Then, Y has a cFUSN distribution
whose form is determined by the parameter values. This model encompasses many
of the competing forms of skew normal distribution that are in the literature
(Table 9.1).

Lin et al. (2009) and Lin (2009) proposed using a mixture of skew normal
distributions for model-based clustering; both the restricted and unrestricted
forms of the skew normal were proposed. The models are fitted using the MCEM
algorithm where a Monte Carlo estimate of the expected complete-data log-
likelihood is used in the E-step. Subsequently, Lee and McLachlan (2014) derived
an analytical form for the E-step which avoids the need for a Monte Carlo estimate.

The `EMMIXskew` package for `R` (Wang et al., 2013) is available for fitting the
mixture of restricted skew normal distributions. It includes constrained versions

Table 9.1 *Examples of skew normal distributions that arise within the cFUSN family.*

Family	Parameters	Example
Restricted Skew Normal	$q = 1$ $\kappa = 0$	Azzalini and Dalla Valle (1996)
Unrestricted Skew Normal	$q = d$ $\kappa = 0$	Sahu et al. (2003)
Extended Skew Normal	$\kappa \neq 0$	Azzalini and Capitanio (1999)

of the scale matrices (Σ_g) that correspond to the EEE, EEI, VVV, VVI and VII models in `mclust`. The package allows for model selection using BIC or ICL. The `mixsmsn` package (Prates et al., 2013) also facilitates fitting mixtures of restricted skew normal distributions and it includes other model selection criteria including DIC and EDC.

The mixture of (restricted) skew normal distributions was fitted to the Old Faithful data using the `EMMIXskew` package (Wang et al., 2013). The model was fitted for $G = 1, 2, \ldots, 10$ and the five available structures for the covariance matrix were used. Model selection was completed using BIC (Figure 9.7) and ICL (Figure 9.8) and in both cases the resulting model was a mixture with $G = 2$ components and a constrained scale matrix; the parameter estimates are given in Equation (9.4). A plot of the resulting mixture density is given in Figure 9.9. The code for fitting the model is given in Listing 9.3.

$$
\begin{aligned}
&\tau_1 = 0.35, &&\tau_2 = 0.65, \\
&\mu_1 = \begin{pmatrix} 1.74 \\ 52.21 \end{pmatrix}, &&\mu_2 = \begin{pmatrix} 4.87 \\ 83.48 \end{pmatrix}, \\
&\Sigma_1 = \begin{pmatrix} 0.013 & 0.000 \\ 0.000 & 30.312 \end{pmatrix}, &&\Sigma_2 = \begin{pmatrix} 0.013 & 0.000 \\ 0.000 & 30.312 \end{pmatrix}, &&(9.4) \\
&\alpha_1 = \begin{pmatrix} 0.373 \\ 2.832 \end{pmatrix}, &&\alpha_2 = \begin{pmatrix} -0.712 \\ -4.329 \end{pmatrix}.
\end{aligned}
$$

The model does an excellent job at clustering data and selects two clusters. The clusters correspond to short waiting/eruption times and long waiting/eruption/-times. The component densities capture the skewness that is visually evident in the data. It is worth noting that the cluster that captures short waiting/eruption times is skewed towards the top right corner of the plot and the cluster that captures the long waiting/eruption times is skewed towards the bottom left corner of the plot. Comparing the BIC values of the mixture of restricted skew normal distributions to the mixture of multivariate distributions shows that the mixture of skew normal distributions is preferred with a BIC difference of 26.

It is worth noting that the extra modeling flexibility yielded by replacing the normal distribution with a skewed alternative also increases the likelihood of finding a local maximum in the likelihood function. This can be seen in the BIC

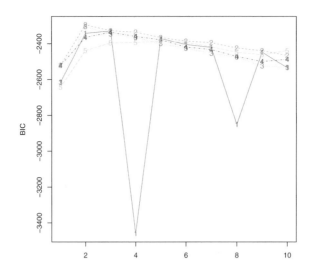

Figure 9.7 The BIC values for the mixtures of restricted multivariate skew normal distributions (five scale matrix structures) fitted using `EMMIXskew` for $G = 1, 2, \ldots, 10$. The low values of BIC for $G = 4$ and $G = 8$ for one model type are due to `EMMIXskew` finding local maxima for those cases.

and ICL plots (Figure 9.7 and Figure 9.8) where all of the model fits for $G = 4$ and $G = 8$ with common scaling matrix have converged to local maxima.

A further example of the phenomenon of converging to a local maximum is shown in Figure 9.10, where the $G = 2$ and general unconstrained scale matrix is fitted to a standardized version of the Old Faithful data. In this case, a local maximum of the log-likelihood was found. The plot shows the local maxima in the likelihood can yield models with dramatically different structure.

It is worth noting that the local maximum found has high density in the region where the data are observed, but it fails to capture the evident cluster structure in the data.

9.3 Skew-t Distribution

The skew-t distribution is a heavy-tailed analogue of the skew normal distribution. Similarly to the skew-normal distribution, there are many competing versions of the model and these are outlined in detail in Lee and McLachlan (2013d), Lee and McLachlan (2016) and the debate in Azzalini et al. (2016) and McLachlan and Lee (2016). A brief summary of three versions is given in Table 9.2.

Examples of the restricted and unrestricted skew-t distributions are given in Figure 9.11 and Figure 9.12 respectively. In these plots, we can see that the type

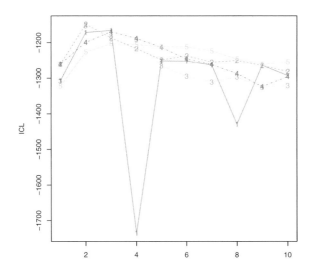

Figure 9.8 The ICL values for the mixtures of restricted multivariate skew normal distributions (five scale matrix structures) fitted using `EMMIXskew` for $G = 1, 2, \ldots, 10$. The low values of ICL for $G = 4$ and $G = 8$ for one model type are due to `EMMIXskew` finding local maxima for those cases.

Table 9.2 *Some skew-t distributions that arise within the cFUST family.*

Family	Parameters	Example
Restricted skew-t	$q = 1$ $\kappa = 0$	Branco and Dey (2001) Azzalini and Capitanio (2003)
Unrestricted skew-t	$q = d$ $\kappa = 0$	Sahu et al. (2003)
Extended skew-t	$\kappa \neq 0$	Arellano-Valle and Genton (2010)

of skew induced by the skewing parameter is very different. Thus, the two models accommodate different cluster shapes.

The restricted skew-t distribution exhibits a skewed shape where the mass of distribution is skewed in a single linear direction, as can be seen in Figure 9.11. By contrast, in the unrestricted skew-t distribution the mass of the distribution is skewed in the direction of the axes, as can be seen in Figure 9.12.

Pyne et al. (2009) proposed using a mixture of (restricted) skew-t distributions for clustering; their work was particularly motivated by the application of mixtures of skew distributions for gating in flow cytometry. The fitting of the mixture of restricted skew-t distributions uses a Monte Carlo EM algorithm. The `EMMIXskew`

Listing 9.3: Clustering the Old Faithful data using the **EMMIXskew** package; this package fits the restricted skew normal distribution

```
# Load the Old Faithful Data
data(faithful)

# Load the EMMIXskew package
library(EMMIXskew)

# Scale the data for ease of comparison with other models
# The results are not invariant to scaling of the data
X <- scale(faithful)

# Fit the Skew Normal Model with G=1,2,...,10
# Fit five different scale matrix structures (EEE,EEI,VVV,VVI,VII)
# Store the BIC and ICL values for each model
BICmat <- matrix(NA,10,5)
ICLmat <- matrix(NA,10,5)
for (g in 1:10)
{
    for (ncov in 1:5)
    {
        fit <- EmSkew(X,g=g,ncov=ncov,distr="msn",nrandom=500,nkmeans
            =500)
        BICmat[g,ncov] <- fit$bic
        ICLmat[g,ncov] <- fit$ICL
    }
}

# Plot the resulting BIC values (minus outputted BIC)
matplot(-BICmat,ylab="BIC",type="l")
matplot(-BICmat,type="p",add=TRUE)

# Plot the resulting ICL values
matplot(ICLmat,ylab="ICL",type="l")
matplot(ICLmat,type="p",add=TRUE)

# Refit the model with the highest BIC
rmsnfaith <- EmSkew(X,g=2,ncov=2,distr="msn",nrandom=500,nkmeans
    =500)

# Examine the fitted model
rmsnfaith

# Plot the resulting fit
EmSkew.contours(faithful,rmsnfaith)
```

package (Wang et al., 2013) can be used to fit mixtures of restricted skew-*t* distributions for model-based clustering.

Lin (2010) proposed a mixture of (unrestricted) skew-*t* distributions for model-based clustering. The fitting of these models was computationally intensive because of the use of a Monte Carlo E-step in the EM algorithm. Recently, Lee and McLachlan (2014) derived analytic calculations to avoid the need for a Monte

Contours of Components using EmSkew: MSN Distribution

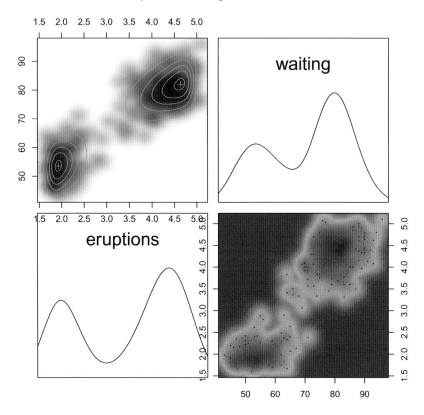

Figure 9.9 The plot of the best fitting mixture of restricted multivariate skew normal distributions, fitted using EMMIXskew. The model has $G = 2$ and constrained diagonal scale matrices Σ_g.

Carlo E-step for model fitting. The EMMIXuskew package (Lee and McLachlan, 2013a) can be used to fit mixtures of unrestricted skew-t distributions for model-based clustering; the package uses BIC for model selection.

9.3.1 Restricted Skew-t Distribution

The mixture of restricted skew-t distributions was fitted to the Old Faithful data using EMMIXskew. The model was fitted for $G = 1, 2, \ldots, 4$ and the five available structures for the covariance matrix were used. The model fitting procedure was unstable for $G > 4$, so these models were not considered. The model with the highest BIC (Figure 9.13) is a mixture with $G = 2$ components and constrained scale matrix; see Equation (9.5). A plot of the resulting fitted mixture density is given in Figure 9.15. The code for fitting the model is given in Listing 9.4.

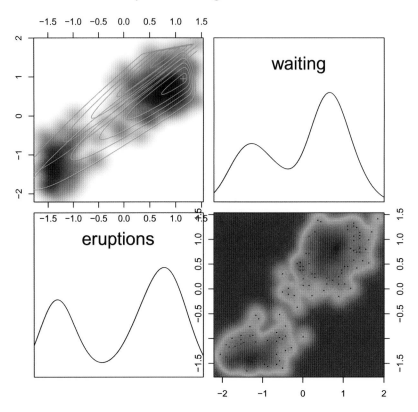

Figure 9.10 The plot of the best fitting mixture of restricted multivariate skew normal distributions, fitted using `EMMIXskew`, to a scaled version of the Old Faithful data. The model has $G = 2$ and general unconstrained scale matrices Σ_g.

$$\begin{aligned}
&\tau_1 = 0.35, && \tau_2 = 0.65, \\
&\mu_1 = \begin{pmatrix} 1.74 \\ 52.21 \end{pmatrix}, && \mu_2 = \begin{pmatrix} 4.87 \\ 83.46 \end{pmatrix}, \\
&\Sigma_1 = \begin{pmatrix} 0.013 & 0.000 \\ 0.000 & 30.046 \end{pmatrix}, && \Sigma_2 = \begin{pmatrix} 0.013 & 0.000 \\ 0.000 & 30.046 \end{pmatrix}, && (9.5) \\
&\alpha_1 = \begin{pmatrix} 0.369 \\ 2.814 \end{pmatrix}, && \alpha_2 = \begin{pmatrix} -0.708 \\ -4.299 \end{pmatrix}, \\
&\nu_1 = 186.3, && \nu_2 = 200.0.
\end{aligned}$$

The model selects the two clusters that are visually evident in the data in the same manner as the mixture of skew normal distributions. The parameter values

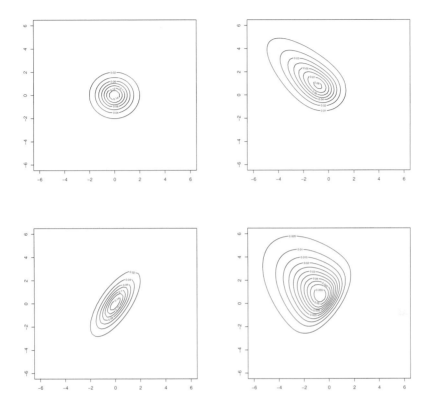

Figure 9.11 Examples of restricted skew-t distributions. The left and right plots in each row have the same location and scale parameters. The left-hand examples have skew parameter $\alpha^T = (0,0)$ and the right-hand examples have skew parameter $\alpha^T = (-2, 2)$; this skewing parameter moves the tail of the distribution towards lower values on the x-axis and high values on the y-axis.

and very high degrees of freedom parameter in each component indicate that the model fit is very similar to a mixture of skew normal distributions. In fact, the BIC for the mixture of skew normal model whose parameters are given in (9.4) is - 2,295, whereas the BIC for the skew-t mixture whose parameters are given in (9.5) is - 2,306, so the skew normal mixture model is preferred.

9.3.2 Unrestricted Skew-t Distribution

A mixture of (unrestricted) skew-t distributions was fitted to the Old Faithful data, for $G = 1, 2, \ldots, 5$, using the `EMMIXuskew` package (Lee and McLachlan, 2013a,b); the code for the model fitting is given in Listing 9.5. The BIC values for the model with differing values of G are given in Figure 9.16 and the model with $G = 2$ achieves the highest BIC. A plot of the best fitting model is given in Figure 9.17 and the parameter estimates for this model are given in (9.6).

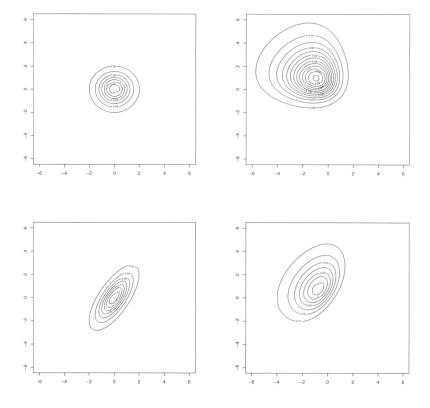

Figure 9.12 Examples of unrestricted skew-t distributions. The left and right plots in each row have the same location and scale parameters. The left-hand examples have skew parameter $\alpha^T = (0,0)$ and the right-hand examples have skew parameter $\alpha^T = (-2,2)$; this skewing parameter moves the tail of the distribution towards lower values on the x-axis and high values on the y-axis.

$$
\begin{aligned}
\tau_1 &= 0.36, & \tau_2 &= 0.64, \\
\mu_1 &= \begin{pmatrix} 1.73 \\ 48.52 \end{pmatrix}, & \mu_2 &= \begin{pmatrix} 4.69 \\ 75.94 \end{pmatrix}, \\
\Sigma_1 &= \begin{pmatrix} 0.0048 & 0.0846 \\ 0.0846 & 13.2039 \end{pmatrix}, & \Sigma_2 &= \begin{pmatrix} 0.074 & 0.835 \\ 0.835 & 25.464 \end{pmatrix}, \\
\alpha_1 &= \begin{pmatrix} 0.37 \\ 7.03 \end{pmatrix}, & \alpha_2 &= \begin{pmatrix} -0.49 \\ 5.01 \end{pmatrix}, \\
\nu_1 &= 18.12, & \nu_2 &= 51.16.
\end{aligned}
\tag{9.6}
$$

The (unrestricted) skew-t mixture model has captured the two evident clusters very well and has modeled the skew within the two clusters very accurately. A comparison of the mixtures of restricted skew-normal distributions, mixture of

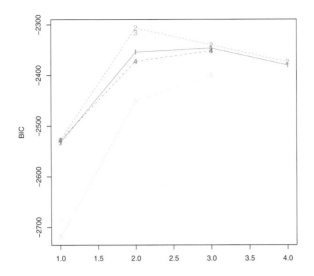

Figure 9.13 The BIC values for the mixtures of restricted multivariate skew-t distributions (five scale matrix structures) fitted using `EMMIXskew` for $G = 1, 2, \ldots, 4$.

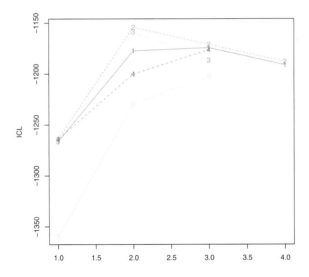

Figure 9.14 The ICL values for the mixtures of restricted multivariate skew-t distributions (five scale matrix structures) fitted using `EMMIXskew` for $G = 1, 2, \ldots, 4$.

Listing 9.4: Clustering the Old Faithful data using the **EMMIXskew** package; this package fits the restricted skew-*t* distribution

```
# Load the Old Faithful Data
data(faithful)

# Load the EMMIXskew package
library(EMMIXskew)

# Fit the Skew t Model with G=1,...,4
# Fit five different scale matrix structures (EEE,EEI,VVV,VVI,VII)
BICmat<-matrix(NA,4,5)
ICLmat<-matrix(NA,4,5)
for (g in 1:4)
{
    for (ncov in 1:5)
    {
        fit<-EmSkew(faithful,g=g,ncov=ncov,distr="mst",nrandom=100,
            nkmeans=100)
        BICmat[g,ncov]<-fit$bic
        ICLmat[g,ncov]<-fit$ICL
    }
}

# Plot the resulting BIC values (minus outputted BIC)
matplot(-BICmat,ylab="BIC",type="l")
matplot(-BICmat,type="p",add=TRUE)

# Plot the resulting BIC values (minus outputted BIC)
matplot(ICLmat,ylab="ICL",type="l")
matplot(ICLmat,type="p",add=TRUE)

# Refit the model with the highest BIC
rmstfaith<-EmSkew(faithful,g=2,ncov=2,distr="mst",nrandom=100,
    nkmeans=100)

# Look at fitted model
rmstfaith

# Plot the resulting fit
EmSkew.contours(faithful,rmstfaith)
```

restricted skew-*t* distributions and the mixture of unrestricted skew-*t* distributions using BIC shows that the restricted skew-normal model has a higher BIC value for the Old Faithful data.

9.4 Box–Cox Transformed Mixtures

The Box–Cox transformation (Box and Cox, 1964) provides a general family of transformations for numerical data, which includes the logarithm and power

Contours of Components using EmSkew: MST Distribution

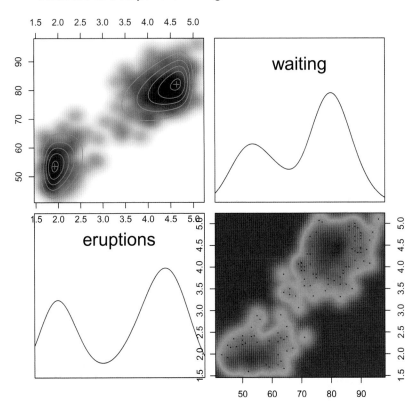

Figure 9.15 The plot of the highest BIC mixture of restricted multivariate skew-t distributions, fitted using `EMMIXskew`. The model has $G = 2$ and constrained diagonal scale matrices, Σ_g, for all g.

transformations as special cases. The transformation has the form

$$g_\delta(y) = \begin{cases} \frac{y^\delta - 1}{\delta} & \text{for } \delta \neq 0 \\ \log(y) & \text{for } \delta = 0 \end{cases}$$

and can be applied to each variable in a data set. The Box–Cox transformation was extended (Bickel and Doksum, 1981) to

$$g_\delta(y) = \frac{\text{sign}(y)|y|^\delta - 1}{\delta} \text{ for } \delta > 0$$

to accommodate negative values.

Lo et al. (2008) proposed using a Box–Cox transformation of the data variables prior to using a finite mixture of multivariate t-distributions for model-based clustering of flow cytometry data. The resulting model, when back transformed

Listing 9.5: Clustering the Old Faithful data using the `EMMIXuskew` package; this
package fits the unrestricted skew-*t* distribution

```
# Load the Old Faithful Data
data(faithful)

# Load the EMMIXuskew package
library(EMMIXuskew)

# Fit the unrestricted Skew t Model with G=1,...,5

BICvec<-rep(NA,5)
for (g in 1:5)
{
    BICvec[g]<-fmmst(g=g,faithful)$bic
}

# Plot the resulting BIC values (minus outputted BIC)
plot(-BICvec,type="l",ylab="BIC",xlab="G")
points(-BICvec)

# Refit the model with the highest BIC
umstfaith<-fmmst(g=2,faithful)

# Examine the fitted model
summary(umstfaith)

# Plot the resulting fit
fmmst.contour.2d(faithful,umstfaith,main=" ")
```

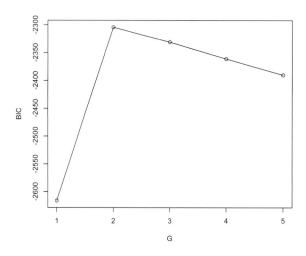

Figure 9.16 The BIC values for the mixtures of unrestricted multivariate
skew-*t* distributions fitted using `EMMIXuskew` for $G = 1, 2, \ldots, 5$

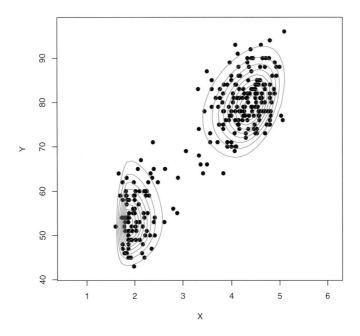

Figure 9.17 The plot of the best fitting mixture of unrestricted multivariate skew-*t* distributions, fitted using `EMMIXuskew`; the model has $G = 2$ components.

to the original data scale, allowed for skewed cluster shapes. The `flowClust` package (Lo et al., 2009) for `Bioconductor` fits the Box–Cox transformed mixture of *t*-distribution model. Lo and Gottardo (2012) outlined an extension of the model where each mixture component is allowed to have its own transformation; this extension greatly extends the range of non-elliptical clusters that the method can accurately model and it is also implemented in the `flowClust` package. The `flowClust` package allows the user to either specify or estimate the Box–Cox parameters and the degrees of freedom of the *t*-distribution component densities (default setting is a *t*-distribution with 4 degrees of freedom).

The Box–Cox transformed mixture models were fitted to the Old Faithful data using `flowClust`. The option of specifying or estimating the transformation parameters and degrees of freedom of the mixture components was also investigated. Code for fitting these models is given in Listing 9.6 and the resulting BIC values are shown in Figure 9.18. A plot of the resulting model is given in Figure 9.19 and the model parameters are given in Equation (9.7).

Listing 9.6: Clustering the Old Faithful data using the `flowClust` package; this package fits the mixture of Box–Cox transformed *t*-distributions

```
# Load the Old Faithful data
data(faithful)

# Load the flowClust package
library(flowClust)

# Fit the Box--Cox transformed multivariate t with G=1,2,...,6
# Fit cases where degrees of freedom fixed, estimated, cluster
    specific (nu.est=0,1,2)
# Fit cases where transformation is not used, estimated, cluster
    specific (trans=0,1,2)

nuest<-c(0,1,2,0,1,2,0,1,2)
lambdaest<-c(0,0,0,1,1,1,2,2,2)
BICmat<-matrix(NA,6,9)
for (j in 1:9){
   BICmat[,j] <-criterion(flowClust(faithful,K=1:6,nu.est=nuest[j],
      trans=lambdaest[j],randomStart=100,level=1))
}

# Fit the best model again
BCfaith<-flowClust(faithful,expName="Old Faithful",K=2,nu.est=1,
    trans=0,randomStart=100,level=1)


# Plot the results
dens<-density(BCfaith, faithful)
plot(dens)
points(faithful,pch=3,col="darkgray")
```

$$\begin{aligned}
\tau_1 &= 0.35, & \tau_2 &= 0.65, \\
\mu_1 &= \begin{pmatrix} 2.01 \\ 54.30 \end{pmatrix}, & \mu_2 &= \begin{pmatrix} 4.30 \\ 79.99 \end{pmatrix}, \\
\Sigma_1 &= \begin{pmatrix} 0.06 & 0.37 \\ 0.37 & 31.07 \end{pmatrix}, & \Sigma_2 &= \begin{pmatrix} 0.15 & 0.83 \\ 0.83 & 32.62 \end{pmatrix}, \\
\nu_1 &= 17.47, & \nu_2 &= 17.47, \\
\delta_1 &= 1 & \delta_2 &= 1.
\end{aligned} \tag{9.7}$$

The final selected and estimated model is a mixture of two multivariate *t*-distributions with no transformation of the waiting time or eruption time variables. The model clusters the data very well but the model fit to each component does not show the non-elliptical structure of the clusters.

9.5 Generalized Hyperbolic Distribution

The multivariate normal inverse Gaussian model arises as a location-scale mixture of multivariate distributions of the form, $Y_i|U_i \sim N(\mu + U_i\beta, U_i\Sigma)$ and $U_i \sim$

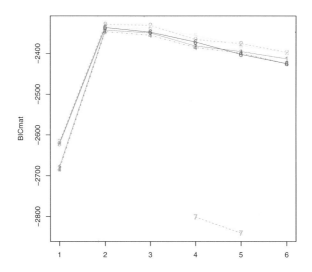

Figure 9.18 The BIC values for the mixtures of Box–Cox transformed t-distributions fitted using `flowClust` for $G = 1, 2, \ldots, 6$.

InverseGaussian$(1, \gamma)$ (Barndorff-Nielsen et al., 1982). That is,

$$f(y_i|u_i, \mu, \beta, \Sigma) = \frac{1}{(2\pi u_i)^{d/2}|\Sigma|^{1/2}} \exp\left[-\frac{1}{2u_i}(y_i - \mu - u_i\beta)^T \Sigma^{-1}(y - \mu - u_i\beta)\right],$$

$$f(u_i) = \frac{1}{\sqrt{2\pi}} \exp(\gamma) u_i^{-3/2} \exp\left[-\frac{1}{2}\left(\frac{1}{u_i} + \gamma^2 u_i\right)\right]. \tag{9.8}$$

The marginal distribution of Y_i is called a multivariate normal inverse Gaussian distribution. This distribution can accommodate heavy-tailed and skewed clusters and all of the model EM algorithm calculations can be written in closed form.

Karlis and Santourian (2009) proposed using a mixture of multivariate normal inverse Gaussian distributions as an alternative model for model-based clustering of non-elliptically shaped clusters. O'Hagan et al. (2016) developed a family of mixtures of multivariate normal inverse Gaussian distributions by using an eigen-decomposition of the scale matrix Σ_g in each component distribution in a way that is analogous to `mclust` and by further allowing the skew parameter β_g and tail weight parameter γ_g to be constrained to be equal or unequal across groups. The mixture of multivariate normal inverse Gaussian distributions model can be represented by a graphical model, as shown in Figure 9.20.

More recently, Browne and McNicholas (2015) extended the mixture of multivariate normal inverse Gaussian distributions by replacing the inverse Gaussian distribution (9.8) with a generalized inverse Gaussian model with parameters $(\omega, 1, \lambda)$; this gives arise to the mixture of generalized hyperbolic distributions

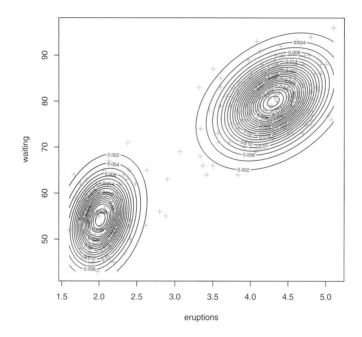

Figure 9.19 The plot of the best fitting mixture of Box–Cox transformed *t*-distributions, fitted using `flowClust`; the model has $G = 2$ components and does not transform the data. The ellipses shown show the 90% contour of each component density and points outside these ellipses are marked as potential outliers.

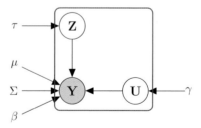

Figure 9.20 Graphical model of the mixture of multivariate normal inverse Gaussian distributions.

(see Blaesild and Jensen, 1981). Thus, the component distribution is of the form,

$$f(y|\mu, \beta, \Sigma, \omega, \lambda) \qquad (9.9)$$

$$= \left[\frac{\omega + \delta(x, \mu|\Sigma)}{\omega + \beta^T \Sigma^{-1} \beta}\right]^{(\lambda - d/2)/2} \frac{K_{\lambda - d/2}\left(\sqrt{[\omega + \beta^T \Sigma^{-1} \beta][\omega + \delta(y, \mu|\Sigma)]}\right)}{(2\pi)^{d/2} |\Sigma|^{1/2} K_\lambda(\omega) \exp\left[-(y - \mu)^T \Sigma^{-1} \beta\right]},$$

where $\delta(y, \mu|\Sigma) = (y - \mu)^T \Sigma^{-1}(y - \mu)$ and $K(\cdot)$ is a modified Bessel function of the second kind. The β parameter controls the skew in the distribution and the parameters (ω, λ, β) contribute to the heavy tailedness of the distribution.

The `MixGHD` package (Tortora et al., 2015a) can be used for fitting mixtures of generalized hyperbolic distributions and extensions. The mixture of generalized hyperbolic distributions was fitted to the Old Faithful data for $G = 1, 2, \ldots, 10$ (Listing 9.7) and the model with the highest BIC is the model with $G = 2$, where the BIC $= -2,306.48$ (Figure 9.21); this is also the highest ICL model with the ICL $= -2,307.55$ (Figure 9.22). The parameter estimates of the model are given in (9.10) and a contour plot of the fitted density is given in Figure 9.23.

Listing 9.7: Clustering the Old Faithful data using the `MixGHD` package; this package fits the generalized hyperbolic distribution

```
# Load the Old Faithful data
data(faithful)

# Load the MixGHD package
library(MixGHD)

# Fit the MGHD models for G=1,2, ..., 10
MGHDfaith<-MGHD(faithful,G=1:10,max.iter=1000,scale=FALSE,modelSel=
    "BIC")
```

$$
\begin{aligned}
&\tau_1 = 0.64, && \tau_2 = 0.36, \\
&\mu_1 = \begin{pmatrix} 5.88 \\ 81.67 \end{pmatrix}, && \mu_2 = \begin{pmatrix} 1.47 \\ 46.36 \end{pmatrix}, \\
&\Sigma_1 = \begin{pmatrix} 0.05 & 0.81 \\ 0.81 & 39.32 \end{pmatrix}, && \Sigma_2 = \begin{pmatrix} 0.01 & -0.55 \\ -0.55 & 37.36 \end{pmatrix}, \\
&\beta_1 = \begin{pmatrix} -1.77 \\ -1.86 \end{pmatrix}, && \beta_2 = \begin{pmatrix} 0.82 \\ 11.83 \end{pmatrix}, \\
&\omega_1 = 20.48, && \omega_2 = 4.62, \\
&\lambda_1 = -2.74, && \lambda_2 = -2.39.
\end{aligned} \qquad (9.10)
$$

The fitted model captures the data structure very well and the estimate density accounts for the skew in both clusters. The BIC for the chosen model is also lower than for the $G = 2$ restricted skew normal distribution. Thus, the extra flexibility afforded by the mixture of generalized hyperbolic distributions may not be needed to accurately model the Old Faithful data.

9.6 Example: Old Faithful Data

A number of models that account for potential non-Gaussian clusters were fitted to the Old Faithful data. These models can account for skew, non-elliptical clusters and heavy tails in the clusters and component densities. In terms of clustering performance, the models performed similarly and captured the two main types of eruptions. Applying the cluster combination approach as described in Chapter 3

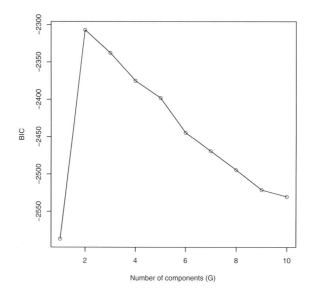

Figure 9.21 The BIC values for the mixture of generalized hyperbolic distributions fitted using `MixGHD` for $G = 1, 2, \ldots, 10$.

to the highest BIC Gaussian mixture model also finds the same data structure, as does the highest ICL Gaussian mixture model. However, in terms of modeling the density of observations within each cluster, the skewed distributions applied to the data yielded better model fit. The best fitting model, in terms of BIC and ICL, was the mixture of restricted skew-normal distributions with $G = 2$ components (Table 9.3).

Table 9.3 *A comparison of the BIC and ICL values for the non-Gaussian mixture models fitted to the Old Faithful data.*

Model	Type	G	BIC	ICL
Multivariate Gaussian	EEE	3	−2314.386	
	VVE	2		−2320.763
Multivariate t	CCCC	3	−2320.68	
	UUUC	2		−2328.35
Restricted skew-normal	**Common diagonal**	**2**	**−2295.02**	**−2297.08**
Restricted skew-t	Common diagonal	2	−2306.24	−2308.43
Unrestricted skew-t		2	−2326.72	−2327.59
Box–Cox transformed t	No transform	2	−2328.53	−2331.70
Generalized hyperbolic		2	−2306.48	−2307.55

Thus, there is little need for models with heavy tails in this application. The restricted skew-normal captures the skew appropriately in this case. However, in

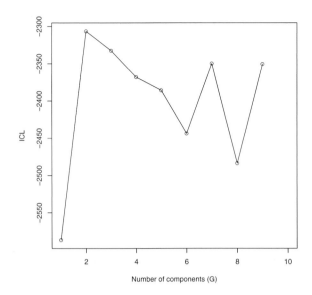

Figure 9.22 The ICL values for the mixture of generalized hyperbolic distributions fitted using `MixGHD` for $G = 1, 2, \ldots, 10$.

many other applications the models with skew and heavy tails may be needed for accurate model-based clustering.

9.7 Example: Flow Cytometry

The non-Gaussian models introduced in this chapter were fitted to the GvHD flow cytometry data and the best fitting models in terms of BIC and ICL were recorded. The data were scaled prior to clustering because many of the clustering algorithms were more stable when run on scaled data. Furthermore, the multivariate Gaussian mixture models within `mclust` were also fitted to the data. The resulting BIC and ICL values are given in Table 9.4.

The model with the highest BIC is the Mixture of Generalized Hyperbolic Distributions (MGHD) with $G = 11$ and a scatter plot of the data with the observations colored by cluster membership is given in Figure 9.24. It is worth noting that the different models gave very different numbers of clusters when BIC is used for model choice, whereas the ICL criterion showed evidence of $G = 4$ or $G = 5$ for all of the models. The restricted skew-t with general variance had the highest ICL value, but the MGHD model a very similar ICL value.

The scatter plot shows that the model-based clustering has found highly over-lapping clusterings of observations. By contrast, the MGHD model with the highest ICL value used $G = 4$ clusters and the result of this clustering is shown in Figure 9.25.

In this case, the clustering shows clearly defined clusters that have little overlap;

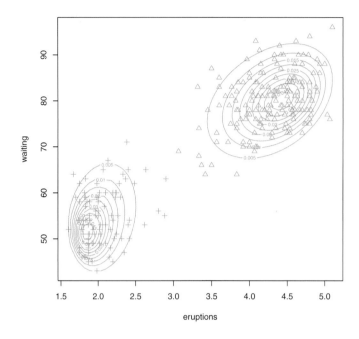

Figure 9.23 The plot of the best fitting mixture of generalized hyperbolic distributions, fitted using `MixGHD`; the model has $G = 2$ components.

this is because of the penalty for overlapping clusters included in the ICL criterion. The clusters found by the various models were not very different when the ICL criterion was used. A cross tabulation of the maximum a posteriori clustering estimates for the MGHD model with $G = 4$ and the (Gaussian) `mclust` VVV model with $G = 4$ shows very little difference in the clustering results (Table 9.5). Thus, the main advantage of the non-Gaussian clustering models, when ICL is used for model choice, is in terms of more accurately modeling the clusters rather than finding the clustering itself.

Further, a cross tabulation between the highest BIC and highest ICL clustering using the MGHD model is shown in Table 9.6. This table shows that the difference between the two clusterings is mainly that the BIC clustering divides the data into smaller groups and that the ICL groups are primarily made up of combinations of those found using BIC. Thus, the main difference is in terms of how separate the clusters are from each other.

9.8 Bibliographic Notes

Stephens (2000a) developed a Bayesian Markov chain Monte Carlo method, based on a birth-death process, for fitting mixtures of multivariate t-distributions with

Table 9.4 *A comparison of the BIC and ICL values for the non-Gaussian mixture models fitted to the GvHD control flow cytometry data.*

Model	Type	G	BIC	ICL
Multivariate Gaussian	VVI	20	−53061.37	
	VVV	4		−54757.84
Multivariate t	UIUC	19	−52951.76	
	UUUU	4		−54710.02
Restricted skew-normal	Diagonal variance	14	−52858.47	
	General variance	5		−53683.46
Restricted skew-t	General variance	7	−52896.12	
	General variance	4		**−53634.42**
Unrestricted skew-t		7	−53240.00	
		4		
Box–Cox transformed t	Common transform	8	−52886.39	
	Common transform	4		−54663.83
Generalized hyperbolic		**11**	**−52790.57**	
		4		−53635.00

Table 9.5 *A cross tabulation of the maximum a posteriori clustering for the MGHD model with $G = 4$ and the* `mclust` *model with $G = 4$.*

	1	2	3	4
1	4271	14	41	0
2	0	1251	36	0
3	1	0	532	1
4	2	0	52	608

unknown number of components (or clusters). This method avoids the need for using BIC to approximate the model probabilities of $G = 1, 2, \ldots$.

Further recent extensions to the mixture of multivariate t-distribution family have been developed in terms of increasing the efficiency of model fitting (Wang and Lin, 2013) and handling missing data (Lin, 2014) in finite mixtures of multivariate t-distributions.

Detailed reviews of the competing versions of the skew normal distribution,

Table 9.6 *A cross tabulation of the maximum a posteriori clustering for the MGHD model when the highest ICL model ($G = 4$) and highest BIC model ($G = 11$) are used.*

	1	2	3	4	5	6	7	8	9	10	11
1	102	115	299	3	0	0	0	4	0	0	11
2	0	26	0	1396	1958	368	562	4	0	11	1
3	0	0	3	0	3	0	0	641	342	298	0
4	0	2	0	1	12	0	0	3	0	0	644

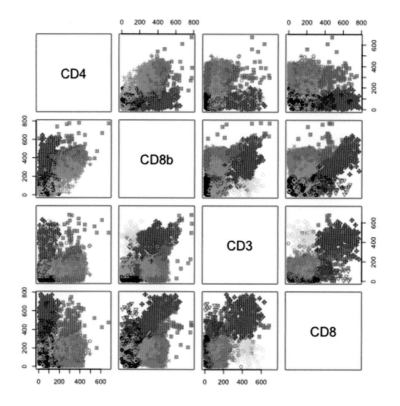

Figure 9.24 A scatter plot of matrix of the GvHD control data with the observations colored by cluster membership for the MGHD model with the highest BIC.

including further versions not described herein, are given by Lee and McLachlan (2013c,d, 2014).

A family of parsimonious models with eigen-decomposed scale matrices was proposed by Vrbik and McNicholas (2014); this work exploits results of Vrbik and McNicholas (2012) to avoid needing a Monte Carlo calculation in the E-step of the EM algorithm. As with the multivariate t-distribution and the skew-normal distribution, the interpretation of the eigen-decomposition of the scale matrix in terms of volume, shape and orientation is not maintained because the component covariance is also a function of the other model parameters.

Lin and Lin (2010) extended the EM algorithm for fitting mixtures of restricted skew normal distributions to accommodate missing data values. Further, Lin and Lin (2011) extended the algorithm for fitting mixtures of unrestricted t-distributions to accommodate missing data. Murray et al. (2017) developed a family of mixtures of unrestricted t-distributions with factor analytic structure for the scaling matrix.

Another family of models recently proposed for model-based clustering of

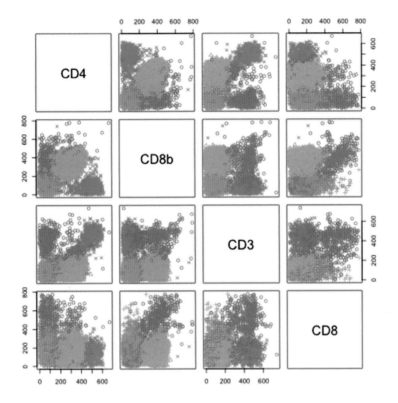

Figure 9.25 A scatter plot of matrix of the GvHD control data with the observations colored by cluster membership for the MGHD model with the highest ICL.

heavy-tailed and non-elliptically clustered data is the mixture of skew-asymmetric Laplace distributions (Franczak et al., 2014), which can be seen as a special case of the mixture of generalized hyperbolic distributions in Section 9.5.

A number of extensions of the mixture of generalized hyperbolic distributions have been proposed, including mixtures of coalesced generalized hyperbolic distribution model (Tortora et al., 2018), mixtures of generalized hyperbolic factor analyzers (Tortora et al., 2015b), mixtures of multiple scaled generalized hyperbolic distribution (Tortora et al., 2014). All of these models are available in the `MixGHD` R package.

Recently, Forbes and Wraith (2014) proposed a model for clustering multivariate data where the tail weight of each variable is allowed to differ; this is in contrast to the models proposed above where the component distributions have effectively fixed degrees of freedom, and thus tail weight, for each variable.

Lee and McLachlan (2016, 2018) have developed a model-based clustering framework based on finite mixtures of canonical fundamental skew-t distributions. This model family includes the models from Sections 9.1—9.3 as special cases.

10

Network Data

Network data arise when the relational data are collected on pairs of entities. In this chapter, models for clustering network data are reviewed and demonstrated on a number of data sets.

Network data are presented in Section 10.1 and some terminology is also introduced. A number of example data sets are described in Section 10.2. The stochastic block model is reviewed in Section 10.3 and it is demonstrated on example data. The mixed membership stochastic block model is introduced in Section 10.4. The latent space model and its extensions are described in Section 10.5. The recently developed stochastic topic block model is introduced in Section 10.6 and the model is fitted to an e-mail data set. The chapter concludes, in Section 10.7, with further topics in network modeling.

10.1 Introduction

A network is any data set that is composed of nodes (or actors) upon which we have relational data recorded for each pair of nodes. Suppose we have a network consisting of n nodes, then the data is of the form of an $n \times n$ adjacency matrix, Y, where y_{ij} is the recorded relationship between node i and node j. The network is said to be undirected if $y_{ij} = y_{ji}$ for all pairs (i, j), otherwise the network is said to be directed. The values taken by y_{ij} can be of any type, but are usually binary, count or continuous quantities. For most of the chapter, we will concentrate on the analysis of binary network data, where $y_{ij} = 1$ if node i and node j are related and $y_{ij} = 0$ otherwise.

An example of an 18-node network of social interactions between monks is shown as a network in Figure 10.2 and as an adjacency matrix in Table 10.1. In these data, $y_{ij} = 1$ if monk i reported monk j as their friend and $y_{ij} = 0$ otherwise. The graph is directed and this can be easily seen in the adjacency matrix Y which is non-symmetric. The adjacency matrix can also be shown as a heatmap plot (Figure 10.1). The layout of the nodes can also be shown using a force-directed layout (Figure 10.2), where the location of the nodes is chosen so that connected nodes tend to be adjacent and non-connected nodes are distant. In this case, the Fruchterman–Reingold force-based algorithm (Fruchterman and Reingold, 1991) is used as implemented in the `network` (Butts et al., 2014) R package.

Many network data sets exhibit structure where subsets of nodes have a tendency to form links with each other. A close inspection of Figure 10.1, Figure 10.2

Table 10.1 *The social interactions of 18 monks as reported in Sampson (1969).*

$$Y = \begin{pmatrix}
0 & 1 & 1 & 0 & 1 & 0 & 1 & 0 & 0 & 0 & 1 & 0 & 0 & 0 & 1 & 0 & 0 & 0 \\
0 & 0 & 1 & 0 & 1 & 1 & 0 & 0 & 1 & 0 & 0 & 0 & 0 & 0 & 1 & 0 & 0 & 0 \\
0 & 1 & 0 & 0 & 0 & 0 & 1 & 1 & 0 & 0 & 0 & 0 & 0 & 1 & 0 & 0 & 0 & 0 \\
0 & 1 & 1 & 0 & 1 & 0 & 0 & 0 & 1 & 0 & 0 & 0 & 0 & 0 & 0 & 0 & 0 & 0 \\
1 & 1 & 0 & 1 & 0 & 1 & 0 & 0 & 0 & 0 & 0 & 0 & 0 & 0 & 0 & 0 & 0 & 0 \\
0 & 1 & 0 & 0 & 1 & 0 & 1 & 0 & 0 & 0 & 1 & 0 & 0 & 1 & 0 & 0 & 0 & 0 \\
1 & 0 & 1 & 1 & 1 & 0 & 0 & 0 & 1 & 1 & 0 & 0 & 0 & 0 & 0 & 0 & 0 & 0 \\
0 & 0 & 0 & 0 & 0 & 0 & 0 & 0 & 1 & 1 & 1 & 0 & 1 & 0 & 0 & 0 & 0 & 0 \\
0 & 1 & 0 & 0 & 0 & 0 & 1 & 1 & 0 & 1 & 1 & 0 & 0 & 0 & 0 & 1 & 0 & 0 \\
0 & 0 & 0 & 0 & 0 & 0 & 0 & 1 & 1 & 0 & 1 & 1 & 1 & 0 & 0 & 0 & 0 & 0 \\
0 & 0 & 0 & 0 & 0 & 1 & 0 & 1 & 1 & 1 & 0 & 1 & 0 & 0 & 0 & 0 & 0 & 0 \\
0 & 1 & 0 & 0 & 0 & 0 & 0 & 1 & 1 & 1 & 1 & 0 & 1 & 0 & 0 & 0 & 0 & 0 \\
0 & 0 & 0 & 0 & 0 & 0 & 1 & 1 & 1 & 1 & 0 & 0 & 0 & 1 & 0 & 0 & 0 & 0 \\
0 & 0 & 0 & 0 & 0 & 0 & 0 & 1 & 1 & 1 & 0 & 1 & 1 & 0 & 0 & 0 & 0 & 0 \\
0 & 1 & 0 & 0 & 0 & 0 & 0 & 0 & 0 & 0 & 0 & 0 & 1 & 0 & 0 & 0 & 0 & 1 \\
0 & 0 & 0 & 0 & 0 & 0 & 0 & 0 & 1 & 1 & 0 & 0 & 0 & 0 & 1 & 0 & 1 & 1 \\
0 & 0 & 0 & 0 & 0 & 0 & 0 & 0 & 0 & 1 & 0 & 0 & 0 & 0 & 1 & 1 & 0 & 1 \\
0 & 0 & 0 & 0 & 0 & 0 & 0 & 0 & 1 & 1 & 0 & 0 & 1 & 0 & 1 & 1 & 1 & 0
\end{pmatrix}$$

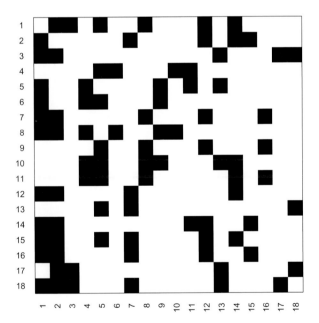

Figure 10.1 The social interactions of 18 monks as reported in Sampson (1969) shown as a heatmap. The dark shading shows the connections between nodes.

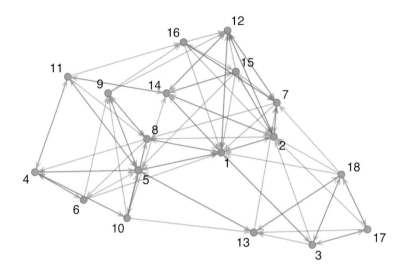

Figure 10.2 The social interactions of 18 monks as reported in Sampson (1969) shown as a network. The layout of nodes in the graph is determined using the Fruchterman–Reingold method.

and Table 10.1 reveals that subsets of nodes $\{1, 2, \ldots, 7\}$, $\{8, 9, \ldots, 14\}$ and $\{15, 16, 17, 18\}$ have a higher tendency to link within subset than across subsets. However, there are some connections across subsets also. In network analysis, such structure is commonly referred to as community structure rather than clustering structure. Whilst in many applications, the terms cluster and community are identical, we will see in Section 10.3 that clusters can be different from communities.

In this chapter, we introduce a number of model-based clustering methods for network data. Emphasis will be given to clustering binary network data sets but many of the methods can be extended to other data types. Many of the models outlined in this chapter are closely related to the model-based clustering methods outlined in earlier chapters. There are many other aspects of network modeling that are not covered in this chapter. Extensive overviews of network analysis and the statistical modeling of network data include Wasserman and Faust (1994), Carrington et al. (2005), Airoldi et al. (2007), Wasserman et al. (2007), Kolaczyk (2009), Goldenberg et al. (2010), Salter-Townshend et al. (2012) and Channarond (2015).

10.2 Example Data

In this section, we introduce example data sets that are used to illustrate the model-based clustering methods for network data.

10.2.1 Sampson's Monk Data

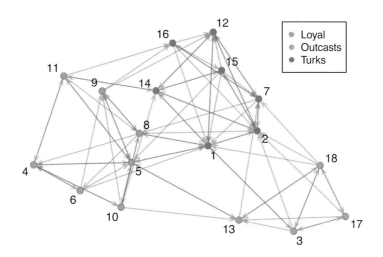

Figure 10.3 The social interactions of 18 monks as reported in Sampson (1969). The nodes are colored according to the grouping given by Sampson.

Sampson (1969) recorded the social interactions among a group of 18 novice monks while resident in a monastery. During his stay, a political "crisis in the cloister" resulted in the expulsion of four monks and the voluntary departure of several others. Of particular interest is the data on positive affect relations ("liking"), in which each monk was asked if they had positive relations to each of the other monks.

The data were gathered at three times to capture changes in group sentiment over time. They represent three time points in the period during which a new cohort entered the monastery near the end of the study but before the major conflict began. Each member ranked only his top three choices on "liking." (Some subjects offered tied ranks for their top four choices.) A tie from monk A to monk B exists if A nominated B as one of his three best friends at that that time point. The most commonly analyzed version of the data aggregates the "liking" data over the three time points. The data are shown in Figure 10.2 and the adjacency matrix of the network is also shown in Table 10.1.

Sampson (1969) grouped the novice monks into three groups which he called "Loyal," "Outcasts" and "Turks"; thus, there are known social clusters in this network. A plot of the data where the nodes are colored by Sampson's grouping is given in Figure 10.3. This plot indicates that there appears to be higher connectivity within Sampson's groups than between them, thus the nodes in the network exhibit clustering.

10.2.2 Zachary's Karate Club

The Zachary's karate club data set, introduced in Chapter 1, consists of the friendship network of 34 members of a university-based karate club. It was collected following observations over a three-year period in which a factional division caused the members of the club to formally separate into two organizations. The network exhibits many of the phenomena observed in social networks, in particular community (clustering) structure and nodes with low and high degree; the data were shown in Figure 1.6.

10.2.3 AIDS Blogs

The AIDS blog data set records the pattern of citation among 146 unique blogs related to AIDS patients and their support networks. The data were originally collected by Gopal (2007) over a randomly selected three-day period in August 2005. The nodes in the network correspond to blogs and a directed edge from one blog to another indicates that the former had a link to the latter in their web page; the data are further described and analyzed in Kolaczyk (2009). A plot of the data is given in Figure 10.4 and this shows that the graph has a strong clustering structure where many nodes connect to hub nodes and some hub nodes connect with each other.

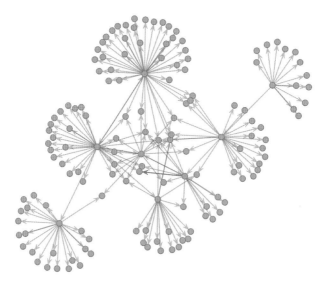

Figure 10.4 The network of AIDS blogs as reported in Kolaczyk (2009).

10.2.4 French Political Blogs

The French political blog data set shows the linking structure in online blogs which commentate on French political issues; the data were collected by Observatoire Presidentielle in October 2006 and are available in the `mixer` R package (Ambroise et al., 2013). In this data set, the nodes represent blogs and edges represent hyperlinks between different blogs; an edge is present if there is one or more links between the blogs. The graph is represented as an undirected graph where the presence of an edge is recorded irrespective of which blog cited which.

This network presents an interesting community structure due to the existence of several political parties and commentators in the French political scene. Seven known political groupings compose this network. The data are shown in Figure 10.5 and some clustering of nodes can be seen in this visualization of the network; the clustering into communities is strongly related to the known political groups.

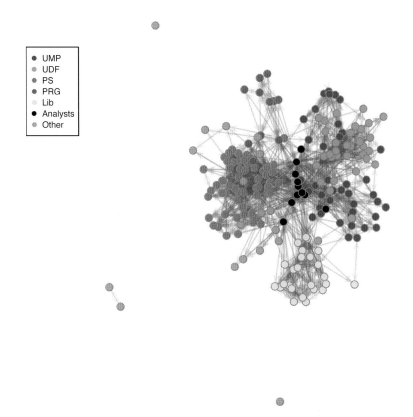

Figure 10.5 The network of French political blogs as reported in the `mixer` package. The nodes have been colored by political grouping and high connectivity between nodes of the same grouping can be seen.

10.2.5 Lazega Lawyers

Lazega (2001) collected a network data set detailing interactions between a set of 71 lawyers in a corporate law firm in the USA. The data include measurements of the advice network, friendship network and co-worker network between the lawyers within the firm. Further covariates associated with each lawyer in the firm are also available including age, seniority, college education and office location. The data can be used to investigate the social processes within the firm such as knowledge sharing and organizational structure. Graphs of the three networks are shown in Figure 10.6, and differences between the three networks can be easily seen.

10.3 Stochastic Block Model

The most fundamental model for clustering nodes in a network is the stochastic block model (Snijders and Nowicki, 1997; Nowicki and Snijders, 2001). This model is similar in flavor to the latent class model (LCA) (Lazarsfeld and Henry, 1968) as described in Chapter 6.

First, we assume that there are G blocks present in the population and that the probability that a node comes from block g is given by τ_g. Let $z_i = (z_{i1}, z_{i2}, \ldots, z_{iG})^T$ be a column indicator vector of block membership for node i, so $z_{ig} = 1$ if node i belongs to block g and $z_{ig} = 0$ otherwise. We also assume that there is a $G \times G$ block-to-block interaction matrix, Θ, where θ_{gh} is the probability that a node in block g is related to a node in block h. Further, we assume that all pairwise interactions are independent.

Thus,

$$\mathbb{P}\{y_{ij} = 1 | \Theta, z_{ig} = 1, z_{jh} = 1\} = \theta_{gh} \tag{10.1}$$
$$= z_i^T \Theta z_j, \tag{10.2}$$

and

$$\mathbb{P}\{y_{ij} = 0 | \Theta, z_{ig} = 1, z_{jh} = 1\} = 1 - \theta_{gh} = 1 - z_i^T \Theta z_j. \tag{10.3}$$

A graphical model for the stochastic block model is shown in Figure 10.7.

The property that the interaction probabilities only depend on the cluster membership of the nodes is called stochastic equivalence (Fienberg and Wasserman, 1981) and this concept has origins in structural equivalence (Lorrain and White, 1971). Further, if the block-to-block interaction matrix, Θ, is diagonally dominant (i.e. $\theta_{gg} > \theta_{gh}$ for $h \neq g$) then the network has assortative mixing and the blocks will correspond to communities; this is because nodes in the same block are more likely to connect to each other. However, if Θ has the property that $\theta_{gg} < \theta_{gh}$, for $h \neq g$, then disassortative mixing is present; in this case, the blocks do not correspond to communities, but they are still valid clusters because the nodes in the block are stochastically equivalent.

Advice Network   Friendship Network

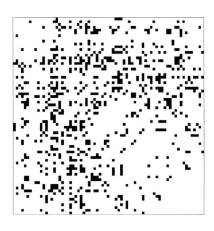

Co-worker Network

Figure 10.6 The three networks (Advice, Friendship and Co-Worker)
recorded in the Lazega lawyer data (Lazega, 2001) represented using
heatmaps.

10.3.1 Inference

The complete-data likelihood for the stochastic block model has the form,

$$L_c(\tau, \Theta) = \mathbb{P}\{Y, Z | \tau, \Theta\} = \mathbb{P}\{Y|Z, \Theta\}\mathbb{P}\{Z|\tau\}$$

$$= \prod_{i,j} \left(z_i^T \Theta z_j\right)^{y_{ij}} \left(1 - z_i^T \Theta z_j\right)^{1-y_{ij}} \times \prod_{i=1}^{n} \prod_{g=1}^{G} \tau_g^{z_{ig}}, \qquad (10.4)$$

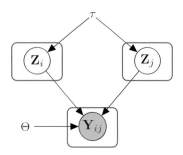

Figure 10.7 Graphical model representation of the stochastic block model (SBM).

where $Z = (z_1, z_2, \ldots, z_n)$ and where the product is taken over all pairs $\{(i,j) : i \neq j\}$ if the network is directed and over pairs $\{(i,j) : i < j\}$ if the network is undirected.

The observed likelihood function does not have a tractable form for even moderate values of n because it involves summing (10.4) over all G^n possible values for Z. Thus, maximum likelihood inference for this model is difficult. However, Snijders and Nowicki (1997) outlines an EM algorithm for fitting stochastic block models when $G = 2$.

Bayesian inference using Markov Chain Monte Carlo (MCMC) has been widely adopted for fitting this model. In order to implement Bayesian inference for the stochastic block model, conjugate priors are used for the model parameters. In particular, the following priors are commonly used $\tau \sim \text{Dirichlet}(\delta)$ and $\theta_{gh} \sim \text{beta}(\alpha, \beta)$ where δ, α and β are the prior hyperparameters. With this prior specification a Gibbs sampler (Gelfand and Smith, 1990) can be used to generate samples from the joint posterior, $p(Z, \tau, \Theta | Y)$, for the model parameters and cluster memberships. MCMC schemes for fitting the stochastic block model are given in Snijders and Nowicki (1997) for the case when $G = 2$ and in Nowicki and Snijders (2001) for any G.

McDaid et al. (2012, 2013) developed a collapsed sampler for the stochastic block model by analytically integrating out the model parameters and thus sampling from the posterior $p(Z|Y)$; this sampler facilitates fitting stochastic block models to large network data sets.

Further efficient algorithms for fitting stochastic block models have been developed including a CEM algorithm (Zanghi et al., 2008) and a variational Bayesian algorithm (Daudin et al., 2008; Latouche et al., 2010); these algorithms are implemented in the `mixer` R package (Ambroise et al., 2013).

Finally, Channarond et al. (2012) exploit the degree distribution of the stochastic block model which can be approximately modeled by a finite mixture of Poisson distributions to find a highly efficient algorithm for approximate inference for the stochastic block model; this algorithm uses only the degree of the nodes to estimate the number of blocks, block membership and block parameters and the resulting estimates are consistent.

Listing 10.1: Fitting a stochastic block model to Sampson's monks data using the `mixer` package; the Bayesian method for fitting the model within this package is used

```
# Load libraries
library(network)

# Load Monk Data
data(sampson, package="ergm")
ySampson <- as.matrix(samplike)
rownames(ySampson) <- 1:18
colnames(ySampson) <- 1:18

# Plot Data
layout <- network.layout.fruchtermanreingold(samplike, layout.par=
    NULL)
plot(samplike, label=1:18, mode="fruchtermanreingold", coord=layout
    )

# Fit the Stochastic Block Model
set.seed(1)
library(mixer)
fitb <- mixer(ySampson, qmin=1, qmax=5, method="variational")

# Choose the best model
mod <- getModel(fitb)
mod

# Compare estimated block membership with Sampson's groups
library(mclust)
z <- t(mod$Taus)
grp <- get.vertex.attribute(samplike, "group")
table(map(z), grp)
```

10.3.2 Application

Sampson's Monks Data

The stochastic block model was fitted to Sampson's monks data from Section 10.2.1 using the `mixer` (Ambroise et al., 2013) R package; a variational EM algorithm was used for model fitting and the model with the highest integrated complete likelihood (ICL) was chosen. The code for model fitting is contained in Listing 10.1.

The resulting model is a model with $G = 3$ blocks and the estimated model parameters are

$$\hat{\tau} = \begin{pmatrix} 0.39 \\ 0.22 \\ 0.39 \end{pmatrix} \text{ and } \hat{\Theta} = \begin{pmatrix} 0.71 & 0.25 & 0.18 \\ 0.04 & 0.83 & 0.07 \\ 0.10 & 0.04 & 0.55 \end{pmatrix}.$$

The fitted model has strong within-block connection probabilities and low between-block connection probabilities. Thus, the model has clustered the monks into subsets with high connectivity. Table 10.2 shows a cross classification of the maximum a posteriori block membership for each monk against the groupings given in Sampson (1969).

Table 10.2 *Cross classification of the maximum a posteriori block membership for each monk against Sampson's grouping of monks.*

	Loyal	Outcasts	Turks
1	0	0	7
2	0	4	0
3	7	0	0

The values show that the model has clustered the monks into the exact same groupings as postulated by Sampson.

Zachary's Karate Data

The stochastic block model was fitted to Zachary's karate data from Section 10.2.2 using the Bayesian method implemented in the `mixer` R package. The model with the highest integrated likelihood variational Bayes (ILvB) criterion (Latouche et al., 2012) was selected and the resulting model is a model with $G = 4$ blocks and the estimated model parameters are

$$\hat{\tau} = \begin{pmatrix} 0.11 \\ 0.36 \\ 0.46 \\ 0.08 \end{pmatrix} \text{ and } \hat{\Theta} = \begin{pmatrix} 0.80 & 0.53 & 0.17 & 0.25 \\ 0.53 & 0.13 & 0.00 & 0.10 \\ 0.17 & 0.00 & 0.09 & 0.72 \\ 0.25 & 0.10 & 0.72 & 0.67 \end{pmatrix}. \quad (10.5)$$

It is worth noting that the block-to-block interaction matrix, Θ, does not have a dominant diagonal. Thus, there is some evidence of disassortative mixing in this network.

Table 10.3 shows a cross tabulation of the maximum a posteriori block memberships with a faction identity proposed by Zachary (1977); the faction identity score is coded as: -2 (strongly Mr. Hi's), -1 (weakly Mr. Hi's), 0 (neutral), $+1$ (weakly John's), and $+2$ (strongly John's). Table 10.3 also shows a cross tabulation of maximum a posteriori block membership against the new club membership after the club split. The results show that the stochastic block model captures the divide in the club and the roles within this divide. The results are easily visualized in Figure 10.8 where the nodes are colored according to maximum a posteriori block membership.

AIDS Blog

The stochastic block model was fitted to the AIDS blog data using the Bayesian method in the `mixer` R package and the resulting model with the highest ILvB criterion has $G = 2$ blocks and the estimated model parameters are

$$\hat{\tau} = \begin{pmatrix} 0.06 \\ 0.94 \end{pmatrix} \text{ and } \hat{\Theta} = \begin{pmatrix} 0.19 & 0.16 \\ 0.00 & 0.00 \end{pmatrix}.$$

Table 10.3 *Cross tabulation of maximum a posteriori block membership against faction identity and against new club membership.*

		Faction Identity			
	−2	−1	0	+1	+2
1	3	0	0	0	0
2	7	5	1	0	0
3	0	0	2	4	10
4	0	0	0	0	2

	Mr. Hi's	John A.'s
1	3	0
2	13	0
3	1	15
4	0	2

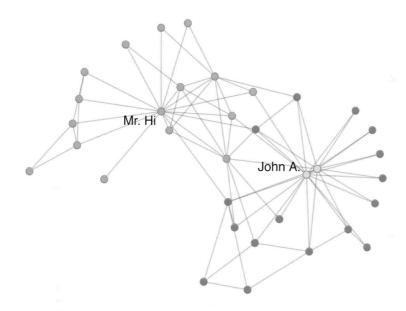

Figure 10.8 A plot of Zachary's karate club network with the nodes colored by maximum a posteriori block membership.

The model picks up the hub nodes and the satellite nodes which reveals the differing roles of different nodes in the network. A plot of the network colored by maximum a posteriori block membership (Figure 10.9) shows the different roles of the nodes in the network. One block of nodes are "hub" nodes and the other are "peripheral" nodes. Thus, the stochastic block model has modeled a clustering structure which differs from the idea of clusters being highly connected components in the network.

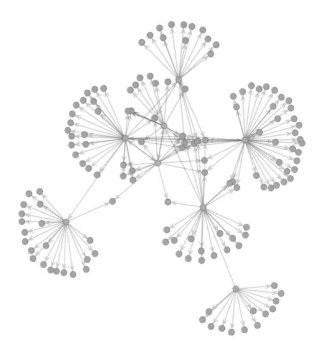

Figure 10.9 A plot of AIDS blog network with the nodes colored by maximum a posteriori block membership.

10.4 Mixed Membership Stochastic Block Model

The mixed membership stochastic block model (MMSB) (Airoldi et al., 2008) is an extension of the stochastic block model (Section 10.3) that allows nodes to have mixed membership to multiple blocks. Thus, the mixed membership stochastic block model is closely related to the latent Dirichlet allocation (Blei et al., 2003) model and the grade of membership model (Erosheva, 2002, 2003; Erosheva et al., 2007) in the same way that the stochastic block model was related to the latent class model.

In this model, as in Sections 10.3, we assume that there are G blocks and any two nodes that belong fully to a block are considered to be stochastically equivalent.

However, in contrast to the stochastic block model, the block membership of a node may change depending on which node in the network it is interacting with. Each node may therefore have multiple block memberships as it links with the other nodes in the network.

Let node i have an individual probability τ_{ig} of belonging to block g, and thus a vector of probabilities τ_i, while interacting with any other node in the network. When node i and node j are interacting, they each have an indicator variable $z_{i \to j}$ and $z_{i \leftarrow j}$ that denote sender and receiver cluster membership for each interaction between nodes i and j respectively; let $\overrightarrow{Z}$ and $\overleftarrow{Z}$ be the matrix of $z_{i \to j}$ and $z_{i \leftarrow j}$ values for all (i, j).

Once cluster membership is accounted for, nodal interaction y_{ij} is modeled, as in Section 10.3, as a Bernoulli random variable with probability $z_{i \to j} \Theta z_{i \leftarrow j}^T$, where Θ is a $G \times G$ matrix of block-to-block interaction probabilities.

Thus the data generation procedure is:

- For each node $i \in \{1, 2, \ldots, n\}$:
  - Draw a G-dimensional mixed membership vector $\tau_i \sim \text{Dirichlet}(\alpha)$, where $\alpha = (\alpha_1, \alpha_2, \ldots, \alpha_G)$.
- For each pair of nodes (i, j) :
  - Draw membership indicator for the initiator, $z_{i \to j} \sim \text{Multinomial}(\tau_i)$
  - Draw membership indicator for the receiver, $z_{i \leftarrow j} \sim \text{Multinomial}(\tau_j)$
  - Sample the value of their interaction, $y_{ij} \sim \text{Bernoulli}(z_{i \to j} \Theta z_{i \leftarrow j}^T)$.

The model can be represented using a graphical model, as shown in Figure 10.10.

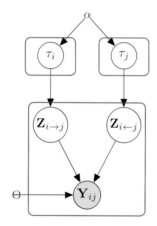

Figure 10.10 Graphical model of mixed membership stochastic block model (MMSBM).

10.4.1 Inference

Fitting the mixed membership stochastic block model (MMSB) can be implemented using Markov Chain Monte Carlo (MCMC) or variational Bayesian methods. The efficiency of the MCMC algorithm can be improved by analytically integrating out the model parameters in the posterior, $p(\overleftarrow{Z}, \overrightarrow{Z}, \Theta, \tau | Y, \alpha)$ and to yield the marginal posterior $p(\overleftarrow{Z}, \overrightarrow{Z} | Y, \alpha)$. A collapsed Gibbs sampler (Liu, 1994) can then be used to sample from the posterior for block memberships; this approach is implemented in the `lda` R package (Chang, 2010). Code for fitting the MMSB using a variational algorithm is publicly available in the user guide for the Infer.Net software (Minka et al., 2010).

Airoldi et al. (2008) propose using a BIC-like criterion for choosing the number of blocks when the network is small and a cross-validated log-likelihood for choosing the number of blocks when the network is large.

10.4.2 Application

Zachary Karate Data

The MMSB model was fitted to Zachary karate data with $G = 4$ blocks to enable comparison with the stochastic block model results of Section 10.3.2. The estimated mixed membership vectors $\hat{\tau}_i$ are shown in a matrix of ternary diagrams in Figure 10.11; this plot was produced using the `compositions` (van den Boogaart, 2009) R package. The ternary diagram in panel (j, k) in Figure 10.13 shows the membership values of block j, block k and versus the other blocks.

The resulting estimated block-to-block interaction matrix and the mean τ_i value are given as

$$\bar{\tau} = \begin{pmatrix} 0.43 \\ 0.31 \\ 0.20 \\ 0.06 \end{pmatrix} \text{ and } \hat{\Theta} = \begin{pmatrix} 0.05 & 0.00 & 0.01 & 0.94 \\ 0.01 & 0.14 & 0.12 & 0.00 \\ 0.00 & 0.11 & 1.00 & 0.25 \\ 1.00 & 0.00 & 0.08 & 1.00 \end{pmatrix};$$

the mean of the τ_i values gives a measure of propensity for nodes to belong to each block.

The fitted model shows higher within-block connection probabilities and lower between-block connection probabilities than the stochastic block model (10.5). This structure is found because nodes are allowed to have mixed membership across blocks, so the probability of a pairwise interaction can be much higher than the value given by Θ if the pair of nodes being considered has a non-negligible probability of belonging to the same block in their mixed membership vector.

White et al. (2012) and White and Murphy (2016a) define the extent of membership (EoM) of a mixed membership vector τ_i as

$$EoM_i = \exp\left(-\sum_{g=1}^{G} \tau_{ig} \log \tau_{ig}\right)$$

which is the exponential of the entropy of τ_i. The EoM value takes values in the range 1 to G and it is a measure of the effective number of blocks that an observation belongs to; previously this quantity has been used to measure the effective number of species in ecological applications.

A histogram of the EoM_i values for each club member is given in Figure 10.12 and this shows that some members exhibit mixed membership across two or more blocks. Finally, we computed the mean τ_i value for nodes from each faction identity (Section 10.2.2) and by new club membership after the split in the club. The results (Table 10.4 and Table 10.5) show that the blocks correspond to the

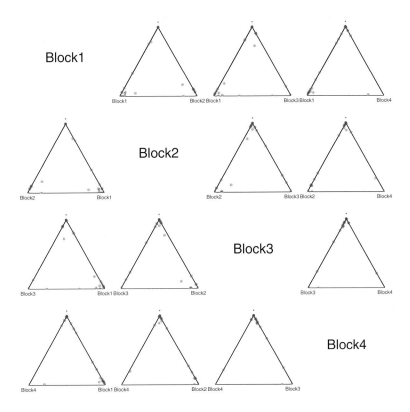

Figure 10.11 A matrix of ternary plots which show the mixed membership values for each node in the karate data. Each ternary diagram shows the mixed membership of a pair of blocks versus the remaining blocks.

split in the club and the role of members within the split and the resulting new clubs that were formed.

Table 10.4 *The mean τ_i value for each faction identity in Zachary's karate data. These values show the correspondence between the estimated blocks and the factions.*

	Block1	Block2	Block3	Block4
−2	0.00	0.47	0.51	0.02
−1	0.21	0.52	0.27	0.00
0	0.70	0.29	0.02	0.00
+1	0.70	0.23	0.04	0.02
+2	0.72	0.11	0.01	0.16

Table 10.5 *The mean τ_i value for the two new clubs formed after the break-up of the club in Zachary's karate data. These values show the correspondence between the estimated blocks and the clubs.*

	Block1	Block2	Block3	Block4
Mr. Hi's	0.12	0.48	0.39	0.01
John A.'s	0.73	0.14	0.01	0.12

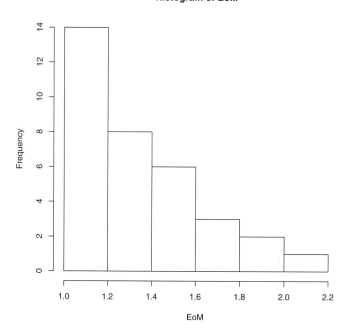

Figure 10.12 A histogram showing the Extent of Membership for each member in Zachary's karate data. The histogram shows that many members have membership in more than one block and one member has mixed membership across more than two blocks.

French Political Blogs

The MMSB model was fitted to the French political blog data with $G = 7$ blocks. The estimated mixed membership vectors $\hat{\tau}_i$ are shown in a matrix of ternary diagrams in Figure 10.13. The resulting block-to-block interaction matrix, Θ, and the mean of the τ_i values are given in Table 10.6.

The fitted model has very low block-to-block interaction probabilities for distinct blocks. Again, this can be explained as being as a result of the extra flexibility given by the mixed membership structure in the model.

A histogram of the EoM_i values for each blog is given in Figure 10.14 and this

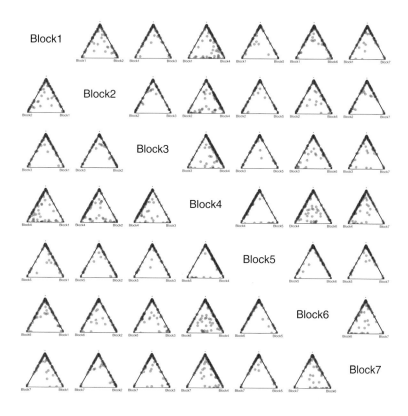

Figure 10.13 A matrix of ternary plots which show the mixed membership values for each node in the French political blog data. Each ternary diagram shows the mixed membership of a pair of blocks versus the remaining blocks.

Table 10.6 *The estimated block-to-block interaction matrix, Θ, for the MMSB model and the mean τ_i value which measures the propensity for a node (blog) to belong to each block.*

$$
\bar{\tau} = \begin{pmatrix} 0.12 \\ 0.15 \\ 0.10 \\ 0.27 \\ 0.12 \\ 0.16 \\ 0.09 \end{pmatrix} \text{ and } \hat{\Theta} = \begin{pmatrix} 1.00 & 0.00 & 0.00 & 0.00 & 0.00 & 0.00 & 0.00 \\ 0.00 & 0.50 & 0.00 & 0.00 & 0.00 & 0.00 & 0.00 \\ 0.00 & 0.00 & 0.40 & 0.00 & 0.00 & 0.01 & 0.04 \\ 0.00 & 0.00 & 0.00 & 0.00 & 0.00 & 0.00 & 0.00 \\ 0.00 & 0.00 & 0.00 & 0.00 & 0.75 & 0.00 & 0.00 \\ 0.00 & 0.00 & 0.01 & 0.00 & 0.00 & 0.08 & 0.88 \\ 0.00 & 0.00 & 0.03 & 0.00 & 0.00 & 0.86 & 0.99 \end{pmatrix}
$$

shows that many blogs exhibit mixed membership across two or more blocks in the MMSB. Finally, we computed the mean τ_i value for nodes from each political grouping outlined in Section 10.2.4. The results (Table 10.7) show that some of the blocks correspond highly with the political groupings, but other blocks correspond less so with the political groupings. For example, the liberals are strongly in Block 5, with some mixed membership with Block 4. By contrast, the Commentators and Analysts are spread across multiple blocks.

Table 10.7 *The mean τ_i value for each political grouping. These values show the correspondence between the estimated blocks and the political groupings.*

	Block1	Block2	Block3	Block4	Block5	Block6	Block7
Commentateurs Analystes	0.28	0.12	0.05	0.24	0.09	0.12	0.11
Other	0.01	0.01	0.39	0.52	0.00	0.07	0.00
Parti Radical de Gauche	0.00	0.00	0.73	0.15	0.00	0.08	0.02
PS	0.01	0.01	0.04	0.25	0.00	0.43	0.26
UDF	0.55	0.02	0.01	0.38	0.00	0.03	0.01
UMP	0.05	0.67	0.01	0.21	0.01	0.02	0.01
Liberaux	0.01	0.00	0.00	0.14	0.84	0.02	0.00

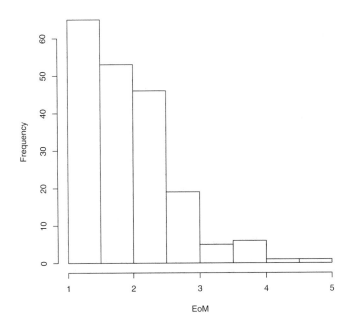

Figure 10.14 A histogram showing the Extent of Membership for each blog in the French political blog data. The histogram shows that many blogs have membership in more than one block and one blog has mixed membership across five blocks.

The code used for this analysis is given in Listing 10.2. It is worth noting that some of the results of the analysis in the `lda` function are highly unstable because

Listing 10.2: Fitting a mixed membership stochastic block model to the French political blog data using the `lda` package

```
# Load the lda package for fitting the MMSB model
library(lda)

# Load the compositions package for the ternary plots
library(compositions)

# Load the network library
library(network)

# Load the data
library(MBCbook)
data(PoliticalBlogs)
yblog<-as.matrix(PoliticalBlogs)

# Set parameters used for model fitting
set.seed <- 1
G <- 7
beta.prior <- list(1, diag(5, G) + 1)
alpha.prior <- 0.1

# Fit the MMSB model
fit <- mmsb.collapsed.gibbs.sampler(yblog, K=G, num.iterations
    =10000, alpha=alpha.prior, beta.prior=beta.prior, burnin=1000L)

# Extract the results and plot them
memberships <- with(fit, t(document_sums) / colSums(document_sums))
colnames(memberships) <- paste("Block", 1:G, sep="")
plot(acomp(memberships))

# Summarize the results
taumean <- apply(memberships,2,mean)
taumean
Theta <- fit$blocks.pos/(fit$blocks.pos+fit$blocks.neg)
Theta

# Further explore the extent of mixed membership
entropy <- function(p){
    p <- p[p>0]
    sum(-p*log(p))
}

EoM <- exp(apply(memberships, 1, entropy))
hist(EoM)
```

some summaries from the MCMC are only taken from the final iteration of the algorithm.

10.5 Latent Space Models

The latent space model (LSM) model (Hoff et al., 2002) assumes that each node has an unknown position, z_i, in a d-dimensional Euclidean latent space. Network edges are assumed to be conditionally independent given the latent positions, and the probability of an edge between nodes i and j is a decreasing function of the distance between z_i and z_j. Thus, in these models the smaller the distance between two nodes in the latent space, the greater their probability of being connected.

In particular, the latent space model assumes that

$$\log\left(\frac{\mathbb{P}\{y_{ij}=1|z_i,z_j,\alpha\}}{\mathbb{P}\{y_{ij}=0|z_i,z_j,\alpha\}}\right) = \alpha - d(z_i,z_j), \qquad (10.6)$$

where $d(z_i,z_j)$ is the Euclidean distance between z_i and z_j and thus

$$\mathbb{P}\{y_{ij}=1|z_i,z_j,\alpha\} = \frac{\exp(\alpha - d(z_i,z_j))}{1 + \exp(\alpha - d(z_i,z_j))}.$$

In the case where additional edge covariates x_{ij} are observed, the latent space models can be extended to include these by adding an extra term to (10.6) to yield

$$\log\left(\frac{\mathbb{P}\{y_{ij}=1|z_i,z_j,x_{ij},\alpha,\boldsymbol{\beta}\}}{\mathbb{P}\{y_{ij}=0|z_i,z_j,x_{ij},\alpha,\boldsymbol{\beta}\}}\right) = \alpha + \beta^T x_{ij} - d(z_i,z_j)$$

$$= \alpha + \sum_{p=1}^{P} \beta_p x_{ijp} - d(z_i,z_j), \qquad (10.7)$$

where $\beta = (\beta_1,\beta_2,\ldots,\beta_P)^T$ are model coefficients for each edge variable and $x_{ij} = (x_{ij1},x_{ij2},\ldots,x_{ijP})^T$ are the edge covariates for the pair of nodes (i,j).

Homophily by attributes is the propensity for nodes with similar attributes to be more (or less) likely to connect with each other. Node covariates can be converted to edge covariates using differences or some other function of the node covariates. Thus, the node covariates can make an edge more (or less) likely to occur in the LSM so that homophily by attributes can be accommodated.

Further, the latent space model assumes that all edges are independent conditional on the latent positions and covariates. Thus,

$$\mathbb{P}\{Y|Z,\mathbf{X},\alpha,\beta\} = \prod_{i=1}^{n}\prod_{\substack{j=1\\j\neq i}}^{n} P(y_{ij}|z_i,z_j,x_{ij},\alpha,\beta).$$

Finally, a model for the latent positions needs to be specified. The original latent space model of Hoff et al. (2002) proposed modeling the z_i values using a multivariate normal distribution. The latent position cluster model (Handcock et al., 2007) extended this model by assuming a finite mixture model for the latent positions in order to model network communities. These and further extensions of the model are described in the subsections that follow.

As an illustrative example, a fit of the LPCM model with $G = 3$ groups was fitted to Sampson's network data and a plot of the fitted model is shown in

Figure 10.15. From this, it can be seen that the model correctly identifies the three groups of monks in the network. Further, the groups are well separated as evidenced by the separate clusters of latent positions and the pie charts which show definitive group membership for each node. Thus, the LPCM clusters the nodes into highly connected clusters and provides a visualization of this clustering. The layout of the nodes in Figure 10.15 has an interpretation in terms of the model edge probabilities which is in contrast to the layout of the same data given in Figure 10.3.

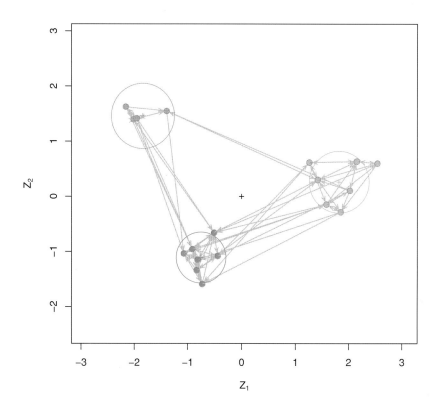

Figure 10.15 Plot of the posterior latent positions for three groups under the Latent Position Cluster Model for Sampson's network data set. The pie charts depict the posterior probability of group membership for each node. The empty circles with a plus symbol at the center show the mean of each group and the 50% contour of the density of latent positions for each group.

10.5.1 The Distance Model and the Projection Model

Hoff et al. (2002) proposed two main latent space models; the distance model and the projection model. The former is the most widely used for its simple interpretation, since it depends directly on the distance between the nodes in a latent (social) space. The most used distance is the Euclidean distance, but any dissimilarity $d_{ij} = d(z_i, z_j)$ may be used, for example Gollini and Murphy (2016) use squared Euclidean distance; we write $d(z_i, z_j) = |z_i - z_j|$ when Euclidean distance is used. This model supposes the network to be approximately symmetric because the probability that i links with j is identical to the probability that j links with i. So the distance model is particularly suitable for undirected graphs or for directed graphs that exhibit strong reciprocity. The distance model has edge probabilities of the form,

$$\log\left(\frac{\mathbb{P}\{y_{ij} = 1|z_i, z_j, x_{ij}, \alpha, \beta\}}{\mathbb{P}\{y_{ij} = 0|z_i, z_j, x_{ij}, \alpha, \beta\}}\right) = \alpha + \beta^T x_{ij} - |z_i - z_j|. \qquad (10.8)$$

The projection model is more suitable for asymmetric networks and it has edge probabilities of the form,

$$\log\left(\frac{\mathbb{P}\{y_{ij} = 1|z_i, z_j, x_{ij}, \alpha, \beta\}}{\mathbb{P}\{y_{ij} = 0|z_i, z_j, x_{ij}, \alpha, \beta\}}\right) = \alpha + \beta^T x_{ij} - |z_i|\cos(\theta_{ij})$$

$$= \alpha + \beta^T x_{ij} - \frac{z_i^T z_j}{|z_j|},$$

where θ_{ij} is the angle between z_i and z_j. Thus, the probability of observing an edge between two nodes i and j depends on the angle between z_i and z_j and the length of z_i in the latent space. If the angle between z_i and z_j is small the probability of having an edge is large, and if the angle is large then the probability of having an edge is small.

10.5.2 The Latent Position Cluster Model

Handcock et al. (2007) proposed the Latent Position Cluster Model (LPCM), a new model which extends the latent space distance models to allow for model-based clustering of the nodes. A Gaussian mixture model with the VII covariance structure is assumed for the latent positions, that is,

$$z_i \sim \sum_{g=1}^{G} \tau_g \text{MVN}_d(\mu_g, \sigma_g^2 \mathbf{I}),$$

where τ_g is the probability that a node belongs to the gth group; this structure allows for clusters of highly connected nodes. The model for the latent positions can be expressed in terms of latent cluster memberships (l_i) and latent positions (z_i) as

$$l_i \sim \text{Multinomial}(\tau_g),$$
$$z_i|(l_{ig} = 1) \sim N(\mu_g, \sigma_g^2 \mathbf{I})$$

and this representation is used in the inference for this model (Section 10.5.5).

A graphical model representation of the latent position cluster model is given in Figure 10.16.

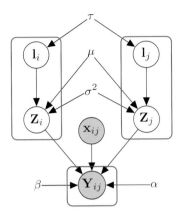

Figure 10.16 Graphical model representation of the latent position cluster model (LPCM).

10.5.3 The Sender and Receiver Random Effects

Krivitsky et al. (2009) proposed an extension of the LPCM to account for *degree heterogeneity*, which is the tendency of some nodes to send and/or receive edges more than others. This method allows the modeling of asymmetric networks within the distance model. In undirected graphs there is only one parameter, δ_i, called the sociality factor. This denotes the propensity of each mode i to form edges with other nodes.

$$\log\left(\frac{\mathbb{P}\{y_{ij}=1|z_i,z_j,\delta_i,\delta_j,x_{ij},\alpha,\beta\}}{\mathbb{P}\{y_{ij}=0|z_i,z_j,\delta_i,\delta_j,x_{ij},\alpha,\beta\}}\right) = \alpha + \beta^T x_{ij} - |z_i - z_j| + \delta_i + \delta_j.$$

In directed graphs the sociality effect in the dyad y_{ij} depends on two parameters, a sender random effect δ_i and a receiver random effect γ_j. Thus,

$$\log\left(\frac{\mathbb{P}\{y_{ij}=1|z_i,z_j,\delta_i,\gamma_j,x_{ij},\alpha,\beta\}}{\mathbb{P}\{y_{ij}=0|z_i,z_j,\delta_i,\gamma_j,x_{ij},\alpha,\beta\}}\right) = \alpha + \beta^T x_{ij} - |z_i - z_j| + \delta_i + \gamma_j,$$

where $\delta_i \sim \mathcal{N}(0,\sigma_\delta^2)$ and $\gamma_i \sim \mathcal{N}(0,\sigma_\gamma^2)$, and σ_δ^2 and σ_γ^2 measure the variance in the propensity to send and receive edges.

A graphical model representation of the latent position cluster model with sender and receiver effects is given in Figure 10.17. In the case of undirected network data, the graphical model is similar but there are no γ_i or γ_j nodes.

10.5.4 The Mixture of Experts Latent Position Cluster Model

Gormley and Murphy (2010a) proposed the mixture of experts latent position

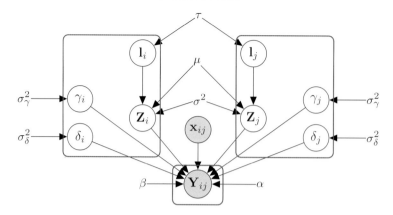

Figure 10.17 Graphical model representation of the latent position cluster model with sender/receiver effects.

cluster model (MoE-LPCM) as an extension of the latent position cluster model using the mixture of experts framework (Chapter 11). This model assumes that the mixing proportions $(\tau_1, \ldots, \tau_G)$ are node specific and can be modeled using a multinomial logistic function of the node covariates $\mathbf{x}_i = (x_{i1}, \ldots, x_{iQ})^T$. The distribution of the latent position z_i conditional on $\mathbf{x}_i$ is assumed to be,

$$z_i \sim \sum_{g=1}^{G} \tau_g(\mathbf{x}_i) \mathrm{MVN}_d(\mu_g, \sigma_g^2 \mathbf{I})$$

where

$$\tau_g(\mathbf{x}_i) = \frac{\exp(\delta_{g0} + \delta_{g1} x_{i1} + \cdots + \delta_{gQ} x_{iQ})}{\sum_{g'=1}^{G} \exp(\delta_{g'0} + \delta_{g'1} x_{i1} + \cdots + \delta_{g'Q} x_{iQ})} = \frac{\exp(\delta_g^T \mathbf{x}_i)}{\sum_{g'=1}^{G} \exp(\delta_{g'}^T \mathbf{x}_i)},$$

and $(\delta_{10}, \ldots, \delta_{1Q}) = (0, 0, \ldots, 0)$.

A graphical model representation of the mixture of experts latent position cluster model is given in Figure 10.18.

10.5.5 Inference

The main approaches for fitting latent space models are maximum likelihood estimation and a Bayesian approach using MCMC or a variational approximation.

The log-likelihood function is a convex function of the distances between latent positions of the nodes but it is not generally a convex function of latent positions. Thus, maximum likelihood estimation for latent space models is computationally challenging. A two-step estimation approach for the latent position cluster model has been proposed by Handcock et al. (2007) where in the first step the maximum likelihood estimates of the latent space model are computed considering no clusters, and in the second step the maximum likelihood estimates of the mixture model are found, conditioned on the estimated latent positions from the first step.

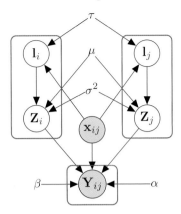

Figure 10.18 Graphical model of mixture of experts latent position cluster model (MoE-LPCM).

Handcock et al. (2007) also developed a Bayesian approach that allows the estimation of all the parameters and the latent position simultaneously using MCMC sampling. This approach usually gives better results than the two-stage MLE, but it is more computationally intensive. Salter-Townshend and Murphy (2009, 2013) developed a variational Bayesian inference routine to approximate the posterior distribution of the parameters in the LPCM. Raftery et al. (2012) propose a likelihood approximation using case-control sampling which greatly improves the efficiency of fitting the LPCM model.

The `latentnet` (Krivitsky and Handcock, 2010, 2008) R package provides two-stage maximum likelihood estimation, minimum Kullback–Leibler and MCMC-based inference estimation for the LPCM for both Euclidean and bilinear latent spaces. `VBLPCM` (Salter-Townshend, 2012) is an R package that performs the variational Bayesian inference of the LPCM for Euclidean latent spaces. `lvm4net` (Gollini, 2015) R package has been developed for fitting latent space models using squared Euclidean distance and where the latent positions are modeled using a single multivariate normal distribution. Also, the `MEclustnet` (Gormley and Murphy, 2018) R package fits the mixture of experts latent position cluster model using an MCMC algorithm.

10.5.6 Application

Zachary's Karate Data Set

The latent position cluster model was fitted to Zachary's karate data with $G = 4$ clusters. A plot of the model fit is given in Figure 10.19 where the posterior mean latent position for each node is shown, the posterior cluster membership is shown using a pie chart and the cluster means and 50% contours for each cluster are also shown. Many of the nodes have high posterior probability of belonging to two components, so it suggests that a model with a lower value of G could be sufficient for these data.

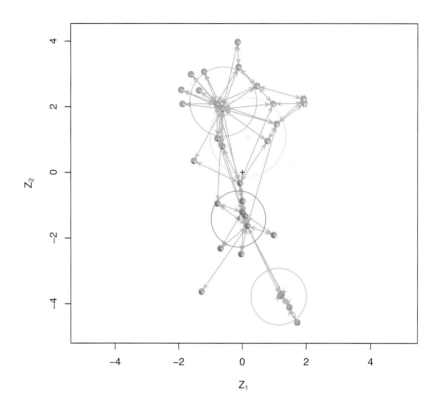

Figure 10.19 The estimated latent positions for each node in Zachary's karate club data for the LPCM model with $G = 4$. Each node is represented by a pie chart which represents the posterior probability of the node belonging to each cluster.

A cross tabulation of the cluster membership versus faction identity and club membership is given in Table 10.8 and it shows that the model has successfully found the division in the karate club, with two components being used to model each side of the divide.

The latent position cluster model was fitted to Zachary's karate data with $G = 2$ clusters but with each actor having a sender and receiver effect. A plot of the model fit is given in Figure 10.20 where the posterior mean latent position for each node is shown, the posterior cluster membership is shown using a pie chart, the size of the plotting symbol shows the magnitude of the receiver effect and the cluster means and 50% contours for each cluster are also shown. The two key actors in the network (Mr. Hi and John A.) have the largest receiver effects.

Table 10.8 *A cross tabulation of the cluster membership versus faction identity and new club membership for the LPCM model with G = 4.*

	−2	−1	0	+1	+2			Mr. Hi's	John A.'s
1	2	0	2	2	10		1	3	13
2	5	1	1	0	0		2	7	0
3	3	4	0	0	0		3	7	0
4	0	0	0	2	2		4	0	4

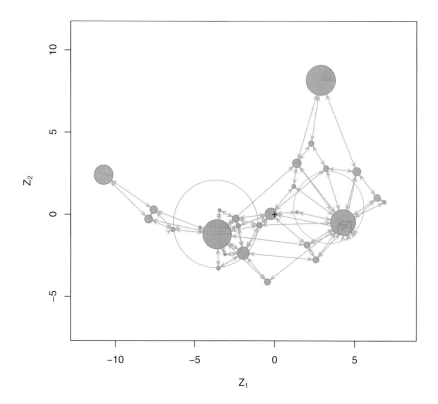

Figure 10.20 The estimated latent positions for each node in Zachary's karate club data for the LPCM model with G = 2 and where each actor has a sender and receiver effect. Mr. Hi's faction is represented in yellow and John A.'s faction is represented in blue.

A cross tabulation of the cluster membership versus faction identity and new club membership (Table 10.9) shows that the model clusters the nodes into the correct groups and further it identifies the most prominent actors in the network.

Table 10.9 *A cross tabulation of the cluster membership versus faction identity and new club membership for the LPCM model with sender and receiver effects and with G = 2.*

	−2	−1	0	+1	+2			Mr. Hi's	John A's
1	10	5	2	2	2		1	17	4
2	0	0	1	2	10		2	0	13

Lazega Lawyers

The latent position cluster model was fitted to the Lazega lawyer friendship data with $G = 2$ clusters and a receiver effect (Listing 10.4). A plot of the estimated latent positions when the model is fitted using the variational Bayesian approach as implemented in the VBLPCM package (Salter-Townshend, 2012) is given in Figure 10.21. For contrast, the estimated latent positions when the model is fitted using the minimum Kullback–Leibler approach as implemented in the latentnet package (Krivitsky and Handcock, 2010) are shown in Figure 10.22.

It can be seen that the two approaches for inference yield similar estimates of the latent positions, but the variational approach gives overly confident estimates of the clustering — very few observations have appreciable uncertainty in their cluster membership in the variational inference model fit.

A cross tabulation of the cluster membership versus seniority is given in Table 10.10 and it shows that the model has found seniority is an important factor in cluster membership.

Table 10.10 *A cross tabulation of lawyer seniority versus cluster membership in the Lazega lawyer friendship data. The left-hand table gives the cross tabulation for the VBLPCM fit and the right-hand table gives the cross tabulation for the latentnet fit.*

	Cluster 1	Cluster 2			Cluster 1	Cluster 2
Associate	36	0		Associate	15	21
Partner	13	20		Partner	25	8

The results show that the latent position cluster model has strong performance at finding clusters of highly connected nodes within the data.

10.6 Stochastic Topic Block Model

The significant and recent increase of interactions between individuals via social media or through electronic communications enables to frequently observe networks with textual edges. It is obviously of strong interest to be able to model and cluster the vertices of those networks using information on both the network structure and the text contents. Techniques able to provide such a clustering would allow a deeper understanding of the studied networks. As a motivating example, Figure 10.23 shows a network made of three "communities" of vertices

Listing 10.3: Fitting a latent position cluster model to the Zachary karate data using the `latentnet` package

```
# Load the relevant libraries
library(latentnet); library(ergm.count)

# Set a color palette for plots
colPalette <- c("#999999","#E69F00","#56B4E9","#009E73","#F0E442","
    #0072B2","#D55E00","#CC79A7")

# Load the karate data
data(zach)

# Fit the model to the data. In this case a 2D latent space and G=4
set.seed(5)
fit <- ergmm(zach~euclidean(d=2, G=4))

# Plot the fitted model
plot(fit, pie=TRUE, cluster.col=colPalette[-1], edge.col="gray")

# Compare the estimated group with faction.id
lab <- get.vertex.attribute(zach, "faction.id")
table(fit$mkl$Z.K, lab)

# Compare the estimated group with club
lab <- get.vertex.attribute(zach, "club")
table(fit$mkl$Z.K, lab)

# Fit a model with sender/receiver effects, 2D latent space and G=2
set.seed(5)
fit <- ergmm(as.network(ykarate)~euclidean(d=2, G=2)+rsender(var=1,
    var.df=3)+rreceiver(var=1, var.df=3))

# Plot the fitted model
plot(fit, pie=TRUE, rand.eff="receiver", cluster.col=colPalette[-1
    ], edge.col="gray")

# Compare the estimated group with faction.id
lab <- get.vertex.attribute(zach, "faction.id")
table(fit$mkl$Z.K, lab)

# Compare the estimated group with club
lab <- get.vertex.attribute(zach, "club")
table(fit$mkl$Z.K, lab)
```

where one of the communities can in fact be split into two separate groups based on the topics of communication between nodes of these groups (see legend of Figure 10.23 for details). Despite the important efforts in both network analysis and text analytics, only a few works have focused on the joint modeling of network vertices and textual edges. The most advanced statistical model able to deal with such data was recently proposed by Bouveyron et al. (2018b) and is called the stochastic topic block model (STBM). In a few words, the STBM model generalizes two popular models for networks and texts respectively, SBM and

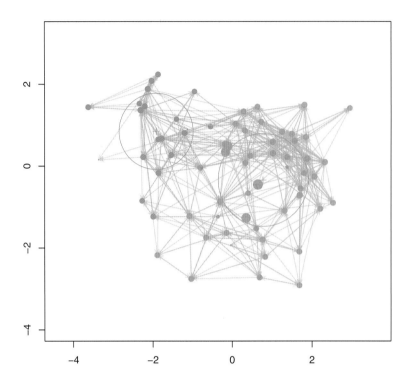

Figure 10.21 Variational Bayesian estimated latent positions for the
Latent Position Cluster Model with $G = 2$ and a receiver effect. The plot
symbol size is proportional to the receiver effect and the symbol is a pie
chart showing the probability of cluster membership.

LDA (latent Dirichlet allocation, Blei et al. (2003), see Chapter 12.2.2), and thus
allows to deal with both data types at once.

10.6.1 Context and Notation

Let us consider a directed network with n vertices, described by its $n \times n$ adjacency
matrix A. If an edge from i to j is present, i.e. $Y_{ij} = 1$, then it is characterized
by a set of D_{ij} documents, denoted $W_{ij} = (W_{ij}^d)_d$. Each document W_{ij}^d is made
of a collection of N_{ij}^d words $W_{ij}^d = (W_{ij}^{dn})_n$. In the directed network scenario, W_{ij}
can model for instance a set of e-mails or text messages sent from actor i to actor
j. In the case of undirected networks, $Y_{ij} = Y_{ji}$ and $W_{ij}^d = W_{ji}^d$ for all i and j.
The set W_{ij}^d of documents can thus model for example books or scientific papers
written by both i and j. In the following, we denote $W = (W_{ij})_{ij}$ the set of all

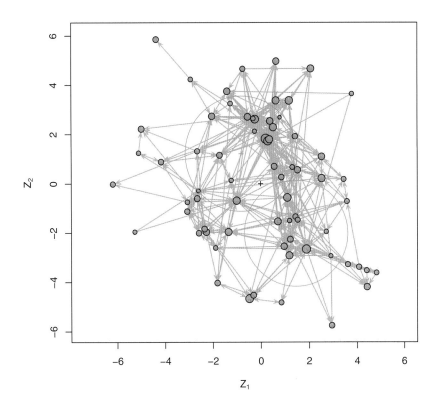

Figure 10.22 Minimum Kullback–Leibler estimated latent positions for the Latent Position Cluster Model with $G = 2$ and a receiver effect. The plot symbol size is proportional to the receiver effect and the symbol is a pie chart showing the probability of cluster membership.

documents exchanged, for all the edges present in the network. The STBM model relies on two concepts at the core of the SBM and LDA models respectively. On the one hand, a generalization of the SBM model would assume that any kind of relationship between two vertices can be explained by their latent clusters only. In the LDA model on the other hand, the main assumption is that words in documents are drawn from a mixture distribution over topics, each document d having its own vector of topic proportions π_d. The STBM model combines these two concepts to introduce a new generative procedure for documents in networks.

10.6.2 The STBM Model

Let us first consider the modeling of the presence of edges between pairs of vertices. The vertices are assumed to be spread into G latent clusters such that $Z_{iq} = 1$ if

Listing 10.4: Fitting a latent position cluster model with $G = 2$ and a receiver effect to the Lazega lawyers friendship data using the VBLPCM package. The results are compared to the latentnet fit. The two non-connected nodes were deleted from the analysis

```
# Load the relevant libraries
library(VBLPCM); library(network); library(MBCbook)

# Set a color palette for plots
colPalette <- c("#999999","#E69F00","#56B4E9","#009E73","#F0E442","
    #0072B2","#D55E00","#CC79A7")

# Load the data
data(Friend)
yfriend <- as.matrix(Friend)

# Identify disconnected nodes (and remove)
ind <- apply(yfriend+t(yfriend),1,sum)!=0
yfriend <- yfriend[ind,][,ind]

# Fit the model to the data. In this case a 2D latent space and G=2
set.seed(1111)
v.start<-vblpcmstart(as.network(yfriend),G=2,model=c("rreceiver"))
v.fit<-vblpcmfit(v.start)

# Plot the fitted model
plot(v.fit, R2=0.03,pie=TRUE, rand.eff="rreceiver", colours=
    colPalette[-1], main="")

# Compare the fitted groups with seniority (for connected nodes
    only)
lab <- get.vertex.attribute(Friend,"seniority")[ind]
table(lab, map(t(v.fit$V_lambda)))

# Fit the same model using latentnet
library(latentnet)
set.seed(5)
fit <- ergmm(as.network(yfriend)~euclidean(d=2, G=2)+rreceiver(var
    =1,var.df=3))

# Plot the fitted model
plot(fit, pie=TRUE, rand.eff="receiver", cluster.col=colPalette[-1
    ], edge.col="gray", main="",  print.formula=FALSE)

# Compare the fitted groups with seniority
table(lab,fit$mkl$Z.K)
```

vertex i belongs to cluster g, and 0 otherwise. The binary vector Z_i is assumed to be drawn from a multinomial distribution

$$Z_i \sim \mathcal{M}\left(1, \tau = (\tau_1, \ldots, \tau_G)\right),$$

where τ denotes the vector of class proportions. Conditionally on Z, an edge from i to j is then sampled from a Bernoulli distribution, depending on their respective

Figure 10.23 A sample network made of three "communities" where one of the communities is made of two topic-specific groups. The left panel only shows the observed (binary) edges in the network. The center panel shows the network with only the partition of edges into three topics (edge colors indicate the majority topics of texts). The right panel shows the network with the clustering of its nodes (vertex colors indicate the groups) and the majority topic of the edges. The latter visualization allows to see the topic-conditional structure of one of the three communities.

clusters

$$Y_{ij}|Z_{ig}Z_{jh} = 1 \sim \mathcal{B}(\theta_{gh}). \tag{10.9}$$

Second, concerning the construction of documents, if an edge is present from vertex i to vertex j, then a set of documents $W_{ij} = (W_{ij}^d)_d$, characterizing the oriented pair (i, j), is assumed to be given. Each pair of clusters (g, h) of vertices is now associated to a vector of topic proportions $\pi_{gh} = (\pi_{ghk})_k$ sampled independently from a Dirichlet distribution

$$\pi_{gh} \sim \mathrm{Dir}\left(\alpha = (\alpha_1, \ldots, \alpha_K)\right),$$

such that $\sum_{k=1}^K \pi_{gh} = 1, \forall (g, h)$. Note that a usual setting for α is $\alpha = 1$ in order to obtain a uniform prior distribution. The nth word W_{ij}^{dn} of documents d in W_{ij} is then associated to a latent topic vector S_{ij}^{dn} assumed to be drawn from a multinomial distribution, depending on the latent vectors Z_i and Z_j

$$S_{ij}^{dn}|\{Z_{ig}Z_{jr}Y_{ij} = 1, \pi\} \sim \mathcal{M}\left(1, \pi_{gh} = (\pi_{gh1}, \ldots, \pi_{ghK})\right). \tag{10.10}$$

Notice that Equations (10.9) and (10.10) are related: they both involve the construction of random variables depending on the cluster assignment of vertices i and j. Thus, if an edge is present ($Y_{ij} = 1$) and if i is in cluster g and j in h, then the word W_{ij}^{dn} is in topic k ($S_{ij}^{dnk} = 1$) with probability π_{ghk}.

Then, given S_{ij}^{dn}, the word W_{ij}^{dn} is assumed to be drawn from a multinomial distribution

$$W_{ij}^{dn}|S_{ij}^{dnk} = 1 \sim \mathcal{M}\left(1, \beta_k = (\beta_{k1}, \ldots, \beta_{kV})\right), \tag{10.11}$$

where V is the number of words in the vocabulary considered and $\sum_{v=1}^{V} \beta_{kv} = 1, \forall k$ as well as $\sum_{v=1}^{V} W_{ij}^{dnv} = 1, \forall (i,j,d,n)$. Therefore, if W_{ij}^{dn} is from topic k, then it is associated to word v of the vocabulary ($W_{ij}^{dnv} = 1$) with probability β_{kv}.

Note that words of different documents d and d' in W_{ij} have the same mixture distribution which only depends on the respective clusters of i and j. Because pairs (g,h) of clusters can have different vectors of topics proportions π_{gh}, the documents they are associated with can have different mixture distribution of words over topics. For instance, most words exchanged from vertices of cluster g to vertices of cluster h can be related to *mathematics* while vertices from g' can discuss with vertices of h' with words related to *cinema* and in some cases to *sport*.

10.6.3 Links with Other Models and Inference

First, it is worth noticing that STBM generalizes both SBM and LDA. Indeed, on the one hand, let us assume that the group memberships (Z) of vertices are known, then it is possible to form G^2 meta-documents which gather all documents of a group. In this context, STBM is equivalent to applying LDA on the G^2 meta-documents. On the other hand, if the number (K) of topics is limited to a unique topic, then STBM is also equivalent to SBM.

Regarding model inference, the authors proposed to use a classification version of a variational EM algorithm (Hathaway, 1986a) in order to exploit a key property of the model. Indeed, the set of latent variables Y allows the decomposition of the full joint distribution in two terms. Consequently, the optimization of the variational bound can also be done alternatively: one part is optimized over S, π, β with Z fixed and the other part is optimized over Z, τ, θ where Z is considered as a binary "parameter" to estimate.

10.6.4 Application: Enron E-mail Network

To illustrate the main features of STBM, we consider here a classical communication network, the Enron data set, which contains all e-mail communications between 149 employees of the famous company from 1999 to 2002. The original data set is available at `https://www.cs.cmu.edu/~./enron/`. Here, we focus on the period September 1 to December 31, 2001. We chose this specific time window because it corresponds to a critical period for the company. The data set considered here contains 20,940 e-mails sent between the $n = 149$ employees. All messages sent between two individuals were merged into a single meta-message. Thus, we end up with a data set of 1,234 directed edges between employees, each edge carrying the text of all messages between two persons.

The C-VEM algorithm for STBM was run on these data, using the SaaS web platform `https://linkage.fr`, for a number G of groups and a number K of topics from 2 to 20. The model with the highest value was $(G,K) = (10,5)$. Figure 10.24 shows the clustering obtained with STBM for ten groups of nodes

Final clustering

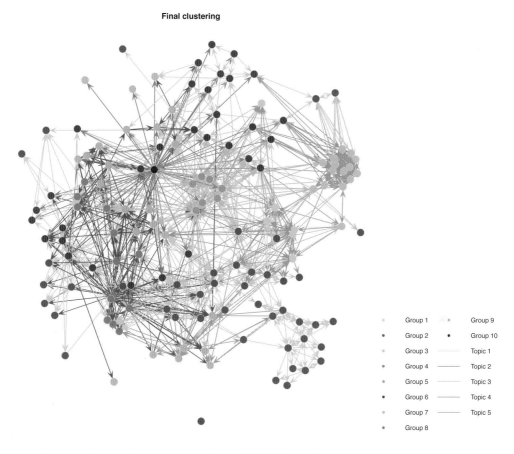

Figure 10.24 Clustering result with STBM on the Enron data set (Sept.-Dec. 2001).

and five topics. Edge colors refer to the majority topics for the communications between the individuals. The found topics can be easily interpreted by looking at the most specific words of each topic, displayed in Figure 10.25. In a few words, we can summarize the found topics as follows:

- Topic 1 seems to refer to the financial and trading activities of Enron,
- Topic 2 is concerned with Enron activities in Afghanistan (Enron and the Bush administration were suspected to work secretly with the Taliban up to a few weeks before the 9/11 attacks),
- Topic 3 contains elements related to the California electricity crisis, in which Enron was involved, and which almost caused the bankruptcy of SCE-corp (Southern California Edison Corporation) early 2001,
- Topic 4 is about usual logistic issues (building equipment, computers, ...),
- Topic 5 refers to technical discussions on gas deliveries (mmBTU is for instance a British thermal unit).

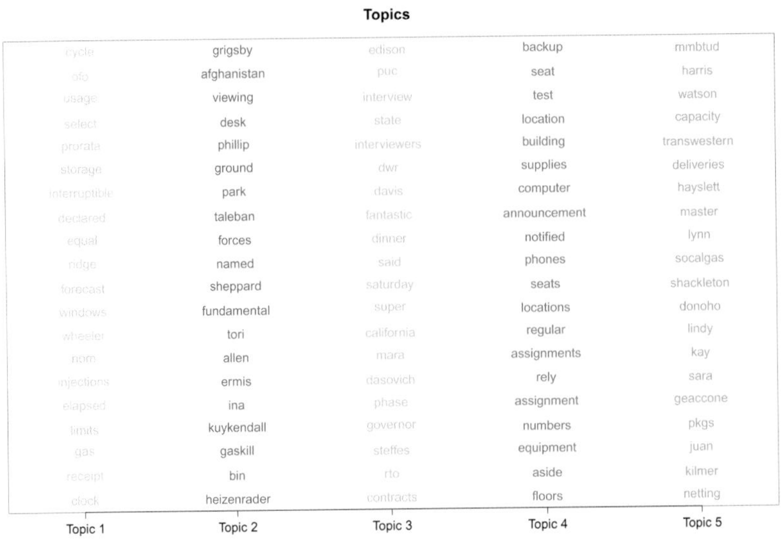

Topics

	Topic 1	Topic 2	Topic 3	Topic 4	Topic 5
	cycle	grigsby	edison	backup	mmbtud
	olb	afghanistan	puc	seat	harris
	usage	viewing	interview	test	watson
	select	desk	state	location	capacity
	prorate	phillip	interviewers	building	transwestern
	storage	ground	dwr	supplies	deliveries
	interruptible	park	davis	computer	hayslett
	declared	taleban	fantastic	announcement	master
	equal	forces	dinner	notified	lynn
	ridge	named	said	phones	socalgas
	forecast	sheppard	saturday	seats	shackleton
	windows	fundamental	super	locations	donoho
	wheeler	tori	california	regular	lindy
	nom	allen	mara	assignments	kay
	injections	ermis	dasovich	rely	sara
	elapsed	ina	phase	assignment	geaccone
	limits	kuykendall	governor	numbers	pkgs
	gas	gaskill	steffes	equipment	juan
	receipt	bin	rto	aside	kilmer
	clock	heizenrader	contracts	floors	netting

Figure 10.25 Most specific words for the five topics found with STBM on the Enron data set.

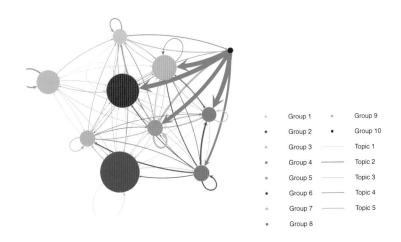

Figure 10.26 Enron data set: summary of connection probabilities between groups (θ, edge widths), group proportions (ρ, node sizes) and most probable topics for group interactions (π, edge colors).

It is worth noticing that two of the five topics recovered by STBM are directly related to the "Enron scandal." Figure 10.26 presents a visual summary of connection probabilities between groups (the estimated θ matrix) and majority topics for group interactions. A few elements deserve to be highlighted in view of this summary. First, group 10 contains a single individual who has a central place in the network and who mostly discusses logistic issues (topic 4) with groups 4, 5, 6 and 7. Second, group 8 is made of six individuals who mainly communicate about Enron activities in Afghanistan (topic 2) between them and with other groups. Finally, groups 4 and 6 seem to be more focused on trading activities (topic 1) whereas groups 1, 3 and 9 are dealing with technical issues on gas deliveries (topic 5).

10.7 Bibliographic Notes

Recent studies have considered the stochastic block model within a broader context. Bickel and Chen (2009) studied the connections between modularity and stochastic block models; Newman (2016) makes further connections between modularity and stochastic block models. Bickel et al. (2011) discusses the asymptotic properties of a class of models of which stochastic block models are a subset. An extension of the stochastic block model that includes node covariates was proposed by Mariadassou et al. (2010). The model allows for the covariates to impact the block membership probabilities and link probabilities directly; inference for this model utilized a variational approach. Further, Zanghi et al. (2010) proposed an extension of the stochastic block model that jointly modeled the network data and node covariates by assuming that the covariates were conditionally Gaussian given the latent block membership of the node; a variational EM algorithm was developed for model inference.

A number of extensions and alternatives to the MMSB model have recently been developed. In particular, Xing et al. (2010) developed a dynamic version of the mixed membership stochastic block model (dMMSB). Chang and Blei (2010) developed a model based upon the MMSB which jointly models network and nodal feature data. White and Murphy (2016b) developed a version of the MMSB that allows the mixed membership vector to depend on node covariates; the model is called the mixed membership of experts stochastic block model.

Latouche et al. (2011) and McDaid and Hurley (2010) developed a model called the overlapping stochastic block model that allows nodes to have full membership of more than one block, thus providing an alternative version of mixed membership; this model was further extended in McDaid et al. (2014).

Salter-Townshend and Murphy (2015) developed a finite mixture of exponential random graph models (ERGM) (Holland and Leinhardt, 1981) to cluster networks with similar structure. The model is fitted using a pseudo-likelihood EM algorithm. They applied the method to cluster ego-networks of nodes in a network to find nodes with similar local structure. A related model called the random subgraph model was developed by Jernite et al. (2014) for a historical application investigating social relations in Merovingian Gaul.

A number of extensions of the network models have been developed for dynamic network data, where the network is observed over time and where edges can be added and removed over the period of observation. Dynamic stochastic block models have been developed by Yang et al. (2011), Xu and Hero (2014), Matias and Miele (2017) and Zreik et al. (2017). A dynamic version of STBM has also been proposed recently by Corneli et al. (2018) for clustering dynamic networks with textual edges. The mixed membership stochastic block model has been extended for dynamic network data by Fu et al. (2009) and Xing et al. (2010). Further, the latent space model for social networks has been extended for dynamic network data by Sarkar and Moore (2005a,b), Sarkar et al. (2007), Sewell and Chen (2015) and Friel et al. (2016). Hanneke et al. (2010) developed a dynamic exponential random graph model but this doesn't account for clustering in the network.

11

Model-based Clustering with Covariates

In this chapter we consider model-based clustering methods that account for external information that is available in the presence of covariates. In this case, each observation has two parts: a response variable y_i on which the clustering is based and covariates x_i which are used to explain the clustering structure.

The mixture of experts model provides a general modeling framework that extends the mixture model to accommodate the presence of covariates.

This modeling approach is demonstrated on three data sets: CO_2 emissions and GNP of different countries (Section 11.1.1), the Australian Institute of Sport data (Section 11.1.2) and the Italian wine data (Section 11.1.3).

The mixture of experts model is described in Section 11.2, methods for fitting the mixture of experts model are outlined in Section 11.2.1, software for mixture of experts modeling is summarized in Section 11.4 and the application of the modeling framework to the three illustrative examples is shown in Section 11.5.

11.1 Examples

11.1.1 CO_2 and Gross National Product

This data set gives the CO_2 emissions and GNP for a number of countries in the year 1996 (Hurn et al., 2003). There is interest in studying the relationship between CO_2 emissions and GNP. A scatter plot of the data (Figure 11.1) shows that the data may follow two different linear regression lines, one that has a large slope value (accounting for TUR, MEX, CAN, AUS, NOR and USA) and one with a slope of almost zero (accounting for the other countries).

Thus, it appears that a suitable model for these data would be a mixture of linear regression models. This model would be able to account for the fact that the relationship between CO_2 and GNP appears to be clustered around two different regression lines; the fit of such a model is shown in Figure 11.2.

11.1.2 Australian Institute of Sport (AIS)

Various physical and hematological measurements were made on 102 male and 100 female athletes at the Australian Institute of Sport; the data are available in Cook and Weisberg (1994) and in the sn package (Azzalini, 2015). The 13 variables recorded in the study are detailed in Table 11.1.

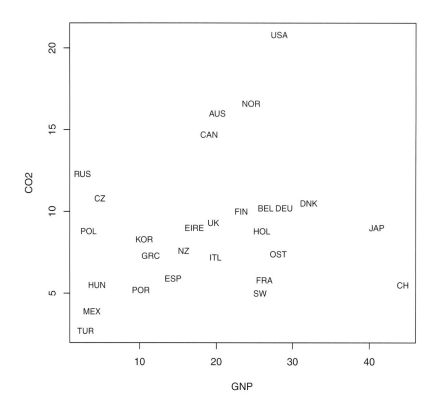

Figure 11.1 Scatter plot of CO_2 emissions and GNP for a number of countries in 1996.

A generalized pairs plot (Emerson and Green, 2014) of the variables, except sport, is given in Figure 11.3 and a generalized pairs plot of the hematological variables only is given in Figure 11.4.

A mixture of experts model can be used to investigate the clustering structure in the hematological measurements of the athletes and investigate how covariates may influence these measurements and the clusters.

11.1.3 Italian Wine

We reconsider the Italian wine data (Forina et al., 1986), which were previously analyzed in Chapters 2, 7 and 8. The wine data record 27 chemical and physical properties of three different types of Italian wine. It was shown in the previous chapters that a model-based clustering of these data captures the three different wine types with high accuracy. Further, there is some evidence that the clustering solution is also capturing the different years (vintages) of the wines. We will

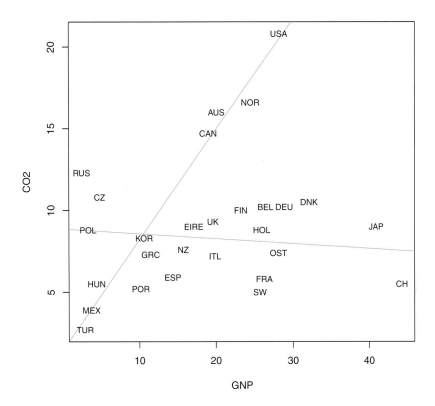

Figure 11.2 Scatter plot of CO_2 emissions and GNP for a number of countries in 1996, with the two linear regression components from a mixture of regression models shown.

reanalyze these data using the mixture of experts framework where the wine type and year are included as covariates, so that the clustering solution can be further investigated.

11.2 Mixture of Experts Model

The mixture of experts model (Jacobs et al., 1991) extends the mixture model by allowing the parameters of the model for observation i to depend on covariates x_i. Thus, the density of each observation, y_i, conditional on the covariates, x_i, is given as:

$$f(y_i|x_i) = \sum_{g=1}^{G} \tau_g(x_i) f(y_i|\theta_g(x_i)). \tag{11.1}$$

Table 11.1 *Australian Institute of Sport data variables. The first five are hematological variables and the others are physical measurements for the athlete.*

Variable	Description
RCC	red cell count
WCC	white cell count
Hc	hematocrit
Hg	hemoglobin
Fe	plasma ferritin concentration
BMI	body mass index
SSF	sum of skin folds
Bfat	body fat percentage
LBM	lean body mass
Ht	height
Wt	weight
sex	a factor with levels: female, male
sport	a factor with levels: Basketball, Field, Gym, Netball, Row, Swim, Track 400m, Tennis, Track Sprint, Water Polo

Thus, both the mixing proportions and the parameters of component densities of the mixture model can depend on the covariates x_i. Bishop (2006, Chapter 14.5) refers to the mixture of experts model as a conditional mixture model because conditional on observing covariate x_i, the distribution of the response variable, y_i, is a finite mixture model. The standard terminology used in the mixture of experts model literature refers to the component densities, $f(y_i|\theta_g(x_i))$, as "experts" and to the mixing proportions, $\tau_g(x_i)$, as "gating networks."

In the original formulation of the mixture of experts model (Jacobs et al., 1991) the mixture proportions (gating network) are modeled using a multinomial logistic model and the mixture components (expert networks) are generalized linear models (McCullagh and Nelder, 1983). Thus,

$$\tau_g(x_i) = \frac{\exp(\beta_g^T x_i)}{\sum_{h=1}^{G} \exp(\beta_h^T x_i)} \text{ and } \theta_g(x_i) = \psi(\gamma_g^T x_i) \qquad (11.2)$$

for some link function $\psi(\cdot)$.

Figure 11.5 gives a graphical model representation of the mixture of experts model. This representation aids the interpretation of the full mixture of experts model (in which all model parameters are functions of covariates (Figure 11.5(d)) and the special cases where some of the model parameters do not depend on the covariates (Figures 11.5(a)–11.5(c)). The four models detailed in Figure 11.5 have the following interpretations (cf. Gormley and Murphy, 2011):

(a) in the *mixture model*, the distribution of y_i depends on the latent cluster membership variable z_i and the distribution of z_i is independent of the covariate x_i; y_i is independent of the covariate x_i conditional on z_i;

(b) in the *expert network mixture of experts model*, the outcome variable distri-

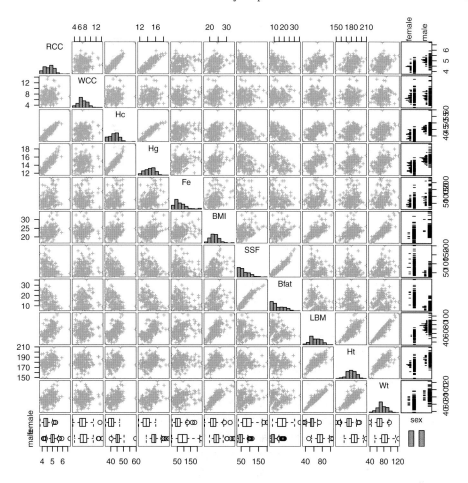

Figure 11.3 A generalized pairs plot of the Australian Institute of Sport data.

bution depends on both the covariate x_i and the latent cluster membership variable z_i; the distribution of the latent variable is independent of the covariate x_i;

(c) in the *gating network mixture of experts model*, the outcome variable distribution depends on the latent cluster membership variable z_i and the distribution of the latent variable depends on the covariate x_i; the distribution of the outcome variable is independent of x_i conditional on z_i;

(d) in the *full mixture of experts model*, the outcome variable distribution depends on both the covariate x_i and on the latent cluster membership variable z_i. Additionally, the distribution of the latent variable z_i depends on the covariate x_i.

The full mixture of experts model has been widely applied and is commonly used as a flexible methodology to estimate the conditional density of y_i given x_i

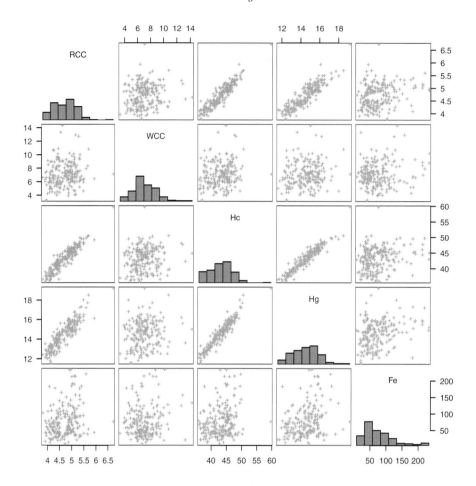

Figure 11.4 A generalized pairs plot of the hematological variables in the Australian Institute of Sport data.

(e.g. Bishop, 2006, Chapter 14.5). In such applications, there is not necessarily a direct correspondence between the mixture components and clusters. Further, the hierarchical mixture of experts model (Jordan and Jacobs, 1994) that extends the mixture of experts model outlined in (11.1) can give a more accurate density estimate with fewer components.

The gating network mixture of experts model (Figure 11.5(c)) closely matches the model-based clustering framework outlined in previous chapters. This model has the functional form,

$$f(y_i|x_i) = \sum_{g=1}^{G} \tau_g(x_i) f(y_i|\theta_g). \tag{11.3}$$

Thus, in this model, the component densities have the same form as a finite mixture model, but we allow the covariates to influence the probability of group

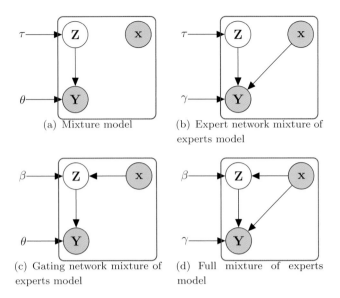

(a) Mixture model   (b) Expert network mixture of experts model

(c) Gating network mixture of experts model   (d) Full mixture of experts model

Figure 11.5 The graphical model representation of the mixture of experts model. The differences between the four special cases are due to the presence or absence of edges between the covariates x and the latent variable z and response variable y.

membership for each observation. This model has an appealing interpretation in terms of clustering because the covariates influence the probability of group membership and thereafter the correspondence between component densities and clusters has the same interpretation as in other model-based clustering methods. In the other formulations of the mixture of experts model, where the covariates enter the component densities, we can find that observations with very different response values can be clustered together because they are being modeled using the same generalized linear model.

11.2.1 Inference

The primary tools for inference for mixture of experts models are maximum likelihood using the EM algorithm (Dempster et al., 1977), Bayesian inference using Markov Chain Monte Carlo (MCMC) (e.g. Diebolt and Robert, 1994; Frühwirth-Schnatter, 2006) or Bayesian inference using variational methods (e.g. Smídl and Quinn, 2006, Chapter 3.3).

Following Section 2.3, we will focus on maximum likelihood using the EM algorithm, where the likelihood is of the form

$$L(\boldsymbol{\beta}, \boldsymbol{\gamma}) = \prod_{i=1}^{N} \sum_{g=1}^{G} \tau_g(x_i) f(y_i | \theta_g(x_i)), \qquad (11.4)$$

where $\tau(x_i)$ and $\theta_g(x_i)$ are defined using (11.2).

We introduce the missing group labels z_i, where $z_i = (z_{i1}, z_{i2}, \ldots, z_{iG})$ and $z_{ig} = 1$ if observation i comes from cluster g and $z_{ig} = 0$ otherwise. Thus, the conditional distribution of (y_i, z_i) given x_i is of the form

$$f(y_i, z_i | x_i) = \prod_{g=1}^{G} [\tau_g(x_i) f(y_i | \theta_g(x_i))]^{z_{ig}} .$$

Thus, the complete-data likelihood is of the form

$$L_c(\boldsymbol{\beta}, \boldsymbol{\gamma}) = \prod_{i=1}^{N} \prod_{g=1}^{G} [\tau_g(x_i) f(y_i | \theta_g(x_i))]^{z_{ig}} \qquad (11.5)$$

and the complete-data log-likelihood has the form

$$\ell_c(\boldsymbol{\beta}, \boldsymbol{\gamma}) = \sum_{i=1}^{N} \sum_{g=1}^{G} z_{ig} \left[\log \tau_g(x_i) + \log f(y_i | \theta_g(x_i)) \right] \qquad (11.6)$$

$$= \underbrace{\sum_{i=1}^{N} \sum_{g=1}^{G} z_{ig} \log \tau_g(x_i)}_{\text{Gating network contribution}} + \underbrace{\sum_{i=1}^{N} \sum_{g=1}^{G} z_{ig} \log f(y_i | \theta_g(x_i))}_{\text{Expert network contribution}} .$$

Therefore, the complete-data log-likelihood decomposes into two separate parts, where one part comes from the gating network and the other part comes from the expert network.

The EM algorithm (Dempster et al., 1977) for the mixture of experts model follows in a similar manner to that for mixture models (as outlined in Section 2.3).

The E-step of the EM algorithm involves computing

$$\hat{z}_{ig}^{(t+1)} = \frac{\hat{\tau}_g^{(t)}(x_i) f(y_i | \hat{\theta}_g^{(t)}(x_i))}{\sum_{h=1}^{G} \hat{\tau}_h^{(t)}(x_i) f(y_i | \hat{\theta}_h^{(t)}(x_i))},$$

where $\hat{\tau}_g^{(t)}(x_i)$ and $\hat{\theta}_g^{(t)}(x_i)$ are the gating and expert network models calculated using the estimates of the parameters $(\hat{\boldsymbol{\beta}}^{(t)}, \hat{\boldsymbol{\gamma}}^{(t)})$ from the tth iteration of the algorithm.

For the M-step of the EM algorithm, the decomposition of the expected complete-data log-likelihood (11.7) into two parts means that these two terms can be maximized separately.

$$\mathbb{E}\ell_c(\boldsymbol{\beta}, \boldsymbol{\gamma}) = \sum_{i=1}^{N} \sum_{g=1}^{G} \hat{z}_{ig}^{(t+1)} \log \tau_g(x_i) + \sum_{i=1}^{N} \sum_{g=1}^{G} \hat{z}_{ig}^{(t+1)} \log f(y_i | \theta_g(x_i)). \qquad (11.7)$$

The first term is of the same functional form as a multinomial logistic regression model and thus methods for fitting such models can be used to maximize this term and find estimates of the parameters in the mixing proportions (gating networks). The second term is of the same form as fitting G separate generalized linear models, and thus the methods for fitting such models can be used to estimate the component density (expert network) parameters. In the special case of the gating network mixture of experts model the second term is identical to the term used to find the component density estimates in a standard mixture model. Thus, the model fitting for mixture of experts models is straightforward, in principle.

11.3 Model Assessment

Model comparison for the mixture of experts models can be implemented in a similar manner to the other model-based clustering methods outlined in previous chapters. The Bayesian Information Criterion (BIC) and Integrated Completed Likelihood (ICL) have been shown to give suitable model selection criteria, both for the number of component densities (and thus clusters) required and for selecting variables to include in the model. Gormley and Murphy (2010a) demonstrate the use of BIC for selecting the appropriate model structure (as described in Section 11.2), the inclusion of covariates and the choice of the number of components. Likewise, the component merging methods outlined in Chapter 3 can be applied to study the correspondence between component densities and clusters; this has been demonstrated in Gormley and Murphy (2011) in the context of clustering voting data using a mixture of experts model.

11.4 Software

A number of tools for fitting mixture of experts models are available in the R programming environment (R Development Core Team, 2010). These include `flexmix` (Leisch, 2004; Grün and Leisch, 2008), `mixtools` (Benaglia et al., 2009) and others.

11.4.1 flexmix

The `flexmix` package (Leisch, 2004; Grün and Leisch, 2008) can accommodate the full mixture of experts model as outlined in Section 11.2 in the case where y_i is a univariate outcome. The package has a similar interface to the generalized linear model (glm) functions within R. The user can specify the form of the general linear model and covariates (if any) are to be used in the component density (expert network) and mixture components (gating network). In the case of multivariate outcome continuous variables, the package has functionality for multivariate Gaussian component distributions (diagonal or non-diagonal). Further, `flexmix` allows to use the local independence assumption for outcome variables (see Chapter 6) for clustering more general multivariate outcomes, including continuous, count, discrete or mixed outcomes. `flexmix` has a modular structure, so a user can extend the range of models to which it can be applied to facilitate more general univariate or multivariate outcomes.

11.4.2 mixtools

The `mixtools` package (Benaglia et al., 2009) can also accommodate the full mixture of experts model as outlined in Section 11.2. The package allows for non parametric estimation of the functional form for the mixing components (gating networks) and for the component densities (expert networks), so it offers some flexibility beyond `flexmix`. However, the multivariate models in `mixtools` use the local independence assumption, so it does not directly offer the facility to model multivariate Gaussian component densities with non-diagonal covariance.

11.4.3 MoEClust

The `MoEClust` package (Murphy and Murphy, 2018a,b) can be used to fit the full mixture of experts model when the expert network model is a univariate or multivariate Gaussian distribution. The software also facilitates using the `mclust` parsimonious covariance structures in the expert network models.

11.4.4 Other

The `poLCA` package (Linzer and Lewis, 2011) can be used to fit a mixture of experts version of the latent class analysis model (Chapter 6) where the mixing coefficients depend on the covariates; that is a gating network mixture of experts model (Figure 11.5(b)). The `mixreg` package (Turner, 2014) can be used to fit a mixture of regression models using the EM algorithm; the `fpc` package (Hennig, 2015a) can also be used to fit mixtures of regression models.

11.5 Results

11.5.1 CO$_2$ and GNP Data

The mixture of experts model was fitted to the CO$_2$ data using flexmix (Leisch, 2004; Grün and Leisch, 2007, 2008) where the component density (expert network) is a Gaussian distribution. Models with different numbers of clusters (expert networks) were considered and the inclusion/exclusion of GNP as a covariate was also considered. An optimal model was found by maximizing BIC. The final model selected had $G = 2$ and GNP entered the model in the component densities; the same model maximized the ICL criterion.

A table of the BIC and ICL values of a number of models are given in Table 11.2, the parameters of the final model are detailed in Table 11.3 and the fitted model was shown previously in Figure 11.2; the code for a single model fit with 50 random starting values is given in Listing 11.1.

The final model which is a mixture of regression models (Quandt and Ramsey, 1978) and this model can also be fitted using the `mixtools` package (Benaglia et al., 2009) as is shown in Listing 11.2. The fitted model is almost identical to that given in Table 11.3 and a plot of the fitted model is given in Figure 11.6.

Table 11.2 *The model BIC and ICL values for the CO_2 data.*

Groups (G)	Expert Covariates	Gating Covariates	Log-Likelihood	BIC	ICL
1			−78.63	−163.92	−163.92
1	GNP		−77.98	−165.96	−165.96
1		GNP	−78.63	−163.92	−163.92
1	GNP	GNP	−77.98	−165.96	−165.96
2			−74.92	−166.50	−167.48
2	GNP		−66.98	**−157.29**	**−160.19**
2		GNP	−74.70	−169.40	−170.15
2	GNP	GNP	−66.34	−159.34	−161.67
3			−74.91	−176.50	−201.46
3	GNP		−63.93	−164.53	−179.36
3		GNP	−70.37	−174.06	−174.41
3	GNP	GNP	−65.12	−173.56	−176.35

Table 11.3 *Estimated parameter of the fitted mixture of experts model to the CO_2 data.*

Parameter	Component 1	Component 2
Proportion	0.756	0.244
Intercept	8.65	1.40
GNP	−0.02	0.68
sigma	2.14	0.85

11.5.2 Australian Institute of Sport

The mixture of experts model was fitted to the AIS data using `flexmix`. The component densities were assumed to be Gaussian and two potential covariance structures were considered, which are equivalent to the VVI and the VVV models in `mclust`. Only the model with the VVI structure allows for covariates in the component densities, but the VVV models achieved the highest BIC amongst all of the models considered. Thus, further analysis was restricted to the models with VVV covariance structure and with the mixing proportions potentially depending on the covariates. A greedy forwards search algorithm was used to explore the model space, starting with modeling the data as a single Gaussian (i.e. $G = 1$) and with no covariates in the model. At each stage of the search, the possibility of increasing G or adding an extra covariate to the mixing proportion (gating network) part of the model was considered. The new model with the greatest positive increase in BIC was used for the next step of the algorithm. If no new model yielded a positive increase in BIC, then the algorithm terminated. Example code of how a single mixture of experts model of this type is fitted is given in Listing 11.3.

The best fitting model found by the greedy forwards search algorithm was a model with $G = 2$ and where sex and BMI are covariates in the mixing proportions

Listing 11.1: Fitting a mixture of experts model to the CO_2 data using the `flexmix` package. The model in the code is a mixture of experts model with $G = 2$, where GNP is a covariate in the expert networks and in the gating network

```
# Load the flexmix package
library(flexmix)

# Load the CO_2 data
data(CO2data, package="mixtools")

# Fit a mixture of experts model with 50 random starting values
# The highest BIC value is stored as bicval
# The best fitting model is stored as bestfit
bicval <- Inf
itermax <- 50
for (iter in 1:itermax)
{
  fit <- flexmix(CO2~GNP, data=CO2data, k=2,
    concomitant = FLXPmultinom(~GNP))

  if (bicval>BIC(fit))
  {
    bicval <- BIC(fit)
    bestfit <- fit
    print(c(iter, bicval))
  }
}

# Explore the fitted model
summary(bestfit)
parameters(bestfit)

# Explore the gating network part of the model
summary(bestfit, which="concomitant")
parameters(bestfit, which="concomitant")
```

(gating network); this model also yielded the highest ICL value. The models found at each iteration of the greedy model search are given in Table 11.4 and the model parameters are given in Table 11.5.

Table 11.4 *The models chosen in the greedy forward model search procedure for the Australian Institute of Sport data.*

Groups (G)	Expert Covariates	Gating Covariates	Log-Likelihood	BIC	ICL
1			−2048.320	−4202.806	−4225.664
2			−2265.879	−4188.115	−4611.382
2		sex	−1938.726	−4100.399	−4103.539
2		sex, BMI	−1931.634	**−4091.524**	**−4091.581**
2		sex, BMI, SSF	−1931.481	−4096.526	−4096.823

Listing 11.2: Fitting a mixture of experts model to the CO_2 data using the `mixtools` package

```
# Set seed of random number generator
set.seed(1)

# Load the mixtools package
library(mixtools)

# Load the CO_2 data
data(CO2data)

# Fit a mixture of experts model with 50 random starting values
# The highest BIC value is stored as bicval
# The best fitting model is stored as bestfit
bicval <- -Inf
itermax <- 50
for (iter in 1:itermax)
{
  G <- 2
  fit <- regmixEM(CO2data$CO2,CO2data$GNP,k=G)
  n <- nrow(CO2data)
  p <- nrow(fit$beta)*G+G+(G-1)
  fitbic <- 2*fit$loglik - log(n)*p
   if (bicval<fitbic)
   {

     bicval <- fitbic
     bestfit <- fit
     print(c(iter,bicval))
   }
}

# Explore the fitted model
summary(bestfit)
plot(bestfit,which=2)
```

The data and clustering results are shown using scatter plot matrices of the hematological variables and the two covariates, using coloring for cluster membership in Figure 11.7.

11.5.3 Italian Wine

Mixture of experts models were fitted to the Italian wine data using the `flexmix` package where a diagonal (VVI) covariance structure was assumed. In this case, we restrict attention only to models where the covariates enter the mixing proportions (gating network); that is, the model as shown in Figure 11.5(b). The chosen model allows for studying the clustering structure in the outcome variables in a manner similar to the approach described in Chapter 2, but where the relationship between the clusters and the covariates is explored using the mixture of experts framework.

Most Probable Component Membership

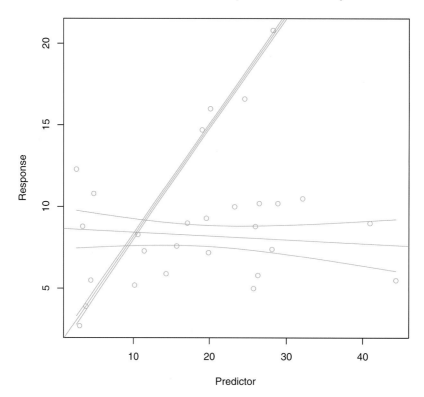

Figure 11.6 A plot of the mixture of regressions model with $G = 2$ fitted to the CO_2 data using `mixtools`. Confidence intervals for the fitted lines are also shown on the plot.

The restriction of the models considered assumes that the outcome variable and the covariates are conditionally independent, given the cluster membership values.

The models were fitted with $G = 1, 2, \ldots, 5$ and for the four models where the type and year covariates were included/excluded. The code used for model fitting is given in Listing 11.4.

The model with the highest BIC and ICL values is the $G = 3$ model with both covariates included (Table 11.6). This indicates that both year and wine type are important determinants of the clustering of the Italian wine data. However, a cross tabulation of the wine type with year (Table 11.7) shows that there is considerable dependence between these two factors.

An examination of the clustering results (Table 11.8) shows that the resulting clustering captures the wine types very accurately. This shows that the model captures the data structure accurately and that the groups are essentially determined by the wine types. It would not be a fair assessment of the classification

Table 11.5 *Estimated parameters of the fitted mixture of experts model to the AIS data. (a) Gives the component mean values and (b) gives the coefficients of the multinomial regression for the component densities.*

(a)	Component 1	Component 2
RCC	5.01	4.39
WCC	7.22	6.98
Hc	45.47	40.41
Hg	15.52	13.49
Fe	95.45	55.95

(b)	Mixing Proportion 1	Mixing Proportion 2
(Intercept)	0.00	145.69
sexmale	0.00	−45.59
BMI	0.00	−5.49

Table 11.6 *The BIC and ICL values for the mixture of experts models with Gaussian component densities and VVI covariance structure, when these models are fitted to the Italian wine data. For each value of G, four different models were considered by including/excluding each covariates. The model with the highest BIC is marked in bold face.*

Groups (G)	Covariates	Log Likelihood	BIC	ICL
1	None	−12730.95	−25741.71	−25741.71
	Type	−12730.95	−25741.71	−25741.71
	Year	−12730.95	−25741.71	−25741.71
	Type, Year	−12730.95	−25741.71	−25741.71
2	None	−12024.37	−24613.55	−24614.65
	Type	−11938.61	−24452.40	−24452.95
	Year	−11975.38	−24520.75	−24520.86
	Type, Year	−11935.57	−24451.50	−24451.87
3	None	−11552.36	−23954.52	−23955.98
	Type	−11374.42	**−23619.38**	**−23619.41**
	Year	−11490.96	−23842.09	−23843.86
	Type, Year	−11372.08	−23625.07	−23625.11
4	None	−11397.45	−23929.72	−23930.49
	Type	−11230.40	−23626.70	−23627.16
	Year	−11362.66	−23875.68	−23878.62
	Type, Year	−11228.46	−23638.36	−23639.33
5	None	−11279.52	−23978.86	−23979.53
	Type	−11147.95	−23757.16	−23760.50
	Year	−11208.73	−23858.00	−23862.04
	Type, Year	−11175.57	−23833.13	−23839.63

Listing 11.3: Fitting a mixture of experts model using the `flexmix` package. The model in the code is a mixture of experts model with $G = 2$, where sex and BMI are covariates in the gating networks and there are no covariates in the expert network

```
# Load the ais data
# Extract the hematological variables in a matrix
library(sn)
data(ais)
hematological <- as.matrix(ais[,3:7])

# Load the flexmix package
library(flexmix)

# Fit a mixture of experts model with 50 random starting values
# The highest BIC value is stored as bicval
# The best fitting model is stored as bestfit
bicval <- Inf
for (iter in 1:50)
{
  fit1 <- flexmix(hematological~1,model=FLXMCmvnorm(diagonal=FALSE)
             ,data=ais,k=2,concomitant=FLXPmultinom(~sex+BMI))
  parameters(fit1)
  if (bicval>BIC(fit1))
  {
    bicval <- BIC(fit1)
    bestfit <- fit1
    print(c(iter,bicval))
  }
}

# Explore the fitted model
summary(bestfit)
clusters(bestfit)
parameters(bestfit)

# Explore the gating network model
summary(bestfit,which="concomitant")
parameters(bestfit,which="concomitant")
```

Table 11.7 *A cross tabulation of the wine type and year (vintage) for the Italian wine data.*

	70	71	72	73	74	75	76	78	79
Barbera	0	0	0	0	9	0	5	29	5
Barolo	0	19	0	20	20	0	0	0	0
Grignolino	9	9	7	9	16	9	12	0	0

performance of the model, because the wine type is part of the model structure and is an input into the prediction.

The resulting model parameters, as given in Table 11.9 and Table 11.10, show the cluster means and gating parameter coefficients, respectively.

Table 11.8 *A cross tabulation of the clustering of the wine samples and the type of wine. The table shows that only one wine sample is misclassified by the mixture of experts model.*

	1	2	3
Barbera	0	0	48
Barolo	0	59	0
Grignolino	70	1	0

Table 11.9 *Component means for the mixture of experts model fitted to the Italian wine data.*

Variable	Component Mean		
	1	2	3
Alcohol	13.71	13.15	12.27
Sugar-free extract	26.70	25.02	24.21
Fixed acidity	76.18	103.43	81.63
Tartaric acid	1.64	2.60	1.90
Malic acid	1.99	3.33	1.95
Uronic acids	0.82	1.17	0.82
pH	3.33	3.26	3.30
Ash	2.45	2.44	2.24
Alkalinity of ash	17.28	21.42	20.11
Potassium	897.22	891.10	859.55
Calcium	64.85	71.94	93.73
Magnesium	106.55	99.31	94.02
Phosphate	429.13	345.08	321.97
Chloride	74.95	41.27	68.82
Total phenols	2.86	1.68	2.23
Flavanoids	2.98	0.78	2.05
Nonflavanoid phenols	0.29	0.45	0.37
Proanthocyanins	1.90	1.15	1.62
Color Intensity	5.47	7.40	3.06
Hue	1.07	0.68	1.05
OD280/OD315 of diluted_wines	3.16	1.68	2.77
OD280/OD315 of flavanoids	3.44	1.89	3.33
Glycerol	9.90	8.86	7.81
2-3-butanediol	756.94	806.48	658.47
Total nitrogen	270.60	210.06	237.00
Proline	1102.04	629.90	514.24
Methanol	110.16	115.54	105.71

Table 11.10 *Gating network parameter values for the mixture of experts model fitted to the Italian wine data.*

Variable	Component		
	1	2	3
(Intercept)	0.00	14.24	−34.69
Grignolino	0.00	−23.39	38.23
Barolo	0.00	−27.50	−36.95

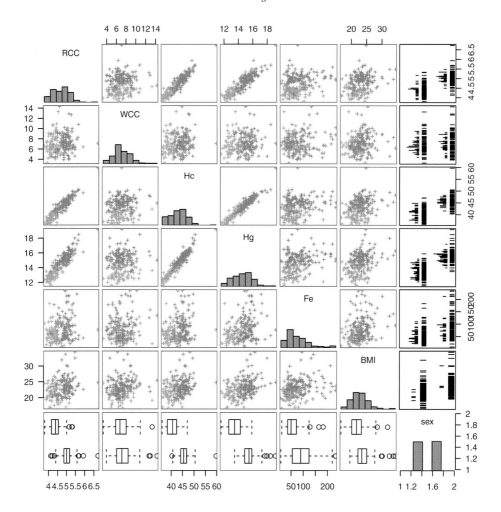

Figure 11.7 A scatter plot matrix of the AIS data, colored by cluster. Only the five hematological variables and the two selected covariates are included in the plot.

11.6 Discussion

This chapter has illustrated the utility of mixture of experts models for model-based clustering where covariates are available. The mixture of experts framework provides a principled method for describing and exploring the clustering found in the outcome variables and to determine which covariates can assist with explaining the clusters found. In the examples considered in this chapter, the covariates entered either the component density or the mixing proportions, but not both. In principle the covariates can enter both parts of the model, but in that case the interpretation of the model and the effect of the covariates becomes more difficult. A model of this type can provide a useful estimation of the conditional density of

Listing 11.4: Fitting a mixture of experts model to the wine data using the `flexmix` package. The model in the code is a mixture of experts model with $G = 3$, where type is a covariate in the gating network

```
# Load the flexmix package
library(flexmix)

# Load the wine data
library(MBCbook)
data(wine27)

# Select the physical and chemical measurements as the outcome
# Select wine type and year as possible covariates
wine <- as.matrix(wine27[,1:27])
type <- wine27[,"Type"]
year <- wine27[,"Year"]

# Fit a mixture of experts model with 50 random starting values
# The highest BIC value is stored as bicval
# The best fitting model is stored as bestfit
bicval <- Inf
itermax <- 50
for (iter in 1:itermax)
{
fit <- flexmix(wine~1,model=FLXMCmvnorm(diagonal=TRUE),k=3,
    concomitant=FLXPmultinom(~type))
  if (bicval>BIC(fit))
  {
      bicval <- BIC(fit)
      bestfit <- fit
      print(c(iter,bicval))
  }
}

# Explore the fitted model
summary(bestfit)
clusters(bestfit)
parameters(bestfit)

# Explore the gating network of the model
summary(bestfit,which="concomitant")
parameters(bestfit,which="concomitant")
```

the outcome given the covariates but as a clustering model it is more difficult to interpret.

11.7 Bibliographic Notes

Mixture of experts models have been employed in a wide range of modeling settings. Peng et al. (1996) use a mixture of experts model and a hierarchical mixture of experts model in speech recognition applications. Thompson et al. (1998) use a mixture of experts model for studying the diagnosis of diabetic

Model-based Clustering with Covariates

patients. Rosen and Tanner (1999) develop a mixture of experts proportional hazards model and analyze a multiple myeloma data set. Celeux et al. (2005) use a mixture of experts random effect model for clustering gene expression profiles. Hurn et al. (2003) use MCMC to fit a mixture of regressions model, which is a special case of the mixture of experts model, but where the mixing proportions don't depend on the covariates x_i. Carvalho and Tanner (2007) use a mixture of experts model for non-linear time-series modeling. Geweke and Keane (2007) use a model similar to the mixture of experts model, but where the gating network has a probit structure, in a number of econometric applications. Souza and Araújo (2014) apply the mixture of experts framework to quality assessment in production systems with multiple operating modes.

The cluster-weighted model family (Gershenfeld, 1997) is very closely related to the mixture of experts model family. In this modeling approach, both the outcome variable and covariates are treated as random and these are jointly modeled using a mixture model. The conditional distribution of the outcome variable, conditional on the covariates, can thus be derived from the joint model. Some key references for this modeling approach are Ingrassia et al. (2012), Ingrassia et al. (2015), Dang et al. (2017) and Punzo and McNicholas (2017). The flexCWM R package (Mazza et al., 2018) is available for fitting a wide range of cluster-weighted models.

Gormley and Murphy (2008, 2010a, 2011) use a mixture of experts model for model-based clustering of ranking data in educational and political science applications. Gormley and Murphy (2010b) extend the latent position cluster model (Handcock et al., 2007) using the mixture of experts model framework to study clustering in social networks. White and Murphy (2016b) extend the mixed membership stochastic block model (Airoldi et al., 2008) using a variant of the gating network mixture of experts model. These two modeling approaches are discussed in more detail in Chapter 10.

Further details on mixture of experts models and their application in model-based clustering are reviewed in McLachlan and Peel (2000, Chapter 5.13), Tanner and Jacobs (2001), Bishop (2006, Chapter 14.5), Gormley and Murphy (2011), Yuksel (2012), Masoudnia and Ebrahimpour (2014) and Gormley and Frühwirth-Schnatter (2018), where extensions to the modeling framework are also discussed.

12

Other Topics

In this final chapter, we discuss model-based clustering of data of types not considered earlier, such as functional data, texts or images. Model-based techniques for co-clustering are also considered.

12.1 Model-based Clustering of Functional Data

Functional data are now increasingly prevalent. The recent emergence of open data systems and their interface with personal devices through APIs (application programming interfaces) allows people to access useful information as varied as weather data and forecasts, university course schedules, real-time bus timetables, and bike availability in bike sharing systems (BSSs). These applications have in common the functional nature of the data they produce. In principle, these data are of infinite dimension and their analysis is therefore more complex than multivariate or even high-dimensional data (see Chapter 8).

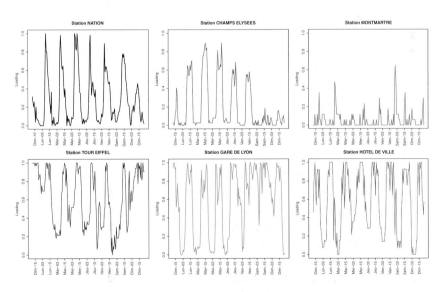

Figure 12.1 Loading profiles of some Vélib stations. A loading value equal to 1 means that the station is full of bikes whereas a value equal to 0 indicates a station without available bikes.

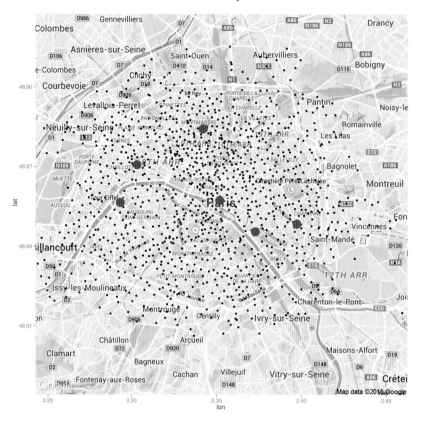

Figure 12.2 Map of the Vélib stations in Paris. The red dots correspond to the stations for which the loading profiles are displayed on Figure 12.1.

As an example, we consider one week of activity of the Paris bike sharing system, Vélib. The data are the loading profiles of the bike stations over one week. The data were collected every hour during the period from Sunday 1 September to Sunday 7 September, 2014, from the open-data API provided by JCDecaux[1]. Each loading profile is the ratio of the number of available bikes to the number of bike docks in the station. This transformation of the raw data allows us to handle the large variability in the station sizes. The data set is therefore made up of 1,189 loading profiles sampled at 181 time points and is available in the `funFEM` package for R. Figure 12.1 shows the loading profiles of some Vélib stations which can be located on the map of Figure 12.2. Note that a loading value of 1 means that the station is full of bikes whereas a value of 0 indicates a station with no available bikes.

Cluster analysis could be useful for understanding the system behavior, but the functional nature of the data and the fact that they live in an infinite dimensional space make the task difficult. The lack of a clear definition of a functional random variable is one reason for this. As a result, most clustering methods for functional

---

[1] The real time data are available at `https://developer.jcdecaux.com/` (with an api key).

data either apply multivariate methods to the discretized curves or use distances for functional data within classical algorithms such as k-means.

12.1.1 Model-based Approaches for Functional Clustering

Recent advances in functional data analysis have helped to develop efficient model-based clustering techniques for functional data. James and Sugar (2003) proposed a method that is particularly effective for sparsely sampled functional data. This method, called fclust, assumes that the basis expansion coefficients of the curves into a spline basis are distributed according to a mixture of Gaussian distributions with cluster-specific means and common variances. The use of a spline basis is convenient when the curves are regular, but is not appropriate for peak-like data such as those that arise in mass spectrometry. For this reason, Giacofci et al. (2013) extended the fclust approach by fitting a Gaussian mixture model to the wavelet decomposition coefficients of the curves. This allows us to deal with a wider range of functional shapes than splines.

The method of Samé et al. (2011) assumes that the curves arise from a mixture of regressions on a basis of polynomial functions, with possible regime changes at each instant of observation. Frühwirth-Schnatter and Kaufmann (2008) have built a specific clustering algorithm based on parametric time series models. Bouveyron and Jacques (2011) extended the high-dimensional data clustering (HDDC) algorithm of Bouveyron et al. (2007a) to the functional case. The resulting model, funHDDC, assumes a parsimonious cluster-specific Gaussian distribution for the basis expansion coefficients. FunHDDC was also recently adapted to the multivariate case by Schmutz et al. (2018).

Jacques and Preda (2013) proposed a model-based clustering method based on an approximation of the notion of density for functional variables. This was extended to multivariate functional data by Jacques and Preda (2014b). These models assume that the functional principal component scores of curves have a Gaussian distribution whose parameters are cluster-specific. The FisherEM algorithm (see Chapter 8) has also been adapted to functional data by Bouveyron et al. (2015).

Some Bayesian approaches have also been proposed. Heard et al. (2006) assumed that the basis expansion coefficients are distributed as a mixture of Gaussians whose variances are modeled by an inverse-gamma distribution. Ray and Mallick (2006) proposed a nonparametric Bayes wavelet model for curve clustering based on a mixture of Dirichlet processes. Jacques and Preda (2014a) provided a review of functional data clustering.

We focus here on three models which share a common way of dealing with the observed curves. They assume that the observed curves $\{x_1, ..., x_n\}$ are independent realizations of an L_2-continuous stochastic process $X = \{X(t)\}_{t \in [0,T]}$ for which the sample paths, i.e. the observed curves, belong to $L_2[0, T]$. In practice, the functional expressions of the observed curves are not known, and we have access only to the discrete observations $x_{ij} = x_i(t_{is})$ at a finite set of ordered times $\{t_{is} : s = 1, \ldots, m_i\}$. It is therefore necessary first to reconstruct the functional form of the data from their discrete observations. A common way to do this is to assume that the curves belong to a finite dimensional space spanned by a basis of functions (see, for example, Ramsay and Silverman, 2005).

Let us therefore consider such a basis $\{\psi_1, \ldots, \psi_p\}$ and assume that the stochastic process X admits the following basis expansion:

$$X(t) = \sum_{j=1}^{p} \gamma_j(X)\psi_j(t), \tag{12.1}$$

where $\gamma = (\gamma_1(X), \ldots, \gamma_p(X))$ is a random vector in $\mathbb{R}^p$, and the number p of basis functions is fixed. The basis expansion of each observed curve $x_i(t) = \sum_{j=1}^{p} \gamma_{ij}\psi_j(t)$ can be estimated by an interpolation procedure (see Escabias et al. (2005), for instance) if the curves are observed without noise or by least squares smoothing if they are observed with error, i.e. $x_i^{obs}(t_{is}) = x_i(t_{is}) + \varepsilon_{is}, s = 1, \ldots, m_i$. An advantage of this representation is the possibility of clustering curves sampled at different times.

12.1.2 The fclust Method

We first focus on the model proposed by James and Sugar (2003) for clustering sparsely sampled functional data. The functional clustering model (FCM) assumes that the observed curves $x_i = (x_i(t_{i1}), \ldots, x_i(t_{in_i}))$ admits the following basis expansion:

$$x_i(t) = \sum_{j=1}^{p} \gamma_{ij}\psi_j(t) + \varepsilon_i,$$

where $\varepsilon_i \sim \mathcal{N}(0, \sigma^2 I)$. The basis coefficients γ_i are further assumed to be as follows:

$$\gamma_i = \mu_{z_i} + \xi_i,$$

where $\xi_i \sim \mathcal{N}(0, \Gamma)$ and z_i indicates the group membership of the curve.

It is assumed that the group means can be decomposed as:

$$\mu_k = \lambda_0 + V\alpha_k,$$

where $\lambda_0 \in \mathbb{R}^p$, $\alpha_k \in \mathbb{R}^d$ with $d = \min(p, G-1)$ and V is a $p \times d$ matrix. To avoid identifiability issues, the authors add some constraints on α_k and V. To summarize, the marginal distribution of X is

$$p(x) = \sum_{g=1}^{G} \pi_g \phi(x; \Psi(\lambda_0 + V\alpha_k), \Psi^T \Gamma \Psi + \sigma^2 I),$$

where π_g is the prior probability of the gth group and $\Psi = (\psi_1(t), \ldots, \psi_p(t))$ is the spline basis matrix for the curve $x(t)$. Figure 12.3 provides a graphical model representation of the model.

Thus, FCM is a parsimonious model in which the complexity is controlled by the value of d. Notice that the parsimony is here only on the group means and not

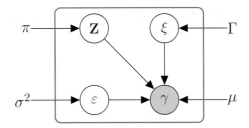

Figure 12.3 Graphical model representation of the fclust model.

Listing 12.1: Clustering of the Vélib data with fclust

```
# Loading libraries and data
library(funcy)
library(funFEM)
data(velib)

# Clustering with fclust
res = funcit(t(velib$data),k = 6,
         methods = "fitfclust")

# Visualization of group means
means = res@models$fitfclust@centers
matplot(means,col=1:res@k,type='l',xaxt='n',lwd=2)
axis(1,at=seq(5,181,6),labels=velib$dates[seq(5,181,6)],las=2)

# Map of the results with ggmap (requires a Google API key)
library(ggmap)
Mymap = get_map(location = 'Paris', zoom = 12,
         maptype = 'terrain')
ggmap(Mymap) + geom_point(data=velib$position,
                 aes(longitude,latitude),
                 colour = I(res@allClusters),
                 size = I(3))

# Map of the results with leaflet (no API key needed)
library(leaflet)
library(RColorBrewer)
df = velib$position
colors = brewer.pal(12,'Paired')
leaflet(df) %>% addTiles() %>%
     addCircleMarkers(color = colors[res@allClusters],
     radius = 5, fillOpacity = 1, stroke = FALSE)
```

on the group covariance matrices as in factor analysis (FA) models (see Chapter 5). Nevertheless, the interpretation of the clustering results is eased by the modeling. The global mean of the curves is $\Psi\lambda_0$, and $\Psi V \alpha_k$ represents the deviation of the group means from the global mean. The model inference is done using an EM algorithm.

The fclust method is implemented in R within the `funcy` package. Listing 12.1 shows the typical call for running `fclust` on the `velib` data set.

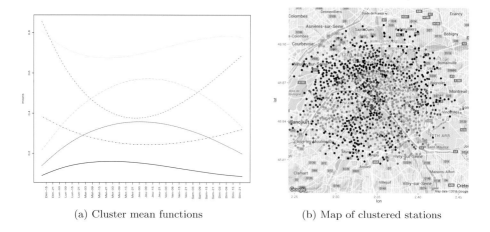

(a) Cluster mean functions   (b) Map of clustered stations

Figure 12.4 Cluster mean functions and map of clustered stations by fclust on the Vélib data set.

Figure 12.4 shows the resulting plots: the left panel displays the cluster mean functions whereas a map of the clustered Vélib stations by fclust is displayed on the right. Although the map of clustered stations may appear meaningful at first glance, fclust is able to recover only some global dynamics of the system here, as the left panel of Figure 12.4 shows. This behavior is related to the chosen modeling of fclust which imposes parsimony over the group means and is not flexible enough in this case, at least for complex dynamics. The method is well adapted to functional data with smoother dynamics, however.

12.1.3 The funFEM Method

The funFEM method (Bouveyron et al., 2015) is based on the discriminative functional mixture (DFM) model, which models the data in terms of a single discriminative functional subspace. With this subspace we can visualize the clustered data more easily and compare the patterns found in different systems.

The DFM model considers a common latent subspace $F[0,T]$ of $L_2[0,T]$, taken to be the most discriminative subspace for the G groups. The subspace $F[0,T]$ is assumed to be spanned by a basis consisting of d basis functions $\{\varphi_j\}_{j=1,\ldots,d}$ in $L_2[0,T]$, with $d < G$ and $d < p$. The assumption $d < G$ is motivated by the fact that a subspace of $d = G - 1$ dimensions is sufficient to discriminate between G groups (Fisher, 1936; Fukunaga, 1990). The basis $\{\varphi_j\}_{j=1,\ldots,d}$ is obtained from $\{\psi_j\}_{j=1,\ldots,p}$ through a linear transformation $\varphi_j = \sum_{\ell=1}^{p} u_{j\ell}\psi_\ell$ such that the $p \times d$ matrix $U = (u_{j\ell})$ is orthogonal. Let $\{\lambda_1, \ldots, \lambda_n\}$ be the latent expansion coefficients of the curves $\{x_1, \ldots, x_n\}$ on the basis $\{\varphi_j\}_{j=1,\ldots,d}$, assumed to be independent realizations of a latent random vector $\Lambda \in \mathbb{R}^d$. The random vectors Γ and Λ are therefore linked through the following linear transformation:

$$\Gamma = U\Lambda + \varepsilon,$$

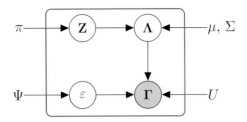

Figure 12.5 Graphical representation of the DFM model.

where $\varepsilon \sim \mathcal{N}(0, \Xi)$ is an independent and random noise term. Conditionally on the group membership variable Z, Λ is also assumed to be distributed according to a multivariate Gaussian density:

$$\Lambda_{|Z=g} \sim \mathcal{N}(\mu_g, \Sigma_g),$$

where μ_g and Σ_g are, respectively, the mean and the covariance matrix of the gth group. With these distributional assumptions, the marginal distribution of Γ is a mixture of Gaussians:

$$p(\gamma) = \sum_{g=1}^{G} \pi_g \phi(\gamma; U\mu_g, U^T \Sigma_g U + \Xi),$$

where ϕ is the standard Gaussian density function, and $\pi_g = P(Z = g)$ is the proportion of the gth group. The noise covariance matrix Ξ is assumed to be such that $\Delta_g = \mathrm{cov}(W^T \Gamma | Z = g) = W^T \Sigma_g W$ has the following form:

$$\Delta_g = \left(\begin{array}{cc} \boxed{\begin{array}{c} \Sigma_g \end{array}} & \mathbf{0} \\ \mathbf{0} & \boxed{\begin{array}{cc} \beta & 0 \\ & \ddots \\ 0 & \beta \end{array}} \end{array} \right) \begin{array}{l} \left.\rule{0pt}{20pt}\right\} \ d \\ \left.\rule{0pt}{20pt}\right\} \ p-d \end{array}$$

where $W = [U, V]$, with V being the orthogonal complement of U. Figure 12.5 provides a graphical summary of the model.

One can recognize this model as an extension of the DLM model (see Chapter 8) to the functional case. We can therefore list the family of 12 DLM models in the functional case too. Similarly to the DLM models, inference for the DFM model has to be done using an algorithm which estimates the discriminative functional subspace separately. The inference algorithm is called funFEM.

Unlike fclust, the parsimony here is focused on the covariance matrices and allows the model to be highly parsimonious. The estimated functional subspace also yields useful visualizations of the clustering results.

Let us now apply the funFEM algorithm to the Vélib data set. The funFEM method is implemented in the `funFEM` package for R. Listing 12.2 shows the necessary code to transform the raw data into curves and then cluster them using

Listing 12.2: Clustering of the Vélib data with funFEM

```
# Loading libraries and data
library(funFEM)
data(velib)

# Transformation of the raw data as curves
basis = create.fourier.basis(c(0, 181), nbasis=25)
fdobj = smooth.basis(1:181,t(velib$data),basis)$fd

# Clustering with funFEM
res = funFEM(fdobj,K=6)

# Visualization of group means
fdmeans = fdobj; fdmeans$coefs = t(res$prms$my)
plot(fdmeans,col=1:res$K,xaxt='n',lwd=2)
axis(1,at=seq(5,181,6),labels=velib$dates[seq(5,181,6)],las=2)

# Map of the results with leaflet
library(leaflet)
library(RColorBrewer)
df = velib$position
colors = brewer.pal(12,'Paired')
leaflet(df) %>% addTiles() %>%
      addCircleMarkers(color = colors[res$cls],
      radius = 5, fillOpacity = 1, stroke = FALSE)
```

funFEM. Here, a basis of Fourier functions is chosen because of the expected daily periodicity of the data.

The two panels of Figure 12.6 correspond respectively to the plot of the group mean functions and the map of the clustered stations. One can recognize some expected patterns: the black stations correspond to stations where Parisians live whereas green stations are located in working places. Stations in tourist areas can also be recognized here and are gathered in the red group. It is also striking that the groups identified are spatially coherent, even though the spatial information was not provided to the algorithm, which provides some further validation of the method.

As we mentioned, funFEM allows us to visualize the projection of the clustered curves in the functional discriminative subspace estimated by the method. Listing 12.3 shows the code to obtain such a visualization and Figure 12.7 is the resulting plot. The stations of Figure 12.1 are highlighted and, unsurprisingly, the touristic bike stations fall in the red cluster, except Montmartre which is in another group due to its elevation.

12.1.4 The funHDDC Method for Multivariate Functional Data

Only a few clustering methods are adapted to multivariate functional data. The term "multivariate functional data" refers to a set of several functions or times series describing the same individual. For instance, a bike station can be represented

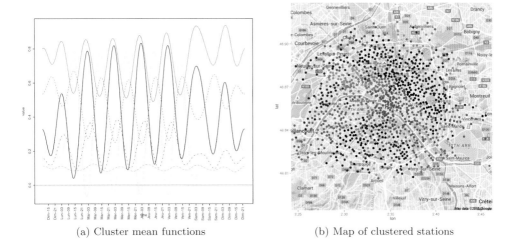

(a) Cluster mean functions   (b) Map of clustered stations

Figure 12.6 Cluster mean functions and map of clustered stations by funFEM on the Vélib data set.

Listing 12.3: Visualization into the functional discriminative subspace of funFEM

```
# Projection into the subspace
fdproj = t(fdobj$coefs) %*% res$U
plot(fdproj,type='p',col=res$cls,pch=19,xlab='Disc axis 1',ylab='
    Disc axis 2',main='Discriminative subspace')

# Add the names of some specific bike stations
sel = c(200,301,384,918,936,1024)
names = c('NATION','CHAMPS ELYSEES','MONTMARTRE','TOUR EIFFEL','
    GARE DE LYON','HOTEL DE VILLE')
text(fdproj[sel,],labels=names)
```

by its loading profile and its percentage of broken docks over the time. Figure 12.8 shows the multivariate functional representations of some Vélib stations.

To date, only the funclust (Jacques and Preda, 2014b) and funHDDC (Schmutz et al., 2018) approaches are able to handle such data. We focus here on the latter, which is the most recent approach and which improves the funclust approach. The funHDDC method is an extension to functional data of the HDDC methodology, originally designed for high-dimensional data (see Chapter 8). The clustering of multivariate functions with funHDDC is made possible by the fact that funHDDC relies on a multivariate version of functional PCA, denoted by MFPCA, proposed by Jacques and Preda (2014b), and which is able to deal with such data. The authors extended the approach proposed by Ramsay and Silverman (2005), which consists of concatenating the observations or basis coefficients into a single vector before performing a functional PCA. The original approach was limited to orthogonal basis functions and MFPCA overcomes this limitation.

Let $\mathbf{X}_1, ..., \mathbf{X}_n$ be an i.i.d. sample of $\mathbf{X} = \boldsymbol{X}(t)_{t \in [0,T]}$, which is a p-variate curve,

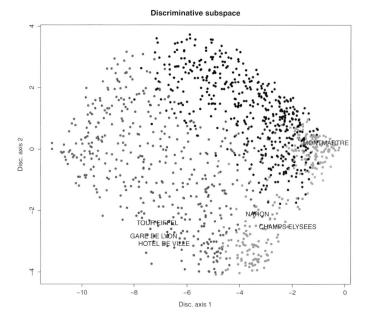

Figure 12.7 Projection of the clustered curves into the functional discriminative subspace estimated by funFEM for the Vélib data.

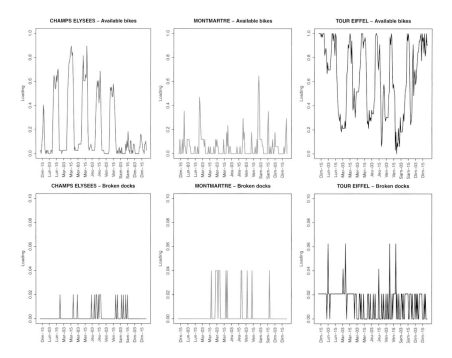

Figure 12.8 Some multivariate curves of the Vélib data set. The top curves are loading profiles of the stations and the bottom curves correspond to the percentage of broken docks in the stations.

i.e. $\mathbf{X} = (X^1, ..., X^p)$. Each curve is assumed to be defined by a linear combination of basis functions $(\phi_r^j(t))_r$:

$$X_i^j(t) = \sum_{r=1}^{R_j} c_{ir}^j(X_i^j)\phi_r^j(t), \tag{12.2}$$

where $(\phi_r^j(t))_{1 \leq r \leq R_j}$ is the basis of functions for the jth component of the multivariate curve and R_j the number of basis functions, $i \in \{1, ..., n\}$, $j \in \{1, ..., p\}$. Let us introduce the following two matrices which gather the coefficients and the basis functions respectively:

$$\mathbf{C} = \begin{pmatrix} c_{11}^1 & \cdots & c_{1R_1}^1 & c_{11}^2 & \cdots & c_{1R_2}^2 & \cdots & c_{11}^p & \cdots & c_{1R_p}^p \\ & & & & \cdots & & & & & \\ c_{n1}^1 & \cdots & c_{nR_1}^1 & c_{n1}^2 & \cdots & c_{nR_2}^2 & \cdots & c_{n1}^p & \cdots & c_{nR_p}^p \end{pmatrix},$$

$$\phi(t) = \begin{pmatrix} \phi_1^1(t) & \cdots & \phi_{R_1}^1(t) & 0 & \cdots & 0 & \cdots & 0 & \cdots & 0 \\ 0 & \cdots & 0 & \phi_1^2(t) & \cdots & \phi_{R_2}^2(t) & \cdots & 0 & \cdots & 0 \\ & & & & \cdots & & & & & \\ 0 & \cdots & 0 & 0 & \cdots & 0 & \cdots & \phi_1^p(t) & \cdots & \phi_{R_p}^p(t) \end{pmatrix}.$$

With these notations, Equation (12.2) can be written in matrix form:

$$\mathbf{X}(t) = \mathbf{C}\phi'(t).$$

The estimation of $\mathbf{C}$ is usually done by least squares smoothing.

The principle of MFPCA reduces to the search for the eigenvalues and eigenfunctions that solve the spectral decomposition of the covariance operator ν:

$$\nu \mathbf{f}_l = \lambda_l \mathbf{f}_l, \forall l \geq 1, \tag{12.3}$$

where λ_l is a set of positive eigenvalues and $\mathbf{f}_l$ is the set of associated multivariate eigenfunctions. An empirical estimator of the covariance operator can be written as:

$$\widehat{\nu}(s, t) = \frac{1}{n}\mathbf{X}'(s)\mathbf{X}(t) = \frac{1}{n}\phi(s)\mathbf{C}'\mathbf{C}\phi'(t).$$

Let us suppose that each principal factor $\mathbf{f}_l$ belongs to the linear space spanned by the matrix ϕ:

$$\mathbf{f}_l(t) = \phi(t)\mathbf{b}_l'$$

where $\mathbf{b}_l = (b_{l11}, ..., b_{l1R_1}, b_{l21}, ..., b_{l2R_2}, ..., b_{lp1}, ..., b_{lpR_p})$. Then the eigenproblem (12.3) becomes:

$$\frac{1}{n}\phi(s)\mathbf{C}'\mathbf{C}\mathbf{W}\mathbf{b}_l' = \lambda_l\phi(s)\mathbf{b}_l', \tag{12.4}$$

where $\mathbf{W} = \int_0^T \phi'(t)\phi(t)$. The MFPCA then reduces to the usual PCA of the matrix $\frac{1}{\sqrt{n}}\mathbf{C}\mathbf{W}^{1/2}$.

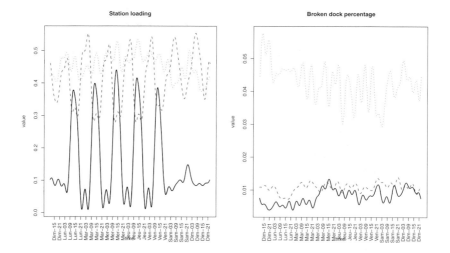

Figure 12.9 Cluster means for the two functional variables obtained with funHDDC on the multivariate Vélib data set.

Thus, each multivariate curve $\boldsymbol{X}_i$ is identified by its score $\boldsymbol{\delta}_i = (\delta_{il})_{l\geq1}$ into the basis of multivariate eigenfunctions $(\boldsymbol{f}_l)_{l\geq1}$:

$$(\delta_{il})_{l\geq1} = \boldsymbol{CW}\boldsymbol{b}_l'.$$

Since those MFPCA scores can be of high dimension, the funHDDC approach then assumes that the scores follow the HD-GMM model of Bouveyron et al. (2007a,b) introduced in Section 8.4.4. In particular, the model assumes that the functional scores follow a mixture distribution:

$$\boldsymbol{\delta} \sim \sum_{g=1}^{G} \tau_g \mathcal{N}(\delta; \mu_g, \Sigma_g),$$

where the covariance matrix Σ_k has a specific structure, encoding the idea that the scores live in low-dimensional functional subspaces, with group-specific intrinsic dimensionalities. Inference for the model above is done using an EM algorithm and the choice of the intrinsic dimensionalities of the group is addressed by model selection.

The funHDDC method is implemented in the `funHDDC` package for R. Listing 12.4 shows the application of funHDDC to the multivariate version of the Vélib data (the `velib2D` data set is available in the `MBCbook` package).

Note that we removed curves with missing values and subsampled the data set before clustering the data with funHDDC. We also asked funHDDC to find only $G = 3$ clusters, in order to make the results easier to interpret. It is of course possible to let the method choose the most appropriate number of clusters through model selection. The two panels of Figure 12.9 correspond to the plots of the group means for each of the two functional variables.

This method identified two clusters (black and red) with a clear temporal

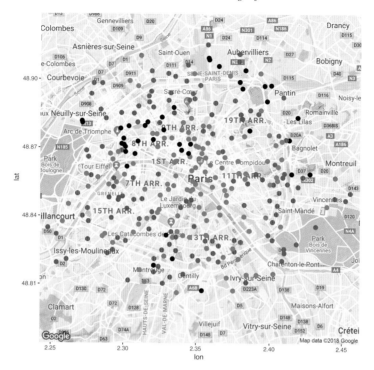

Figure 12.10 Map of clustered stations obtained with funHDDC on the multivariate Vélib data set.

dynamic in terms of station loading, corresponding to the daily commute, and with a low percentage of broken docks. The red cluster also has regular activity during the weekend. The third group (green) has behavior close to that of the red cluster in terms of station loading but clearly differs from the two other clusters in terms of number of broken docks. We can see here the interest of working with several functional variables to provide a deeper analysis of the data.

Finally, Figure 12.10 shows the map of the clustered stations. The map allows us, for instance, to locate the stations of the green cluster which mainly consists of stations within the touristic area. This may explain the higher level of broken docks.

12.2 Model-based Clustering of Texts

Text data has greatly expanded recently, and continues to do so, including with the expansion of social networks such as Twitter and Reddit. Indeed, most new documents are now available in electronic format. There are also large-scale efforts to digitize old documents and make them available to everyone through the Internet, such as Google Books and Gallica. Faced with this mass of texts, clustering is a useful tool to automatically characterize a set of documents. Applications of text clustering range from marketing to sentiment analysis.

Listing 12.4: Clustering of the multivariate Vélib data with funHDDC

```
# Loading libraries and data
library(funHDDC)
library(MBCbook)
data(velib2D)

# Removing NAs and subsampling
ind = which(rowSums(is.na(velib2D$brokenDocks))==0)
X = velib2D$availableBikes[ind,]
Y = velib2D$brokenDocks[ind,]

# Transformation of the raw data as curves
basis = create.fourier.basis(c(0, 181), nbasis=51)
fd1 = smooth.basis(1:181,t(X),basis)$fd
fd2 = smooth.basis(1:181,t(Y),basis)$fd
fdobj = list(fd1,fd2)

# Clustering the multivariate functions with funclust
res = funHDDC(fdobj,K=3)

# Map of the results with leaflet
library(leaflet)
library(RColorBrewer)
df = velib2D$position[ind,]
colors = brewer.pal(12,'Paired')
leaflet(df) %>% addTiles() %>%
  addCircleMarkers(color = colors[res$class],
                   radius = 5, fillOpacity = 1, stroke = FALSE)
```

12.2.1 *Statistical Models for Texts*

A statistical model for texts appeared at the end of the last century with an early model described by Papadimitriou et al. (1998) for latent semantic indexing (LSI) (Deerwester et al., 1990). LSI allows users to recover linguistic notions such as synonymy and polysemy from "term frequency - inverse document frequency" (tf-idf) data. Hofmann (1999) proposed an alternative model for LSI, called probabilistic latent semantic analysis (pLSI), which models each word within a document using a mixture model. In pLSI, each mixture component is modeled by a multinomial random variable and the latent groups can be viewed as "topics". Thus, each word is generated from a single topic and different words in a document can be generated from different topics. pLSI can also be viewed as an extension of the mixture of unigrams, proposed by Nigam et al. (2000). Criticisms of pLSI include that it has no model at the document level and that it may suffer from overfitting.

The model which concentrates all of the desired features was proposed by Blei et al. (2003) and is called latent Dirichlet allocation (LDA[2]). The LDA model has rapidly become a standard tool in statistical text analytics and is even used in different scientific fields such as image analysis (Lazebnik et al., 2006) and

---

[2] Do not confuse with linear discriminant analysis (see Chapter 4), which has the same acronym.

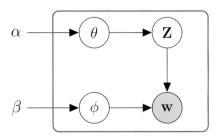

Figure 12.11 Graphical representation of the latent Dirichlet allocation (LDA) model.

transportation research (Côme et al., 2014). The idea of LDA is that documents are represented by random mixtures over latent topics, where each topic is characterized by a distribution over words. LDA is therefore similar to pLSI except that the topic distribution in LDA has a Dirichlet prior. We focus hereafter on the LDA model. Blei (2012) gives an overview of probabilistic topic models.

12.2.2 Latent Dirichlet Allocation

Let us consider a set of D documents $\{W_1, ..., W_D\}$ where each document W_d contains N_d words, i.e. $W_d = \{w_{d1}, ..., w_{dN_d}\}$. As in the usual clustering context, the aim is to cluster the documents and their words into G groups, usually called "topics" in text mining. Notice, however, that LDA is a mixed-membership model (such as the MMSBM model presented in Chapter 10.4) in the sense that each word can have different meanings (and topics) in different documents.

The generative model of LDA for these D documents is as follows. The topic proportion θ_d for the document W_d is assumed have a Dirichlet distribution, namely

$$\theta_d \sim Dir(\alpha),$$

where $\alpha = (\alpha_1, ..., \alpha_G)$ is a common prior over topics for all documents. The word distribution ϕ_g, for topic $g \in \{1, ..., G\}$, is taken to be

$$\phi_g \sim Dir(\beta),$$

where $\beta = (\beta_1, ..., \beta_V)$ and V is the size of the vocabulary. Then the topic z_{id} of the ith word in document d is drawn from a multinomial distribution:

$$z_{id} \sim \mathcal{M}(1, \theta_d).$$

Given the topic z_{id}, the ith word in document d is also drawn from a multinomial distribution, namely

$$w_{id} \sim \mathcal{M}(1, \phi_{z_{id}}).$$

Note that the distribution of words over the topics is a mixture of multinomial

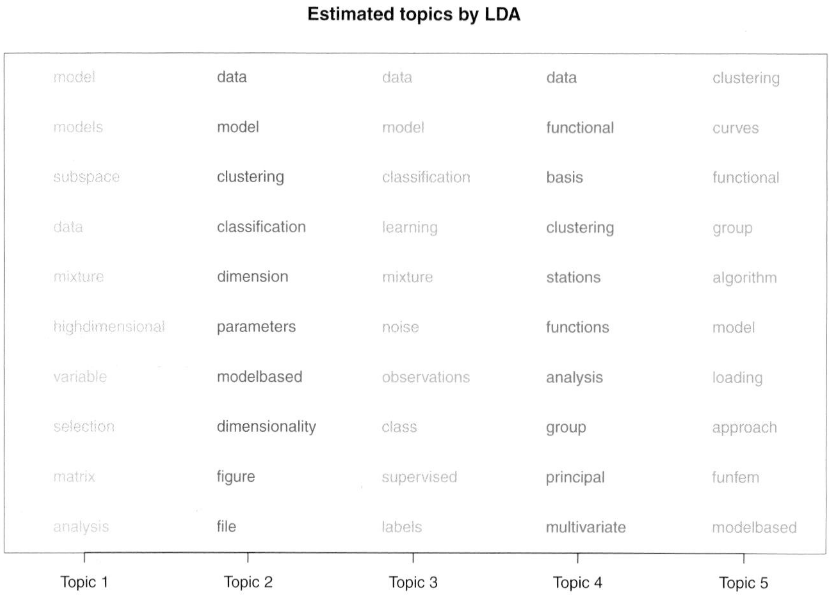

Estimated topics by LDA

Topic 1	Topic 2	Topic 3	Topic 4	Topic 5
model	data	data	data	clustering
models	model	model	functional	curves
subspace	clustering	classification	basis	functional
data	classification	learning	clustering	group
mixture	dimension	mixture	stations	algorithm
highdimensional	parameters	noise	functions	model
variable	modelbased	observations	analysis	loading
selection	dimensionality	class	group	approach
matrix	figure	supervised	principal	funfem
analysis	file	labels	multivariate	modelbased

Figure 12.12 Most representative words in the five estimated topics by latent Dirichlet allocation on Chapters 5, 9 and 12 of this book.

distributions:

$$w_{id}|\theta \sim \sum_{g=1}^{G} \theta_{dg}\mathcal{M}(1, \phi_g).$$

Figure 12.11 provides a graphical summary of the model. Inference for the LDA model can be done using a variational Bayesian expectation-maximization (VBEM) algorithm, as in the original paper (Blei et al., 2003). Alternative possible inference procedures include MCMC or expectation-propagation (EP) algorithms.

12.2.3 Application to Text Clustering

As an illustration, we now apply the LDA function of the R package topicmodels to three chapters of this book (of course, the three texts that are used here can be replaced by any text file). Listing 12.5 illustrates the use of the LDA function on those texts. Some preprocessing of the data is necessary: stemming, and removing stopwords, punctuation and numbers. Figure 12.12 shows the most representative words in the five estimated topics by LDA.

The estimated topics are meaningful in the sense that we can easily recognize a "subspace clustering" topic (Topic 1), a "supervised classification" topic (Topic 3) and a "functional clustering" topic (Topic 5). The two other topics are more blurry, which is not surprising here since we use chapters on very close themes.

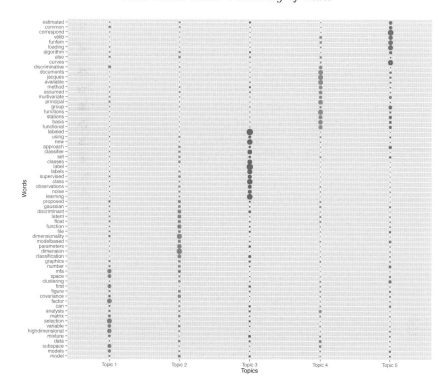

Figure 12.13 Word weights for the five estimated topics by latent Dirichlet allocation on Chapters 5, 9 and 12 of this book.

Original image   Segmented image

Figure 12.14 Color image segmentation with Gaussian mixtures: (left) 800×533 original image, (right) segmented image in $G = 4$ elements.

Figure 12.13 shows word weights (formally the ϕ values) for the five estimated topics. One can see here that some words are specific to certain topics whereas others are used in almost every topic. For instance, the words "Gaussian" and "model-based" have significant weights in several topics, which seems quite natural given the nature of this book.

Listing 12.5: Latent Dirichlet allocation on some chapters of this book

```
# Loading libraries
library(topicmodels)

# Loading data and lexicalization
# (the 3 texts that are used here can be replaced by any text file)
A = scan('MBC-chapter5.txt',what='character')
B = scan('MBC-chapter9.txt',what='character')
C = scan('MBC-chapter12.txt',what='character')
Docs = c(paste(A,collapse=' '),paste(B,collapse=' '),
    paste(C,collapse=' '))
x <- Corpus(VectorSource(Docs))
dtm <- DocumentTermMatrix(x,control=list(steming=TRUE,
    stopwords=TRUE,minWordLength=3,removeNumbers=TRUE,
    removePunctuation=TRUE))

# Running LDA
K = 5; W = 10
out = LDA(dtm,K)

# Visualization of the results
image(x=1:K,y=1:W,z=matrix(0,K,W),col=0,xlab='',ylab='',xaxt='n',
    yaxt='n',main='Estimated topics by LDA')
axis(1,at=1:K,labels=paste('Topic',1:K))
text(expand.grid(x=0.75:(K-0.25), y=1:W), labels=t(terms(out,W))[,W
    :1], col=rep((1:K),W), pos=4)
```

12.3 Model-based Clustering for Image Analysis

In image analysis, model-based clustering is also a common tool for image segmentation, image denoising or inpainting (restoration of damaged images). The following sections present some basics of image analysis with mixture modeling.

12.3.1 Image Segmentation

Let us first consider image segmentation, which is one of the oldest problems in image analysis. It has many applications, including in medical imaging (diagnosis, study of anatomical structure, surgery planning, tumor location), recognition tasks (fingerprint recognition, iris recognition) and object detection (pedestrian detection, face detection, location of objects in satellite images). Classical techniques for image segmentation include thresholding methods, compression-based methods, histogram-based methods, edge detection, region-growing methods and clustering.

Although earlier image segmentation techniques were naturally based on k-means, mixture models were also rapidly adopted in the image analysis field. Let us consider all pixels of a color (RGB) image, such as $x_i \in [0, 255]^3$, denoting the red, green and blue (RGB) intensities for the ith pixel of the image. A basic application of model-based clustering for image segmentation would be to cluster all 3-dimensional pixels into G groups. Note that the choice of the number of groups G depends on the purpose of the segmentation. This could be, for example,

Listing 12.6: Image segmentation with Gaussian mixtures

```
# Loading libraries
library(jpeg)
library(MBCbook)

# Loading of the image
# (the image can be downloaded on the book website
#  or replaced by any JPEG image)
Im = readJPEG('CubaBeach.jpg')
X = as.data.frame(apply(Im,3,rbind))

# Clustering
ind = sample(nrow(X),10000)
out = mixmodCluster(X[ind,],4)
res = mixmodPredict(X,out@bestResult)
P = res@proba

# Going back to the image format
G = out@bestResult@nbCluster
nr = nrow(Im); nc = ncol(Im)
ImP = array(NA,c(nr,nc,G))
for (i in 1:nc){
  ImP[,i,] = P[((i-1)*nr+1):(i*nr),]
}

# Display of posterior probabilities as images
imshow(Im[,,1],main='Original image')
par(mfrow=c(2,2))
for (g in 1:G) imshow(ImP[,,g],main=paste('Posterior proba. for
    group',g))

# Export the segmented image
ImCl = apply(ImP,c(1,2),function(x){return(
    out@bestResult@parameters@mean[which.max(x),])})
ImCl = aperm(ImCl,c(2,3,1))
writeJPEG(ImCl,'CubaBeach-Clustered.jpg',quality=0.95)
```

to identify different elements which are present in the image, or to compress the image. In the first case, the number of groups will correspond to the number of elements to recognize (usually a small number) whereas, in the second case, it will depend on the desired level of compression (usually a high number).

Listing 12.6 provides some R code for image segmentation with the goal of recognizing the different elements (sky, sand, sea and wood) of the picture shown on the left panel of Figure 12.14 (picture taken at Cayo Jutias, Cuba). As the number of elements is fairly clear here, we run Rmixmod with $G = 4$ on the three-dimensional pixels of the image. Since the image is large (800×533 pixels), Rmixmod was run on a subset of 10,000 pixels with $G = 4$ and the whole set was then classified using the estimated model. This strategy avoids running the EM algorithm with 426,400 observations, which could be prohibitive in terms of computing time on some computers in current use.

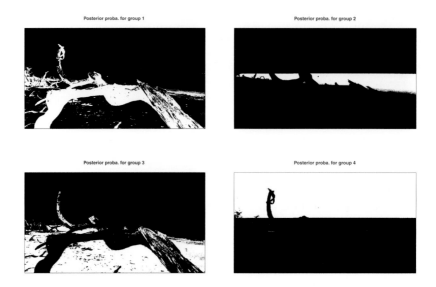

Figure 12.15 Color image segmentation with Gaussian mixtures: posterior probabilities as gray levels (black = 0, white = 1) for the $G = 4$ groups.

The right panel of Figure 12.14 shows the segmented image. Figure 12.15 shows the posterior probabilities that each pixel belongs to each of the $G = 4$ groups as gray levels (black = 0, white = 1), The image is globally well segmented: the sea, sand and sky are almost perfectly recovered and only the wood parts are sometimes confused with the sand. Let us recall that this segmentation has been obtained without taking any spatial information into account.

12.3.2 Image Denoising

The denoising of images is also a central problem in image analysis since all recording devices (both film and digital) can be affected by noise. Common approaches for noise reduction in images include smoothing filters, anisotropic diffusion (based on partial differential equations), wavelet transform and statistical techniques. It is possible to get satisfying denoising results using model-based clustering.

It is first necessary to decompose the image into small images (called patches), for instance of size 8×8 pixels. Thus, the original image of Figure 12.14, which has 800×533 pixels, is transformed into a set of 419,760 patches which are stored as 64-dimensional vectors. The idea of the cluster-based denoising is to form G groups of patches and to denoise each patch using the cluster means.

The model assumed for the patches is as follows: let X be the (unobserved) original patch and Y its noisy observation. Let us assume that both variables are related as follows:

$$Y = \phi(X) + \varepsilon,$$

where $\phi(X) = UX$ is a linear degradation operator and ε a centered Gaussian noise with variance $\sigma^2 I_d$. In most recent work in image denoising, ϕ and σ are assumed to be known. Usually, ϕ is assumed to be the identity operator. Assuming that the distribution of X is a mixture of G Gaussians:

$$X \sim \sum_{g=1}^{G} \tau_g \mathcal{N}(\mu_g, \Sigma_g),$$

we can show that the conditional expectation of X given $Y = y$, which minimizes the expected mean squared error, is as follows:

$$E[X|Y = y] = \sum_{g=1}^{G} P(Z = g|Y = y)E[X|Z = g, Y = y].$$

Making use of the conditional expectation formula for a partitioned Gaussian vector, $E[X|Z = g, Y = y]$ can be written as:

$$E[X|Z = g, Y = y] = \mu_g + \Sigma_g U^T \left(U\Sigma_g U^T + \sigma^2 I_d \right)^{-1} \left(y - U\mu_g \right).$$

We finally obtain:

$$E[X|Y = y] = \sum_{g=1}^{G} P(Z = g|Y = y) \left[\mu_g + \Sigma_g U^T \left(U\Sigma_g U^T + \sigma^2 I_d \right)^{-1} \left(y - U\mu_g \right) \right].$$

Once all patches have been denoised, the image pixels can be reconstructed by averaging over all the patches they belong to. Listing 12.7 provides a simple implementation of this denoising methodology. Notice that, here again, only a subsample of the patches is used to estimate the mixture model, so as to accelerate the computations. The number G of groups has been set to 40 and the model has also been fixed to a spherical Gaussian one in this simple application. However, both the number of groups and the clustering model could be chosen using any model selection criterion.

Figure 12.16 first presents a few group means as images. The patch means capture the wide range of textures and forms present in the image. Figure 12.17 shows the denoising result of the above approach on a noisy version of the Cuban beach picture. The left panel of the figure presents the original black-and-white image, its noisy version and the denoising result. The bottom row shows a zoom of a specific part of the above full-size images. Clearly, although the simple procedure that we applied here allows a significant denoising, the problem remains difficult. The difficulty is here mainly related to the dimension of the patches, here $d = 64$.

It is therefore natural to replace the simple Gaussian mixture model with a model designed for the high dimensional context here. Figure 12.18 shows the denoising that we can reach when using the HD-GMM model of HDDC (see Chapter 8.4.4) in place of the spherical Gaussian. This yields an excellent denoising of the image. This model is known as the HDMI (high-dimensional model for

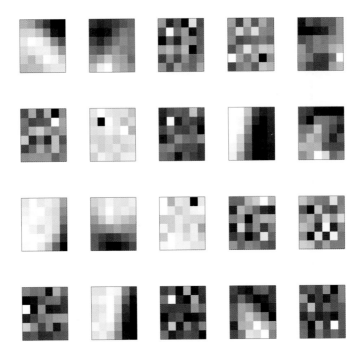

Figure 12.16 Image denoising with spherical Gaussian mixtures: visualization of the group means as 8×8 images.

image denoising) model (Houdard et al., 2019) and is currently the state-of-the-art in patch-based image denoising. It is worth noticing that the end-to-end statistical model of HDMI allows it to estimate the noise level in an unsupervised manner.

12.3.3 Inpainting Damaged Images

Inpainting refers to the process of reconstructing the damaged or missing parts of an image. The applications of inpainting are numerous in everyday life: image restoration, image manipulation, special effects, panoramas, video editing. Once again, mixture modeling is an efficient tool for image inpainting.

The model is the same as the one used above for image denoising, except that the linear transformation $\phi(\cdot)$ of the patch X_i is:

$$\phi(X_i) = \mathrm{diag}(u_i)X_i,$$

where the vector u_i is binary and encodes the observation or the absence of the pixels of the patch X_i. Notice that the mask $U = \{u_1, ..., u_n\}$ is assumed to be observed since it is, in most cases, easy to identify the damaged pixels. This model corresponds to the one used by Yu et al. (2012) in their piecewise linear estimator (PLE) approach.

The initialization phase of the EM algorithm has to be specific in this context.

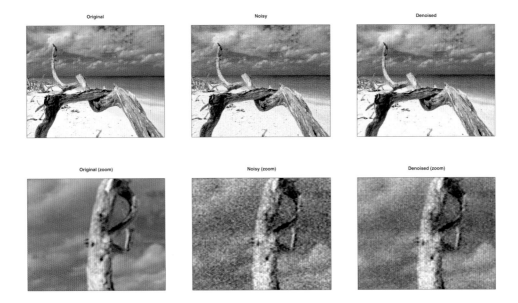

Figure 12.17 Image denoising with spherical Gaussian mixtures: from left to right, the three panels present respectively the original BW image, its noisy version and the denoising result. The bottom row shows a zoom on a specific part of the above images.

Yu et al. (2012) proposed initializing the cluster means with artificial patches supposed to describe the different types of patches that one can find in natural images well.

Figure 12.19 shows the Cuban beach picture (left panel) where 20% of the pixels have been damaged by overprinting a text (center panel). Here, we used the E-PLE implementation of Wang (2013), which is available on the IPOL website at `http://demo.ipol.im/demo/54/`. As one can observe, the result is rather impressive and differences from the original images are hard to find.

12.4 Model-based Co-clustering

Block clustering or co-clustering methods aim to cluster both the rows and the columns of a large array of data in the same analysis. These methods are useful for summarizing large data sets by recovering a hidden block structure. Since more and more huge data sets are available, more and more block clustering methods have been proposed. The checkerboard structure involved in block clustering is attractive for summarizing large data sets with realistic and suggestive blocks. Since block clustering methods treat the rows and the columns of the data array symmetrically, they are especially useful for analyzing tables where this symmetry is natural, such as contingency tables or binary tables. Block clustering methods do not make sense if the variables are not homogeneous.

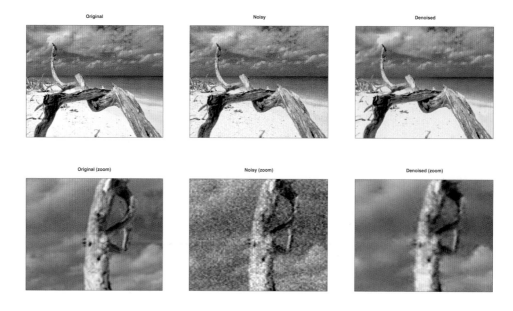

Figure 12.18 Image denoising with HD-GMM: as above, from left to right, the three panels present respectively the original BW image, its noisy version and the denoising result. The bottom row shows a zoom on a specific part of the above images.

Original image   Damaged image   Reconstructed image

Figure 12.19 Image inpainting with S-PLE (Wang, 2013): from left to right, the three panels present respectively the original image, its damaged version and the inpainting result.

The review papers by Madeira and Oliveira (2004) and Brault and Lomet (2015) described extensive lists of block clustering methods. There are two kinds of block clustering methods: deterministic approaches (see for instance Govaert, 1977; Banerjee et al., 2007), and model-based approaches. The model-based view has been implemented through the maximum likelihood methodology (Govaert and Nadif, 2008) and through Bayesian inference (Meeds and Roweis, 2007; Wyse and Friel, 2012). The book of Govaert and Nadif (2014) gives a readable presentation of the model-based methods for co-clustering and is centered around the Latent Block Model.

Listing 12.7: Image denoising with Gaussian mixtures

```
# Loading libraries
library(jpeg)
library(MBCbook)

# Loading and addition of noise on an image
Im = 255 * readJPEG('CubaBeach.jpg')[,,1]
sigma = 20
ImNoise = Im + matrix(rnorm(nrow(Im)*ncol(Im),0,sigma),nrow=nrow(Im
    ))
ImNoise[ImNoise>255] = 255; ImNoise[ImNoise<0] = 0
par(mfrow=c(1,2))
imshow(Im,main='Original')
imshow(ImNoise,main='Noisy')

# Decomposition of the image as 8x8 patches and clustering
# (it may take several minutes to run!)
X = as.data.frame(imageToPatch(ImNoise,8))
ind = sample(nrow(X), 10000)
out = mixmodCluster(X[ind,],40,model=mixmodGaussianModel(family=c("
    spherical")))
res = mixmodPredict(X,out@bestResult)

# Display group means as small images
par(mfrow=c(4,5))
mu = out@bestResult@parameters@mean
for (g in 1:40) imshow(t(matrix(mu[g,64:1],ncol=8,byrow=FALSE)))

# Final denoising and image reconstruction (it may take several
    minutes to run!)
Xdenoised = denoisePatches(X,out,P = res@proba,sigma = sigma)
ImRec = reconstructImage(Xdenoised,nrow(Im),ncol(Im))
ImRec[ImRec>255] = 255; ImRec[ImRec<0] = 0
par(mfrow=c(1,3)); imshow(Im,main='Original')
imshow(ImNoise,main='Noisy')
imshow(ImRec,main='Denoised')
```

12.4.1 The Latent Block Model

The Latent Block Model (LBM) is structured as follows. A population of n observations described by d variables of the same nature is available. The assumption that the variables are of the same nature is necessary for this approach. It is needed to ensure that decomposing the data set in terms of a block structure makes sense.

Let $y = (y_{ij}, i = 1, \ldots, n; j = 1, \ldots, d)$ be the data matrix. It is assumed that there is a partition into G row clusters $z = (z_{ig}; i = 1, \ldots, n; g = 1, \ldots, G)$ and a partition into L column clusters $w = (w_{j\ell}; j = 1, \ldots, d; \ell = 1, \ldots, L)$. The z_{ig}s (respectively $w_{j\ell}$s) are binary indicators of row i (respectively column j) belonging to row cluster g (respectively column cluster ℓ), such that the random variables y_{ij} are conditionally independent knowing z and w with parameterized density

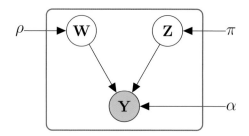

Figure 12.20 Graphical representation of the latent block model (LBM).

$\varphi(y_{ij}; \theta_{g\ell})^{z_{ig}w_{j\ell}}$. Thus, the conditional density of y knowing z and w is

$$f(y|z, w; \theta) = \prod_{i,j,g,\ell} \varphi(y_{ij}; \theta_{g\ell})^{z_{ig}w_{j\ell}}.$$

Moreover, it is assumed that the row and column labels are independent: $p(z, w) = p(z)p(w)$ with $p(z) = \prod_{ig} \pi_g^{z_{ig}}$ and $p(w) = \prod_{j\ell} \rho_\ell^{w_{j\ell}}$, where $(\pi_g = P(z_{ig} = 1), g = 1, \ldots, G)$ and $(\rho_\ell = P(w_{j\ell} = 1), \ell = 1, \ldots, L)$ are the mixing proportions. Hence, the marginal density of y is a mixture density

$$f(y; \eta) = \sum_{(z,w) \in \mathcal{Z} \times \mathcal{W}} p(z; \pi)p(w; \rho)f(y|z, w; \theta),$$

where $\mathcal{Z}$ and $\mathcal{W}$ denote the sets of possible row labels z and column labels w, and $\eta = (\pi, \rho, \theta)$, with

$$\pi = (\pi_1, \ldots, \pi_G), \rho = (\rho_1, \ldots, \rho_L) \text{ and } \theta = (\theta_{g\ell}; g = 1, \ldots, G; \ell = 1, \ldots, L).$$

The density of y parameterized by η is

$$f(y; \eta) = \sum_{(z,w) \in \mathcal{Z} \times \mathcal{W}} \prod_{i,g} \pi_g^{z_{ig}} \prod_{j,\ell} \rho_\ell^{w_{j\ell}} \prod_{i,j,g,\ell} \varphi(y_{ij}; \alpha_{g\ell})^{z_{ig}w_{j\ell}}. \tag{12.5}$$

The LBM involves a double missing data structure, namely z and w, which makes statistical inference more difficult than for the standard mixture model.

12.4.2 Estimating LBM Parameters

With G and L fixed, the likelihood of the model is

$$\mathcal{L}(\theta) = \sum_{(z,w) \in \mathcal{Z} \times \mathcal{W}} p(z; \theta)p(w; \theta)f(y|z, w; \theta)$$

$$= \sum_{(z,w) \in \mathcal{Z} \times \mathcal{W}} \prod_{i,g} \pi_k^{z_{ig}} \prod_{j,\ell} \rho_\ell^{w_{j\ell}} \prod_{i,j,g,\ell} \varphi(y_{ij}; \theta_{g\ell})^{z_{ig}w_{j\ell}}.$$

Even for small tables, computing this likelihood (or its logarithm) is almost impossible. Indeed, for $G = L = 2$, $n = 30$ and $d = 6$, it involves the computation of $G^n \times L^d \approx 10^{12}$ terms.

Similarly, deriving the maximum likelihood estimator with the EM algorithm is challenging. The E-step requires the computation of the joint conditional distributions of the missing labels

$$e_{i,j,g,\ell}^{(c)} = P(z_{ig}w_{j\ell} = 1|y;\theta^{(c)}),$$

for $i = 1, \ldots, n, j = 1, \ldots, d, g = 1, \ldots, G, \ell = 1, \ldots, L$ and $\theta^{(c)}$ being a current value of the model parameters. Thus, the E-step requires computing too many terms that cannot be factorized as for a standard mixture, due to the dependence of the row and column labels conditionally on the observations. Computing penalized model selection criteria such as BIC is therefore also challenging.

The Variational EM Algorithm

To tackle this problem, Govaert and Nadif (2008) proposed using a variational approximation of the EM algorithm by assuming that the conditional joint distribution of the labels knowing the observations factorizes as $q_{zw}^{(c)}(z, w) = q_z^{(c)}(z)q_w^{(c)}(w)$. This VEM algorithm has been shown to provide reasonable estimates of the LBM model in different contexts with continuous, binary, categorical or counting data matrices (Govaert and Nadif, 2008). In this case

$$e_{i,j,g,\ell}^{(c)} \simeq s_{ig}^{(c)}t_{j\ell}^{(c)},$$

where $s_{ig}^{(c)} = P(z_{ig}|y;\theta^{(c)})$ and $t_{j\ell}^{(c)} = P(w_{j\ell} = 1|y;\theta^{(c)})$.

However, the VEM algorithms have some drawbacks that are worth noting. Like most variational approximation algorithms, VEM appears to be quite sensitive to starting values. It also has a tendency to provide solutions with empty clusters and it can only compute a lower bound of the maximum likelihood, called the *free energy*.

The Stochastic EM Algorithm

A possible way to attenuate the dependence of VEM to its initial values is to use stochastic versions of EM which do not stop at the first encountered fixed point of EM (McLachlan and Krishnan, 1997, Chapter 6). The basic idea of these stochastic EM algorithms is to incorporate a stochastic step between the E- and M-steps where the missing data are simulated according to their conditional distribution knowing the observed data and a current estimate of the model parameters. For the LBM, it is not possible to simulate the missing labels z and w in a single exercise, and a Gibbs sampling scheme is required to simulate the pair (z, w). The SEM-Gibbs algorithm is a simple adaptation to the LBM of the standard SEM algorithm of Celeux and Diebolt (1985).

The formulae of the M-steps for VEM and SEM-Gibbs are essentially the same, except that the probabilities of the labels z_i and w_i, conditional on the observations y, are replaced by binary indicator values. While VEM is based on the variational approximation of the LBM, SEM-Gibbs uses no approximation, but runs a Gibbs sampler to simulate the unknown labels with their conditional distribution knowing the observations and a current estimate of the parameters.

Hence, SEM-Gibbs does not increase the log-likelihood at each iteration, but it generates an irreducible Markov chain with a unique stationary distribution which is expected to be concentrated about the ML parameter estimate. Thus a natural estimate of θ derived from SEM-Gibbs is the mean $\bar{\theta}$ of $(\theta^{(c)}; c = B+1, \ldots, B+C)$ obtained after a burn-in period of length B.

Like every stochastic algorithm for estimating mixture models, SEM-Gibbs is theoretically subject to label switching (see Frühwirth-Schnatter, 2006, Section 3.5.5). However, this possible drawback of SEM-Gibbs does not occur in most practical situations. Numerical experiments presented in Keribin et al. (2015) for binary data show that SEM-Gibbs is far less sensitive to starting values than VEM. Those results led them to advocate initializing the VEM algorithm with the SEM-Gibbs mean parameter estimate $\bar{\theta}$ to get a good approximation of the ML estimate for the latent block model.

Gibbs Sampling and Variational Bayes Algorithm

If SEM-Gibbs is insensitive to its initial values, it is not useful to avoid solutions with empty clusters. One purpose of Bayesian inference in statistics is to regularize ML estimates in poorly posed settings. In the LBM setting, Bayesian inference could be thought of as useful to avoid empty cluster solutions and thus to attenuate the "empty cluster" problem. In particular, for the LBM on categorical data or contingency tables, it is possible to consider proper and independent weakly informative prior distributions for the mixing proportions π and ρ and the parameter θ as a product of $G \times L$ non-informative priors on each multinomial parameter $\alpha_{g\ell} = (\alpha_{g\ell}^h)_{h=1,\ldots,H}$.

For instance when the data are categorical, it is possible to consider proper and independent weakly informative prior distributions for the mixing proportions π and ρ, and for parameter α as a product of $G \times L$ weakly informative priors on each multinomial parameter $\alpha_{g\ell} = (\alpha_{g\ell}^h)_{h=1,\ldots,H}$:

$$\pi \sim \mathcal{D}(a, \ldots, a), \quad \rho \sim \mathcal{D}(a, \ldots, a), \quad \alpha_{k\ell} \sim \mathcal{D}(b, \ldots, b),$$

where $\mathcal{D}(v, \ldots, v)$ denotes a Dirichlet distribution with parameter v. Thus a Gibbs sampling iteration is:

1. simulation of $z^{(c+1)}$ according to $p(z|y, w^{(c)}; \theta^{(c)})$ as in SEM-Gibbs,
2. simulation of $w^{(c+1)}$ according to $p(w|y, z^{(c+1)}; \theta^{(c)})$ as in SEM-Gibbs,
3. simulation of $\pi^{(c+1)}$ according to

$$\mathcal{D}(a + z_{.1}^{(c+1)}, \ldots, a + z_{.G}^{(c+1)}),$$

where $z_{.g} = \sum_i z_{ig}$ denotes the number of lines in line cluster g,
4. simulation of $\rho^{(c+1)}$ according to

$$\mathcal{D}(a + w_{.1}^{(c+1)}, \ldots, a + w_{.L}^{(c+1)}),$$

where and $w_{.\ell} = \sum_j w_{j\ell}$ denotes the number of columns in column cluster ℓ,

5 simulation of $\alpha_{g\ell}^{(c+1)}$ according to

$$\mathcal{D}(b + N_{g\ell}^{1\ (c+1)}, \ldots, b + N_{g\ell}^{r\ (c+1)})$$

for $g = 1, \ldots, G; \ell = 1, \ldots, L$ and with

$$N_{g\ell}^{h\ (c+1)} = \sum_{i,j} z_{ig}^{(c+1)} w_{j\ell}^{(c+1)} y_{ij}^h. \tag{12.6}$$

As Gibbs sampling explores the whole distribution, it is subject to label switching. However, it can be avoided by using the identifiability conditions given in Keribin et al. (2015). Then, since Bayesian inference is used here in a regularization perspective, the model parameter can be estimated by maximizing the posterior density $p(\theta|y)$, with a variational Bayes EM algorithm initiated with the Gibbs sampler solution. This leads to the MAP estimate

$$\widehat{\theta}_{MAP} = \arg\max_\theta p(\theta|y).$$

The V-Bayes algorithm is as follows for homogeneous categorical variables with r levels:

- the E-step relies on the computation of the conditional expectation of the complete log-likelihood $E\left(\log p(y, z, w|\theta)|y, \theta^{(c)}\right)$,
- the M-step consists in the estimation of the posterior modes:

$$\pi_k^{(c+1)} = \frac{a - 1 + s_{.k}^{(c+1)}}{n + G(a - 1)}, \quad \rho_\ell^{(c+1)} = \frac{a - 1 + t_{.\ell}^{(c+1)}}{d + L(a - 1)},$$

$$\alpha_{g\ell}^{h\ (c+1)} = \frac{b - 1 + \sum_{i,j} s_{ig}^{(c+1)} t_{j\ell}^{(c+1)} y_{ij}^h}{r(b - 1) + s_{.k}^{(c+1)} t_{.\ell}^{(c+1)}},$$

where $s_{ik}^{(c+1)}$ and $t_{j\ell}^{(c+1)}$ are the current conditional probabilities of the labels and $s_{.k}^{(c+1)}$ and $t_{.\ell}^{(c+1)}$, their sum over the lines and columns respectively.

The hyperparameters a and b act as regularization parameters, and the choice for a and b is important. Following the analysis of Frühwirth-Schnatter (2011a), Keribin et al. (2015) propose taking $a = 4$ for moderate dimensions ($g < 8$) and $a = 16$ for larger dimensions to avoid empty clusters.

12.4.3 Model Selection

It is hard use the BIC criterion to select a model since the likelihood of the LBM is not available. But exact values of ICL can be computed for categorical data or contingency tables since conjugate proper non-informative prior distributions are available in these contexts. For instance with categorical data, it is (see Keribin et al., 2015)

$$\begin{aligned}
\mathrm{ICL}(g,m) = &\log\Gamma(Ga) + \log\Gamma(La) - (L+G)\log\Gamma(a)\\
&+ LG(\log\Gamma(rb) - r\log\Gamma(b))\\
&- \log\Gamma(n+Ga) - \log\Gamma(d+La)\\
&+ \sum_g \log\Gamma(z_{.g}+a) + \sum_\ell \log\Gamma(w_{.\ell}+a)\\
&+ \sum_{g,\ell}\Big[\Big(\sum_h \log\Gamma(N_{g\ell}^h+b)\Big)\\
&- \log\Gamma(z_{.g}w_{.\ell}+rb)\Big],
\end{aligned}$$

where

$$(\hat{z},\hat{w}) = \arg\max_{(z,w)} p(z,w|y;\hat{\theta}),$$

with $\hat{\theta}$ being the estimate of the LBM parameter computed from the V-Bayes algorithm initialized by the Gibbs sampler. The maximizing partition $(\hat{z},\hat{w})$ is obtained with a MAP rule after the last V-Bayes step. It requires an alternating optimization algorithm, which repeats until convergence the two following steps:

- $z^{(c)} = \arg\max_z p(z|y,w^{(c-1)};G,L,\hat{\theta})$,
- $w^{(c)} = \arg\max_w p(w|z^{(c)};G,L,\hat{\theta})$.

12.4.4 An Illustration

We now apply the `blockcluster` package for R for the co-clustering of the Amazon Fine Foods data set, which is available in the `MBCbook` package under the name `amazonFineFoods`. The data set can be freely downloaded at the following address: `https://snap.stanford.edu/data/web-FineFoods.html`. A time horizon of 10 years is considered, up to October 2012. The number of reported reviews is 568,464 and, in the original data set, each row corresponds to one review. Some additional information is reported for each review: the user/product numerical identifiers, a summary of the review and a rating attributed to the product by the user.

The original data set was preprocessed as follows. To focus on the most meaningful part of the data, we only considered the users reviewing more than 20 times and the products being reviewed more than 50 times. The resulting binary matrix indicates whether a user has noted and reviewed a product or not. The left panel of Figure 12.21 displays the raw binary matrix. Listing 12.8 presents the code for running the `coclusterBinary` function of the `blockcluster` package on the `amazonFineFoods` data, with a fixed choice of numbers of row and column clusters.

The left panel of Figure 12.21 displays the $(5,8)$ co-clustering solution proposed by `blockcluster`. It turns out that the data have a clear pattern in terms of blocks of users and products. Indeed, some groups of users have different behaviors

Listing 12.8: Co-clustering of the `classic3` data set with `blockcluster`

```
# Loading libraries
library(blockcluster)
library(MBCbook)
data(amazonFineFoods)

# Co-clustering with blockcluster
X = as.matrix(amazonFineFoods)
out = coclusterBinary(X,nbcocluster = c(5,8))

# Plotting original and reorganized matrices
image(amazonFineFoods,main='Original data matrix',
      xaxt='n',yaxt='n')
image(amazonFineFoods[order(out@rowclass),order(out@colclass)],
      main='Data matrix sorted by groups',
      xaxt='n',yaxt='n')
```

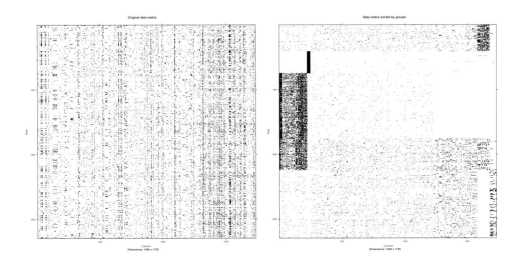

Figure 12.21 Co-clustering of the `amazonFineFoods` data: raw binary matrix (left) and reorganized matrix according to row and column clusters found by `blockcluster`.

regarding the groups of products. Interestingly, the second group of users only reviewed and graded the second group of products. It turns out that all products from this group are from the same brand. In addition, when looking at the grades and reviews provided by those users on those products, they always provide good grades and almost the same positive review. Given these elements, it seems likely that the brand paid users to positively evaluate their products.

12.5 Bibliographic Notes

Clustering of Functional Data

The analysis of functional data is an active topic currently, motivated by the fact that many systems produce functional or time series data. Among the recent contributions to this domain, we can cite the robust version of funHDDC, obtained by trimming, proposed by Rivera-García et al. (2018), the functional k-means algorithm of Yamamoto and Hwang (2017) based on subspace separation to cluster the functions, and the funLBM model that Bouveyron et al. (2018a) introduced for the co-clustering of multivariate functional data. Frühwirth-Schnatter (2011b) also proposed a survey of model-based clustering techniques for time series that can be used for panel data analysis.

Clustering of Texts

The LDA model has become a very popular model and it has been extended in a variety of situations. Among those extensions, the relational topic model (RTM) (Chang and Blei, 2009) models the links between documents as binary random variables conditioned on their contents. The itopic model (Sun et al., 2009) extends RTM to weighted networks. LDA was also extended to the dynamic case by Blei and Lafferty (2006) in order to study the evolution of the use of the topics along the time in a corpus of documents. Note that a limitation of LDA would be the inability to take into account possible topic correlations. This is due to the use of the Dirichlet distribution to model the variability among the topic proportions. To overcome this limitation, the correlated topic model (CTM) has been recently developed by Blei and Lafferty (2005). Let us finally notice that the LDA model has been combined recently with other models to handle complex data structures. For instance, Bouveyron et al. (2018b) proposed the STBM model (see Chapter 10), which encompasses both the stochastic block model (SBM) and LDA to cluster networks with textual edges. The reader may refer to Blei (2012) for an overview on probabilistic topic models.

Clustering for Image Analysis

Model-based clustering has been used for a range of other image analysis applications. One such application is the identification of features in spatial image data. It was the basis of a method for identifying ice floes from satellite images of Arctic regions using model-based clustering of pixel locations centered around principal curves, rather than around the usual straight lines generated by multivariate Gaussian model-based clustering (Banfield and Raftery, 1992). A similar method was used to estimate seismic fault lines from catalogs of the spatial locations of earthquakes (Stanford and Raftery, 2000). Implementation issues were discussed by Fraley and Raftery (2006).

Model-based clustering has also been used for a range of image segmentation problems. For example, it was used to detect the active regions in microarray images, taking account of artefacts such as scratches, blank spots and holes in the middle of active areas, or donut holes (Li et al., 2005). It has also been used for detecting regions of interest, or possible tumors, in dynamic breast MRI (similar to functional MRI) (Forbes et al., 2006).

One issue with model-based clustering for image analysis is that the objects to be clustered are often the pixels or voxels in the image, and so the sample size can become very large. This can cause computational problems for model-based clustering, and several approaches for resolving them have been proposed. For example, Wehrens et al. (2004) investigated subsampling approaches where the clustering model is estimated by the EM algorithm on a random subsample of the pixels, and then inference for the full image is carried out by a single E-step. They found that a random subsample of 2,000 pixels was adequate in most cases.

However, when some clusters were very small, with few pixels or objects, this approach might miss these clusters, and Fraley et al. (2005) described an incremental model-based clustering method to get around this problem. Another approach to this problem is via model-based cluster trees (Murtagh et al., 2005).

Latent Block Models for Co-clustering

Interest in methods for co-clustering has been growing in recent years because of the increasing presence of massive data sets and there are numerous recent publications on the subject. Most of them are extensions of the LBM model to specific data types, such as counting data (Govaert and Nadif, 2010), real data (Lomet, 2012), categorical data (Keribin et al., 2015), ordinal data (Jacques and Biernacki, 2018) and functional data (Bouveyron et al., 2018a).

Bergé et al. (2019) extended the LBM model to text data. They proposed the latent topic block model (LTBM), which takes into account the text content for forming the blocks while recovering the topics present in the texts. As a practical example, the LTBM may be used to co-cluster matrices of reviews of products written by customers. Several new inference procedures have also been explored for LBM, including Bayesian inference (Wyse and Friel, 2012; Keribin et al., 2015) and greedy search approaches (Wyse et al., 2017). A recent algorithm, known as *Largest Gaps*, of Brault and Channarond (2016) allows one to perform clustering and model selection in LBM, using only the marginals (i.e. the sum of the entries in the rows and columns) of the incidence matrix. The approach greatly reduces the computational complexity of the estimation procedure. However, the algorithm is only consistent under certain assumptions concerning the degree distributions of rows and columns.

List of R Packages

We list below the R packages that are used in the code listings throughout this monograph. The package version at the printing date of the book is also indicated (within brackets) to ensure the proper functioning of the code.

Chapter 2

- mclust (5.4.1)

Chapter 3

- Rmixmod (2.1.2)
- prabclus (2.2–6)
- covRobust (1.1–3)
- tclust (1.4–1)
- fpc (2.1–11.1)

Chapter 4

- klaR (0.6–14)
- rrcov (1.4–7)

Chapter 5

- robustDA (1.1)
- adaptDA (1.0.1)

Chapter 6

- FactoMineR (1.4.1)
- clustMD (1.2.1)
- HTSCluster (2.0.8)

Chapter 7

- clustvarsel (2.3.3)
- SelvarMix (1.0)
- LCAvarsel (1.1)

Chapter 8

- HDclassif (2.1.0)
- mvtnorm (1.0–8)
- pls (2.7–0)
- pgmm (1.2.3)
- EMMIXmfa (2.0.7)
- FisherEM (1.5.1)

Chapter 9

- teigen (2.2.2)
- EMMIXskew (1.0.3)
- EMMIXuskew (0.11–6)
- flowClust (3.20.0)
- MixGHD (2.2)

Chapter 10

- network (1.13.0.1)
- mixer (1.9)
- lda (1.4.2)
- compositions (1.40–2)
- latentnet (2.9.0)
- ergm.count (3.3.0)
- VBLPCM (2.4.5)

Chapter 11

- flexmix (2.3–14)
- mixtools (1.1.0)
- sn (1.5–3)

Chapter 12

- funcy (1.0.0)
- funFEM (1.1)
- funHDDC (2.2.0)
- topicmodels (0.2–7)
- jpeg (0.1–8)
- blockcluster (4.2.6)

Bibliography

Ackerson, G. A., and Fu, K. S. 1970. On state estimation in switching environments. *IEEE Transactions on Automatic Control*, **15**, 10–17. 108

Adanson, M. 1757. *Histoire Naturelle du Sénégal. Coquillages. Avec la relation abregée d'un voyage fait en ce pays, pendant les années 1749, 50, 51, 52 et 53*. Paris: Bauche. 2

Adanson, M. 1763. *Familles de Plantes*. Paris: Vincent. 2

Agresti, A. 2002. *Categorical Data Analysis*. 2nd edn. New York: Wiley. 163, 166, 185, 191

Ahlquist, J. S., and Breunig, C. 2012. Model-based clustering and typologies in the social sciences. *Political Analysis*, **20**, 92–112. 77, 78

Airoldi, E. M., Blei, D. M., Fienberg, S. E., Goldberg, A., Xing, E. P., and Zheng, A. X. 2007. *Statistical Network Analysis: Models, Issues and New Directions*. Lecture Notes in Computer Science, vol. 4503. Berlin: Springer. 294

Airoldi, E. M., Blei, D. M., Fienberg, S. E., and Xing, E. P. 2008. Mixed-membership stochastic blockmodels. *Journal of Machine Learning Research*, **9**, 1981–2014. 304, 306, 350

Aitchison, J., and Aitken, C. G. G. 1976. Multivariate binary discrimination by the kernel method. *Biometrika*, **63**, 413–420. 123, 169

Akaike, H. 1974. A new look at the statistical model identification. *IEEE Transactions on Automatic Control*, **19**, 716–723. 133

Allman, E. S., Matias, C., and Rhodes, J. A. 2009. Identifiability of parameters in latent structure models with many observed variables. *The Annals of Statistics*, **37**(6A), 3099–3132. 167

Ambroise, C., Grasseau, G., Hoebeke, M., Latouche, P., Miele, V., Picard, F., and LAPACK authors. 2013. *mixer: Random graph clustering*. R package version 1.7. 297, 300, 301

Anderlucci, L. 2012. *Comparing Different Approaches for Clustering Categorical Data*. Ph.D. thesis, Università di Bologna. 172

Anderlucci, L., and Hennig, C. 2012. Comparing different approaches for clustering categorical data. *Quaderni di Statistica*, **14**, 1–4. 167, 172

Anderson, E. 1935. The irises of the Gaspe peninsula. *Bulletin of the American Iris Society*, **59**, 2–5. 5, 154

Anderson, T. W. 2003. *An Introduction to Multivariate Statistical Analysis*. 3rd edn. New York: Wiley. 110

Andrews, J. L., and McNicholas, P. D. 2011a. Extending mixtures of multivariate t-factor analyzers. *Statistics and Computing*, **21**(3), 361–373. 257, 261

Andrews, J. L., and McNicholas, P. D. 2011b. Mixtures of modified t-factor analyzers for model-based clustering, classification, and discriminant analysis. *Journal of Statistical Planning and Inference*, **141**(4), 1479–1486. 261

Andrews, J. L., and McNicholas, P. D. 2012. Model-based clustering, classification, and discriminant analysis via mixtures of multivariate t-distributions: the tEIGEN family. *Statistics and Computing*, **22**(5), 1021–1029. 261

Andrews, J. L., McNicholas, P. D., and Subedi, S. 2011. Model-based classification via mixtures of multivariate t-distributions. *Computational Statistics and Data Analysis*, **55**(1), 520–529. 261

Andrews, J. L., Wickins, J. R., Boers, N. M., and McNicholas, P. D. 2015. *teigen: Model-based clustering and classification with the multivariate t-distribution.* R package version 2.1.0. 261

Andrews, J. L., Wickins, J. R., Boers, N. M., and McNicholas, P. D. 2018. teigen: An R package for model-based clustering and classification via the multivariate t distribution. *Journal of Statistical Software*, **83**(7), 1–32. 261

Arellano-Valle, R. B., and Genton, M. G. 2005. On fundamental skew distributions. *Journal of Multivariate Analysis*, **96**, 93–116. 267

Arellano-Valle, R. B., and Genton, M. G. 2010. Multivariate extended skew-*t* distributions and related families. *Metron*, **68**, 201–234. 271

Azzalini, A. 2014. *The Skew-Normal and Related Families.* Institute of Mathematical Statistics Monographs. Cambridge University Press. 267

Azzalini, A. 2015. *The R package sn: The Skew-Normal and Skew-t distributions.* R package version 1.3-0. 331

Azzalini, A., and Bowman, A. W. 1990. A look at some data on the Old Faithful geyser. *Journal of the Royal Statistical Society. Series C (Applied Statistics)*, **39**, 357–365. 7

Azzalini, A., and Capitanio, A. 1999. Statistical applications of the multivariate skew normal distribution. *Journal of the Royal Statistical Society. Series B (Statistical Methodology)*, **61**(3), 579–602. 267, 269

Azzalini, A., and Capitanio, A. 2003. Distributions generated by perturbation of symmetry with emphasis on a multivariate skew *t* distribution. *Journal of the Royal Statistical Society. Series B (Statistical Methodology)*, **65**, 367–389. 271

Azzalini, A., and Dalla Valle, A. 1996. The multivariate skew-normal distribution. *Biometrika*, **83**(4), 715–726. 267, 269

Azzalini, A., Browne, R. P., Genton, M. G., and McNicholas, P. D. 2016. On nomenclature for, and the relative merits of, two formulations of skew distributions. *Statistics and Probability Letters*, **110**, 201–206. 267, 270

Baek, J., McLachlan, G. J., and Flack, L. 2009. Mixtures of factor analyzers with common factor loadings: Applications to the clustering and visualisation of high-dimensional data. *IEEE Transactions on Pattern Analysis and Machine Intelligence*, **32**(7), 1298–1309. 238, 241, 246, 257

Banerjee, A., Dhillon, I., Ghosh, J., Merugu, S., and Modha, D. S. 2007. A generalized maximum entropy approach to Bregman co-clustering and matrix approximation. *Journal of Machine Learning Research*, **8**, 1919–1986. 374

Banfield, J. D., and Raftery, A. E. 1989. *Model-based Gaussian and non-Gaussian clustering.* Technical Report 186. Department of Statistics, University of Washington. 76

Banfield, J. D., and Raftery, A. E. 1992. Ice floe identification in satellite images using mathematical morphology and clustering about principal curves. *Journal of the American Statistical Association*, **8**, 7–16. 382

Banfield, J. D., and Raftery, A. E. 1993. Model-based Gaussian and non-Gaussian clustering. *Biometrics*, **49**, 803–821. 6, 20, 34, 76, 105, 237

Barker, M., and Rayens, W. 2003. Partial least squares for discrimination. *Journal of Chemometrics*, **17**(3), 166–173. 233

Barndorff-Nielsen, O., Kent, J., and Sørensen, M. 1982. Normal variance-mean mixtures and *z* distributions. *International Statistical Review*, **50**, 145–159. 283

Bashir, S., and Carter, E. 2005. High breakdown mixture discriminant analysis. *Journal of Multivariate Analysis*, **93**(1), 102–111. 161

Baudry, J. P., Raftery, A. E., Celeux, G., Lo, K., and Gottardo, R. 2010. Combining mixture components for clustering. *Journal of Computational and Graphical Statistics*, **19**, 332–353. 100, 101, 103, 108

Baudry, J.-P., Maugis, C., and Michel, B. 2012. Slope heuristics: overview and implementation. *Statistics and Computing*, **22**, 455–470. 194

Bellman, R. 1957. *Dynamic Programming.* Princeton University Press. 217, 221

Benaglia, T., Chauveau, D., Hunter, D. R., and Young, D. 2009. mixtools: An R package for analyzing finite mixture models. *Journal of Statistical Software*, **32**(6), 1–29. 339, 340

Bensmail, H., and Celeux, G. 1996. Regularized Gaussian discriminant analysis through eigenvalue decomposition. *Journal of the American Statistical Association*, **91**, 1743–1748. 6, 115, 238

Bensmail, H., and Meulman, J. J. 2003. Model-based clustering with noise: Bayesian inference and estimation. *Journal of Classification*, **20**, 49–76. 107

Bensmail, H., Celeux, G., Raftery, A. E., and Robert, C. P. 1997. Inference in model-based cluster analysis. *Statistics and Computing*, **7**, 1–10. 24, 77, 107

Bensmail, H., Golek, J., Moody, M. M., Semmes, J. O., and Haoudi, A. 2005. A novel approach to clustering proteomics data using Bayesian fast Fourier transform. *Bioinformatics*, **21**, 2210–2224. 107

Benzecri, J.-P. 1973. *L'analyse des données*. Paris: Dunod. 172

Bergé, L., Bouveyron, C., and Girard, S. 2016. *HDclassif: High Dimensional Supervised Classification and Clustering*. R package version 2.0.2. 12

Bergé, L., Bouveyron, C., Corneli, M., and Latouche, P. 2019. The latent topic block model for the co-clustering of textual interaction data. *Computational Statistics and Data Analysis*, **in press**. 383

Bhattacharya, S., and McNicholas, P. D. 2014. A LASSO-penalized BIC for mixture model selection. *Advances in Data Analysis and Classification*, **8**, 45–61. 107

Bickel, P. J., and Chen, A. 2009. A nonparametric view of network models and Newman-Girvan and other modularities. *Proceedings of the National Academy of Sciences*, **106**(50), 21068–21073. 329

Bickel, P. J., and Doksum, K. A. 1981. An analysis of transformations revisited. *Journal of the American Statistical Association*, **76**, 296–311. 279

Bickel, P. J., Chen, A., and Levina, E. 2011. The method of moments and degree distributions for network models. *Annals of Statistics*, **39**(5), 2280–2301. 329

Biernacki, C., Celeux, G., and Govaert, G. 1999. An improvement of the NEC criterion for assessing the number of clusters in a mixture model. *Pattern Recognition Letters*, **20**, 267–272. 77

Biernacki, C., Celeux, G., and Govaert, G. 2000. Assessing a mixture model for clustering with the integrated complete likelihood. *IEEE Transactions on Pattern Analysis and Machine Intelligence*, **22**, 719–725. 54, 55, 172, 173

Biernacki, C., Celeux, G., and Govaert, G. 2003. Choosing starting values for the EM algorithm for getting the highest likelihood in multivariate Gaussian mixture models. *Computational Statistics and Data Analysis*, **41**, 561–575. 31, 36, 37, 38, 175

Biernacki, C., Celeux, G., Govaert, G., and Langrognet, F. 2006. Model-based cluster and discriminant analysis with the Mixmod software. *Computational Statistics and Data Analysis*, **51**, 587–600. 198

Biernacki, C., Celeux, G., and Govaert, G. 2010. Exact and Monte Carlo calculations of integrated likelihoods for the latent class model. *Journal of Statistical Planning and Inference*, **140**(11), 2991–3002. 174

Binder, D. A. 1978. Bayesian cluster analysis. *Biometrika*, **65**, 31–38. 76

Bishop, C. M. 2006. *Pattern Recognition and Machine Learning*. Springer. 111, 334, 336, 350

Blaesild, P., and Jensen, J. L. 1981. Multivariate distributions of hyperbolic type. Pages 45–66 of: Taillie, C., Patil, G. P., and Baldessari, B. A. (eds.), *Statistical Distributions in Scientific Work: Volume 4 — Models, Structures, and Characterizations*. Dordrecht: Springer Netherlands. 284

Blashfeld, R. K., and Aldenderfer, M. S. 1988. The methods and problems of cluster analysis. Chap. 14, pages 447–474 of: Nesselroade, J. R., and Cattell, R. B. (eds.), *Handbook of Multivariate Experimental Psychology*. New York: Plenum Press. 3

Blei, D. M. 2012. Probabilistic topic models. *Communications of the ACM*, **55**(4), 77–84. 365, 382

Blei, D. M., and Lafferty, J. D. 2005. Correlated topic models. Pages 147–154 of: *Proceedings of the 18th International Conference on Neural Information Processing Systems*. NIPS'05. Cambridge, MA, USA: MIT Press. 382

Blei, D. M., and Lafferty, J. D. 2006. Dynamic topic models. Pages 113–120 of: *Proceedings of the 23rd International Conference on Machine Learning*. ICML '06. New York, NY, USA: ACM. 382

Blei, D. M., Ng, A. Y., and Jordan, M. I. 2003. Latent Dirichlet allocation. *Journal of Machine Learning Research*, **3**, 993–1022. 304, 322, 364, 366

Bock, H.-H. 1986. Loglinear models and entropy clustering methods for qualitative data. Pages 18–26 of: *Classification as a tool of research. Proceedings of the 9th Annual Conference of the Gesellschaft für Klassifikation*. North Holland. 166, 198

Boser, B., Guyon, I. M., and Vapnik, V. 1992. A training algorithm for optimal margin classifiers. Pages 144–152 of: *Proceedings of the Fifth Annual Workshop on Computational Learning Theory*. COLT '92. New York, NY, USA: ACM. 5

Bouchard, G., and Celeux, G. 2006. Selection of generative model in classification. *IEEE Transactions on Pattern Analysis and Machine Intelligence*, **28**, 544–564. 133, 140, 141

Bouchard, G., and Triggs, B. 2004. The tradeoff between generative and discriminative classifiers. Pages 721–729 of: *16th IASC International Symposium on Computational Statistics (COMPSTAT'04)*. 111

Bouveyron, C. 2014. Adaptive mixture discriminant analysis for supervised learning with unobserved classes. *Journal of Classification*, **31**(1), 49–84. 157, 158, 159, 160

Bouveyron, C., and Brunet, C. 2011. On the estimation of the latent discriminative subspace in the Fisher-EM algorithm. *Journal de la Société Française de Statistique*, **152**(3), 98–115. 253

Bouveyron, C., and Brunet, C. 2012a. *Discriminative variable selection for clustering with the sparse Fisher-EM algorithm*. Tech. rept. Preprint HAL 00685183. Laboratoire SAMM, Université Paris 1 Panthéon-Sorbonne. 254, 255, 256

Bouveyron, C., and Brunet, C. 2012b. Simultaneous model-based clustering and visualization in the Fisher discriminative subspace. *Statistics and Computing*, **22**(1), 301–324. 251, 252, 253

Bouveyron, C., and Brunet, C. 2012c. Theoretical and practical considerations on the convergence properties of the Fisher-EM algorithm. *Journal of Multivariate Analysis*, **109**, 29–41. 253

Bouveyron, C., and Brunet-Saumard, C. 2014. Model-based clustering of high-dimensional data: A review. *Computational Statistics and Data Analysis*, **71**, 52–78. 257

Bouveyron, C., and Girard, S. 2009. Robust supervised classification with mixture models: Learning from data with uncertain labels. *Pattern Recognition*, **42**(11), 2649–2658. 150, 151, 156

Bouveyron, C., and Jacques, J. 2011. Model-based clustering of time series in group-specific functional subspaces. *Advances in Data Analysis and Classification*, **5**(4), 281–300. 353

Bouveyron, C., Girard, S., and Schmid, C. 2007a. High-dimensional data clustering. *Computational Statistics and Data Analysis*, **52**(1), 502–519. 247, 249, 353, 362

Bouveyron, C., Girard, S., and Schmid, C. 2007b. High dimensional discriminant analysis. *Communications in Statistics: Theory and Methods*, **36**(14), 2607–2623. 247, 362

Bouveyron, C., Celeux, G., and Girard, S. 2011. Intrinsic dimension estimation by maximum likelihood in isotropic probabilistic PCA. *Pattern Recognition Letters*, **32**(14), 1706–1713. 249

Bouveyron, C., Côme, E., and Jacques, J. 2015. The discriminative functional mixture model for a comparative analysis of bike sharing systems. *Annals of Applied Statistics*, **9**(4), 1726–1760. 10, 165, 353, 356

Bouveyron, C., Bozzi, L., Jacques, J., and Jollois, F.-X. 2018a. The functional latent block model for the co-clustering of electricity consumption curves. *Journal of the Royal Statistical Society. Series C (Applied Statistics)*, 897–915. 382, 383

Bouveyron, C., Latouche, P., and Zreik, R. 2018b. The stochastic topic block model for the clustering of networks with textual edges. *Statistics and Computing*, **28**, 11–31. 321, 382

Box, G. E. P., and Cox, D. R. 1964. An analysis of transformations. (with Discussion). *Journal of the Royal Statistical Society. Series B (Methodological)*, **26**, 211–252. 278

Boyles, R. A. 1983. On the convergence of the EM algorithm. *Journal of the Royal Statistical Society. Series B (Methodological)*, **45**, 47–50. 23

Branco, M. D., and Dey, D. K. 2001. A general class of multivariate skew-elliptical distributions. *Journal of Multivariate Analysis*, **79**(1), 99–113. 271

Brand, M. 1999. Structure discovery in conditional probability models via an entropic prior and parameter extinction. *Neural Computation*, **11**, 1155–1182. 107

Brault, V., and Channarond, A. 2016. Fast and consistent algorithm for the latent block model. *arXiv preprint arXiv:1610.09005*. 383

Brault, V., and Lomet, A. 2015. Methods for co-clustering: A review. *Journal de la Société Française de Statistique*, **156**, 27–51. 374

Breiman, L. 2001. Random forests. *Machine Learning*, **45**, 5–32. 109

Breiman, L., Friedman, J., Ohlsen, R., and Stone, C. 1984. *Classification and Regression Trees*. New York: Wadsworth. 109

Bretagnolle, V. 2007. *Personal communication*. Source: Museum. 123

Brinkman, R. R., Gasparetto, M., Lee, S.-J. J., Ribickas, A. J., Perkins, J., Janssen, W., Smiley, R., and Smith, C. 2007. High-content flow cytometry and temporal data analysis for defining a cellular signature of graft-versus-host disease. *Biology of Blood and Marrow Transplantation*, **13**, 691–700. 259

Brodley, C., and Friedl, M. 1999. Identifying mislabeled training data. *Journal of Artificial Intelligence Research*, **11**, 131–167. 146

Browne, R. P., and McNicholas, P. D. 2015. A mixture of generalized hyperbolic distributions. *Canadian Journal of Statistics*, **43**(2), 176–198. 283

Bruneau, P., Gelgon, M., and Picarougne, F. 2010. Parsimonious reduction of Gaussian mixture models with a variational-Bayes approach. *Pattern Recognition*, **43**, 850–858. 108

Butts, C. T., Handcock, M. S., and Hunter, D. R. 2014. *network: Classes for Relational Data*. Irvine, CA. R package version 1.10.2. 292

Byar, D. P., and Green, S. B. 1980. The choice of treatment for cancer patients based on covariate information: application to prostate cancer. *Bulletin du Cancer*, **67**, 477–490. 187

Byers, S. D., and Raftery, A. E. 1998. Nearest neighbor clutter removal for estimating features in spatial point processes. *Journal of the American Statistical Association*, **93**, 577–584. 82, 106

Campbell, J. G., Fraley, C., Murtagh, F., and Raftery, A. E. 1997. Linear flaw detection in woven textiles using model-based clustering. *Pattern Recognition Letters*, **18**, 1539–1548. 53

Campbell, J. G., Fraley, C., Stanford, D. C., Murtagh, F., and Raftery, A. E. 1999. Model-based methods for real-time textile fault detection. *International Journal of Imaging Systems and Technology*, **10**, 339–346. 53

Carreira-Perpiñán, M. Á., and Renals, S. 2000. Practical identifiability of finite mixtures of multivariate Bernoulli distributions. *Neural Computation*, **12**(1), 141–152. 167

Carrington, P. J., Scott, J., and Wasserman, S. 2005. *Models and Methods in Social Network Analysis*. Cambridge University Press. 294

Carvalho, A. X., and Tanner, M. A. 2007. Modelling nonlinear count time series with local mixtures of Poisson autoregressions. *Computational Statistics and Data Analysis*, **51**(11), 5266–5294. 350

Cattell, R. B. 1944. A note on correlation clusters and cluster search methods. *Psychometrika*, **9**, 169–184. 2

Cattell, R. B. 1966. The scree test for the number of factors. *Multivariate Behavioral Research*, **1**(2), 145–276. 249

Celeux, G., and Diebolt, J. 1985. Stochastic versions of the EM algorithm. *Computational Statistics Quarterly*, **2**, 73–82. 377

Celeux, G., and Govaert, G. 1991. Clustering criteria for discrete data and latent class models. *Journal of Classification*, **8**(2), 157–176. 168, 172

Celeux, G., and Govaert, G. 1992. A classification EM algorithm for clustering and two stochastic versions. *Computational Statistics and Data Analysis*, **14**, 315–332. 34

Celeux, G., and Govaert, G. 1993. Comparison of the mixture and the classification maximum likelihood in cluster analysis. *Journal of Statistical Computation and Simulation*, **47**, 127–146. 34

Celeux, G., and Govaert, G. 1995. Gaussian parsimonious clustering models. *Pattern Recognition*, **28**, 781–793. 25, 76, 171, 237, 248

Celeux, G., and Mkhadri, A. 1992. Discrete regularized discriminant analysis. *Statistics and Computing*, **2**(3), 143–151. 6

Celeux, G., and Robert, C. 1993. Une histoire de discrétisation (with discussion). *Revue de Modulad*, **11**, 7–42. 186

Celeux, G., and Soromenho, G. 1996. An entropy criterion for assessing the number of clusters in a mixture model. *Journal of Classification*, **13**(2), 195–212. 77

Celeux, G., Hurn, M., and Robert, C. P. 2000. Computational and inferential difficulties with mixture posterior distributions. *Journal of the American Statistical Association*, **95**, 957–970. 107, 183

Celeux, G., Chrétien, S., Forbes, F., and Mkhadri, A. 2001. A component-wise EM algorithm for mixtures. *Journal of Computational and Graphical Statistics*, **10**, 697–712. 200

Celeux, G., Martin, O., and Lavergne, C. 2005. Mixture of linear mixed models for clustering gene expression profiles from repeated microarray experiments. *Statistical Modelling*, **5**, 243–267. 350

Celeux, G., Martin-Magniette, M.-L., Maugis-Rabusseau, C., and Raftery, A. E. 2011. Letter to the editor. *Journal of the American Statistical Association*, **105**, 383. 201

Celeux, G., Martin-Magniette, M. L., Maugis-Rabusseau, C., and Raftery, A. E. 2014. Comparing model selection and regularization approaches to variable selection in model-based clustering. *Journal de la Société Française de Statistique*, **155**, 57–71. 77

Celeux, G., Frühwirth-Schnatter, S., and Robert, C. P. (eds.). 2018a. *Handbook of Mixture Analysis*. Chapman & Hall/CRC. 14

Celeux, G., Maugis, C., and Sedki, M. 2018b. Variable selection in model-based clustering and discriminant analysis with a regularization approach. *Advances in Data Analysis and Classification*, To appear. 202, 209

Cerioli, A., Garcia-Escudero, L. A., Mayo-Iscar, A., and Riani, M. 2018. Finding the number of normal groups in model-based clustering via constrained likelihoods. *Journal of Computational and Graphical Statistics*, **27**(2), 404–416. 107

Chang, J. 2010. *lda: Collapsed Gibbs sampling methods for topic models*. R package version 1.2.1. 305

Chang, J., and Blei, D. M. 2009. Relational topic models for document networks. Pages 81–88 of: *Proceedings of the Twelfth International Conference on Artificial Intelligence and Statistics, AISTATS 2009, Clearwater Beach, Florida, USA, April 16-18, 2009*. 382

Chang, J., and Blei, D. M. 2010. Hierarchical relational models for document networks. *Annals of Applied Statistics*, **4**(1), 124–150. 329

Chang, W. C. 1983. On using principal component before separating a mixture of two multivariate normal distributions. *Journal of the Royal Statistical Society. Series C (Applied Statistics)*, **32**(3), 267–275. 230

Channarond, A. 2015. Random graph models: an overview of modeling approaches. *Journal de la Société Française de Statistique*, **156**(3), 56–94. 294

Channarond, A., Daudin, J.-J., and Robin, S. 2012. Classification and estimation in the stochastic blockmodel based on the empirical degrees. *Electronic Journal of Statistics*, **6**, 2574–2601. 300

Cheeseman, P., and Stutz, J. 1995. Bayesian classification (AutoClass): Theory and results. Pages 153–180 of: Fayyad, U., Piatesky-Shapiro, G., Smyth, P., and Uthurusamy, R. (eds.), *Advances in Knowledge Discovery and Data Mining*. AAAI Press. 77

Chen, J., and Tan, X. 2009. Inference for multivariate normal mixtures. *Journal of Multivariate Analysis*, **100**, 1367–1383. 107

Chen, T., Zhang, N. L., Liu, T. F., Wang, Y., and Poon, L. K. M. 2012. Model-based multidimensional clustering of categorical data. *Artificial Intelligence*, **176**, 2246–2279. 198

Chi, E. C., and Lange, K. 2014. Stable estimation of a covariance matrix guided by nuclear norm penalties. *Computational Statistics and Data Analysis*, **80**, 117–128. 107

Chow, C. 1970. On optimum recognition error and reject tradeoff. *IEEE Transactions on Information Theory*, **16**(1), 41–46. 161

Ciuperca, G., Ridolfi, A., and Idier, J. 2003. Penalized maximum likelihood estimator for normal mixtures. *Scandinavian Journal of Statistics*, **30**, 45–59. 107

Collins, L. M., and Lanza, S. T. 2013. *Latent Class and Latent Transition Analysis: With Applications in the Social, Behavioral, and Health Sciences*. New York: Wiley. 197

Côme, E., and Oukhellou, L. 2014. Model-based count series clustering for bike sharing system usage mining: A case study with the Vélib system of Paris. *ACM Transactions on Intelligent Systems and Technology*, **5**(3), 39:1–39:21. 194

Côme, E., Randriamanamihaga, A., Oukhellou, L., and Aknin, P. 2014. Spatio-temporal analysis of dynamic origin-destination data using latent Dirichlet allocation. Application to the Vélib bike sharing system of Paris. In: *Proceedings of 93rd Annual Meeting of the Transportation Research Board*. 365

Cook, R. D., and Weisberg, S. 1994. *An Introduction to Regression Graphics*. New York: John Wiley & Sons. 331

Coretto, P., and Hennig, C. 2010. A simulation study to compare robust clustering methods based on mixtures. *Advances in Data Analysis and Classification*, **4**, 111–135. 106

Coretto, P., and Hennig, C. 2011. Maximum likelihood estimation of heterogeneous mixtures of gaussian and uniform distributions. *Journal of Statistical Planning and Inference*, **141**, 462–473. 106

Corneli, M., Bouveyron, C., Latouche, P., and Rossi, F. 2018. The dynamic stochastic topic block model for dynamic networks with textual edges. *Statistics and Computing*, In press. 330

Cortes, C., and Vapnik, V. 1995. Support-vector networks. *Machine Learning*, **20**(3), 273–297. 5

Cox, D. R. 1958. The regression analysis of binary sequences. *Journal of the Royal Statistical Society. Series B (Methodological)*, 215–242. 5

Czekanowski, J. 1909. Zur differential-diagnose der Neadertalgruppe. *Korrespondenz-Blatt der Deutschen Gesellschaft für Anthropologie, Ethnologie, und Urgeschichte*, **40**, 44–47. 2

Czekanowski, J. 1911. Objectiv kriterien in der ethnologie. *Korrespondenz-Blatt der Deutschen Gesellschaft für Anthropologie, Ethnologie, und Urgeschichte*, **47**, 1–5. 2

Dang, U. J, Punzo, A., McNicholas, P. D., Ingrassia, S., and Browne, R. P. 2017. Multivariate response and parsimony for Gaussian cluster-weighted models. *Journal of Classification*, **34**(1), 4–34. 350

Das Gupta, S. 1973. Theories and methods in classification: a review. Pages 77–137 of: Cacoullos, T. (ed.), *Discriminant Analysis and Applications*. Elsevier. 6

Dasarathy, B. 1980. Nosing around the neighbourhood: a new system structure and classification rule for recognition in partially exposed environments. *IEEE Transactions on Pattern Analysis and Machine Intelligence*, **2**, 67–71. 161

Dasgupta, A., and Raftery, A. E. 1998. Detecting features in spatial point processes with clutter via model-based clustering. *Journal of the American Statistical Association*, **93**, 294–302. 53, 105

Dasgupta, D., and Nino, F. 2000. A comparison of negative and positive selection algorithms in novel pattern detection. Pages 125–130 of: *IEEE International Conference on Systems, Man and Cybernetics.* 161

Daudin, J.-J., Picard, F., and Robin, S. 2008. A mixture model for random graphs. *Statistics and Computing,* **18**, 173–183. 300

Day, N. E. 1969. Estimating the components of a mixture of two normal distributions. *Biometrika,* **56**, 463–474. 76, 92, 106

Dean, N., and Raftery, A. E. 2005. Normal-uniform mixture differential gene expression detection for cDNA microarrays. *BMC Bioinformatics,* **6**, article 173. 106

Dean, N., and Raftery, A. E. 2010. Latent class analysis variable selection. *Annals of the Institute of Statistical Mathematics,* **62**, 11–35. 212, 214

Deerwester, S., Dumais, S., Furnas, G., Landauer, T., and Harshman, R. 1990. Indexing by latent semantic analysis. *Journal of the American Society for Information Science,* **41**(6), 391. 364

Defays, D. 1978. An efficient algorithm for a complete link method. *Computer Journal,* **20**, 364–366. 33

Dellaportas, P. 1998. Bayesian classification of neolithic tools. *Journal of the Royal Statistical Society. Series C (Applied Statistics),* **47**, 279–297. 107

Dempster, A. P., Laird, N. M., and Rubin, D. B. 1977. Maximum likelihood for incomplete data via the EM algorithm (with discussion). *Journal of the Royal Statistical Society, Series B,* **39**, 1–38. 4, 23, 94, 337, 338

Devos, O., Ruckebusch, C., Durand, A., Duponchel, L., and Huvenne, J.-P. 2009. Support vector machines (SVM) in near infrared (NIR) spectroscopy: Focus on parameters optimization and model interpretation. *Chemometrics and Intelligent Laboratory Systems,* **96**, 27–33. 11, 221

Diebolt, J., and Robert, C. P. 1994. Estimation of finite mixture distributions through Bayesian sampling. *Journal of the Royal Statistical Society, Series B,* **56**(2), 363–375. 337

Donoho, D. 2000. High-dimensional data analysis: The curses and blessings of dimensionality. In: *Math Challenges of the 21st Century.* American Mathematical Society. 222

Dowe, D. L. 2008. Foreword re C. S. Wallace. *The Computer Journal,* **51**(5), 523–560. 198

Driver, H. E., and Kroeber, A. L. 1932. Quantitative expression of cultural relationships. *University of California Publications in Archaeology and Ethnology,* **31**, 211–216. 2

Duda, R., Hart, P., and Stork, D. 2000. *Pattern Classification.* New York: John Wiley & Sons. 233

Edwards, A. W. F., and Cavalli-Sforza, L. L. 1965. A method for cluster analysis. *Biometrics,* **21**, 362–375. 76

Efron, B., and Tibshirani, R. 1997. Improvements on cross-validation: the .632+ bootstrap method. *Journal of the American Statistical Association,* **92**, 648–560. 119

Emerson, J. W., and Green, W. A. 2014. *gpairs: The Generalized Pairs Plot.* R package version 1.2. 332

Erosheva, E. A. 2002. *Grade of membership and latent structure models with application to disability survey data.* Ph.D. thesis, Department of Statistics, Carnegie Mellon University. 304

Erosheva, E. A. 2003. Bayesian estimation of the Grade of Membership model. Pages 501–510 of: Bernardo, J., Bayarri, M., Berger, J., Dawid, A., Heckerman, D., Smith, A., and West, M. (eds.), *Bayesian Statistics, 7.* UK: Oxford University Press. 304

Erosheva, E. A., Fienberg, S. E., and Joutard, C. 2007. Describing disability through individual-level mixture models for multivariate binary data. *The Annals of Applied Statistics,* **1**(2), 502–537. 304

Escabias, M., Aguilera, A. M., and Valderrama, M. J. 2005. Modeling environmental data by functional principal component logistic regression. *Environmetrics,* **16**, 95–107. 354

Evans, K., Love, T., and Thurston, S. W. 2015. Outlier identification in model-based cluster analysis. *Journal of Classification,* **32**, 63–84. 106

Everitt, B. S. 1993. *Cluster Analysis.* 3rd edn. London: Edward Arnold. 14

Everitt, B. S., and Hand, D. J. 1981. *Finite Mixture Distributions.* London: Chapman & Hall. Monographs on Applied Probability and Statistics. 14

Fienberg, S. E., and Wasserman, S. 1981. Discussion of "An exponential family of probability distributions for directed graphs" by Holland and Leinhardt. *Journal of the American Statistical Association,* **76**(373), 54–57. 298

Figueiredo, M. A. T., and Jain, A. K. 2002. Unsupervised learning of finite mixture models. *IEEE Transactions on Pattern Analysis and Machine Intelligence,* **24**, 381–396. 107

Fisher, R. A. 1936. The use of multiple measurements in taxonomic problems. *Annals of Eugenics,* **7**, 179–188. 5, 6, 154, 231, 237, 252, 356

Fisher, R. A. 1938. The statistical utilization of multiple measurements. *Annals of Human Genetics,* **8**(4), 376–386. 5

Foley, D. H., and Sammon, J. W. 1975. An optimal set of discriminant vectors. *IEEE Transactions on Computers,* **24**, 281–289. 253

Fop, M., and Murphy, T. B. 2018. Variable selection methods for model-based clustering. *Statistics Surveys,* **12**, 18–65. 216

Fop, M., Smart, K., and Murphy, T. B. 2017. Variable selection for latent class analysis with application to low back pain diagnosis. *Annals of Applied Statistics,* **11**(4), 2080–2110. 213, 214

Fop, M., Scrucca, L., and Murphy, T. B. 2018. Model-based clustering with sparse covariance matrices. *Statistics and Computing,* To appear. 77

Forbes, F., and Wraith, D. 2014. A new family of multivariate heavy-tailed distributions with variable marginal amounts of tailweight: Application to robust clustering. *Statistics and Computing,* **24**(6), 971–984. 291

Forbes, F., Peyrard, N., Fraley, C., Georgian-Smith, D., Goldhaber, D. M., and Raftery, A. E. 2006. Model-based region-of-interest selection in dynamic breast MRI. *Journal of Computer Assisted Tomography,* **30**, 675–687. 383

Forina, M., Armanino, C., Castino, M., and Ubigli, M. 1986. Multivariate data analysis as a discriminating method of the origin of wines. *Vitis,* **25**, 189–201. 8, 60, 332

Fraiman, R., Justel, A., and Svarc, M. 2008. Selection of variables for cluster analysis and classification rules. *Journal of the American Statistical Association,* **103**, 1294–1303. 215

Fraley, C., and Raftery, A. E. 1998. How many clusters? Which clustering method? - Answers via model-based cluster analysis. *Computer Journal,* **41**, 578–588. 53, 78, 105

Fraley, C., and Raftery, A. E. 1999. MCLUST: Software for model-based cluster analysis. *Journal of Classification,* **16**, 297–306. 76

Fraley, C., and Raftery, A. E. 2002. Model-based clustering, discriminant analysis and density estimation. *Journal of the American Statistical Association,* **97**, 611–631. 78, 105, 126, 248, 261

Fraley, C., and Raftery, A. E. 2003. Enhanced model-based clustering, density estimation and discriminant analysis software: MCLUST. *Journal of Classification,* **20**, 263–286. 94, 106

Fraley, C., and Raftery, A. E. 2005. *Bayesian Regularization for Normal Mixture Estimation and Model-based Clustering.* Technical Report 486. Department of Statistics, University of Washington. 95

Fraley, C., and Raftery, A. E. 2006. Some applications of model-based clustering in chemistry. *R News,* **6**, 17–23. 77, 382

Fraley, C., and Raftery, A. E. 2007a. Bayesian regularization for normal mixture estimation and model-based clustering. *Journal of Classification,* **24**, 155–181. 94, 95, 107

Fraley, C., and Raftery, A. E. 2007b. Model-based methods of classification: Using the mclust software in chemometrics. *Journal of Statistical Software,* **18**, paper i06. 76

Fraley, C., Raftery, A. E., and Wehrens, R. 2005. Incremental model-based clustering for large datasets with small clusters. *Journal of Computational and Graphical Statistics,* **14**, 520–546. 383

Franczak, B. C., Browne, R. P., and McNicholas, P. D. 2014. Mixtures of shifted asymmetric Laplace distributions. *IEEE Transactions on Pattern Analysis and Machine Intelligence*, **36**(6), 1149–1157. 291

Frénay, B., and Verleysen, M. 2014. Classification in the presence of label noise: a survey. *IEEE Transactions on Neural Networks and Learning Systems*, **25**(5), 845–869. 160

Friedman, H. P., and Rubin, J. 1967. On some invariant criteria for grouping data. *Journal of the American Statistical Association*, **62**, 1159–1178. 76

Friedman, J. 1989. Regularized discriminant analysis. *Journal of the American Statistical Association*, **84**, 165–175. 115, 233, 236

Friel, N., Rastelli, R., Wyse, J., and Raftery, A. E. 2016. Interlocking directorates in Irish companies using a latent space model for bipartite networks. *Proceedings of the National Academy of Sciences*, **113**(24), 6629–6634. 330

Fritz, H., García-Escudero, L. A, and Mayo-Iscar, A. 2012. tclust: An R package for a trimming approach to cluster analysis. *Journal of Statistical Software*, **47**(12), 1–26. 106

Fruchterman, T. M. J., and Reingold, E. M. 1991. Graph drawing by force-directed placement. *Software - Practice and Experience*, **21**(11), 1129–1164. 292

Frühwirth-Schnatter, S. 2006. *Finite Mixture and Markov Switching Models*. Springer Series in Statistics. New York: Springer-Verlag. 14, 174, 337, 378

Frühwirth-Schnatter, S. 2011a. Dealing with label switching under model uncertainty. Pages 213–240 of: Mengersen, K. L., Robert, C., and Titterington, D. M. (eds.), *Mixtures: Estimation and Applications*. Wiley. 181, 184, 185, 379

Frühwirth-Schnatter, S. 2011b. Panel data analysis: a survey on model-based clustering of time series. *Advances in Data Analysis and Classification*, **5**(4), 251–280. 382

Frühwirth-Schnatter, S., and Kaufmann, S. 2008. Model-based clustering of multiple time series. *Journal of Business and Economic Statistics*, **26**, 78–89. 353

Fu, W., Song, L., and Xing, E. P. 2009. Dynamic mixed membership blockmodel for evolving networks. Pages 329–336 of: *Proceedings of the 26th Annual International Conference on Machine Learning*. ICML '09. New York, NY, USA: ACM. 330

Fukunaga, K. 1990. *Introduction to Statistical Pattern Recognition*. San Diego: Academic Press. 233, 252, 356

Fukunaga, K. 1999. Statistical pattern recognition. Pages 33–60 of: Chen, C. H., Pau, L. F., and Wang, P. S. P. (eds.), *Handbook Of Pattern Recognition And Computer Vision*. World Scientific. 6

Galimberti, G., Manisi, A., and Soffritti, G. 2017. Modelling the role of variable in model-based cluster analysis. *Statistics and Computing*, **28**, 146–169. 216

Gallegos, M. T., and Ritter, G. 2005. A robust method for cluster analysis. *Annals of Statistics*, 347–380. 88, 90

Gallegos, M. T., and Ritter, G. 2009a. Trimmed ML estimation of contaminated mixtures. *Sankhyā A*, **71**, 164–220. 106

Gallegos, M. T., and Ritter, G. 2009b. Trimming algorithms for clustering contaminated grouped data and their robustness. *Advances in Data Analysis and Classification*, **3**, 135–167. 106

Gamberger, D., Lavrac, N., and Groselj, C. 1999. Experiments with noise filtering in a medical domain. Pages 143–151 of: *Proceedings of the Sixteenth International Conference on Machine Learning*. ICML '99. San Francisco, CA, USA: Morgan Kaufmann Publishers Inc. 161

García-Escudero, L. A., Gordaliza, A., Matrán, C., and Mayo-Iscar, A. 2008. A general trimming approach to robust cluster analysis. *Annals of Statistics*, **36**, 1324–1345. 88, 90, 91, 93, 106

García-Escudero, L. A., Gordaliza, A., Matrán, C., and Mayo-Iscar, A. 2010. A review of robust clustering methods. *Advances in Data Analysis and Classification*, **4**, 89–109. 106

García-Escudero, L. A., Gordaliza, A., Matrán, C., and Mayo-Iscar, A. 2011. Exploring the number of groups in robust model-based clustering. *Statistics and Computing*, **21**, 585–599. 106

García-Escudero, L. A., Gordaliza, A., Matrán, C., and Mayo-Iscar, A. 2015. Avoiding spurious local maximizers in mixture modeling. *Statistics and Computing*, **25**, 619–633. 107

Gates, G. 1972. The reduced nearest neighbor rule. *IEEE Transactions on Information Theory*, **18**(3), 431–433. 161

Gelfand, A. E., and Smith, A. F. M. 1990. Sampling-based approaches to calculating marginal densities. *Journal of the American Statistical Association*, **85**(410), 398–409. 300

Gelman, A., Carlin, J. B., Stern, H. S., Dunson, D. B., Vehtari, A., and Rubin, D. B. 2013. *Bayesian Data Analysis*. 3rd edn. London: Chapman and Hall. 94

Gershenfeld, N. 1997. Nonlinear inference and cluster-weighted modeling. *Annals of the New York Academy of Sciences*, **808**(1), 18–24. 350

Geweke, J., and Keane, M. 2007. Smoothly mixing regressions. *Journal of Econometrics*, **136**(1), 252–290. 350

Ghahramani, Z., and Hinton, G. E. 1997. *The EM algorithm for factor analyzers*. Tech. rept. University of Toronto. 238, 240, 244, 246, 257

Giacofci, M., Lambert-Lacroix, S., Marot, G., and Picard, F. 2013. Wavelet-based clustering for mixed-effects functional models in high dimension. *Biometrics*, **69**, 31–40. 353

Goldenberg, A., Zheng, A. X., Fienberg, S. E., and Airoldi, E. M. 2010. A survey of statistical network models. *Foundations and Trends in Machine Learning*, **2**, 129–233. 294

Gollini, I. 2015. *lvm4net: Latent Variable Models for Networks*. R package version 0.2. 317

Gollini, I., and Murphy, T. B. 2014. Mixture of latent trait analyzers for model-based clustering of categorical data. *Statistics and Computing*, **24**, 569–588. 166, 167, 198

Gollini, I., and Murphy, T. B. 2016. Joint modelling of multiple network views. *Journal of Computational and Graphical Statistics*, **25**(1), 246–265. 314

Goodman, L. A. 1974. Exploratory latent structure models using both identifiable and unidentifiable models. *Biometrika*, **61**, 215–231. 166, 167

Gopal, S. 2007. The evolving social geography of blogs. Pages 275–293 of: Miller, H. J. (ed.), *Societies and Cities in the Age of Instant Access*. The GeoJournal Library, vol. 88. Springer Netherlands. 296

Gordon, A. D. 1999. *Classification*. 2nd edn. Boca Raton: Chapman & Hall/CRC. 14

Gormley, I. C., and Frühwirth-Schnatter, S. 2018. Mixtures of experts. Chap. 12, pages 279–316 of: Frühwirth-Schnatter, S., Celeux, G., and Robert, C. P. (eds.), *Handbook of Mixture Analysis*. CRC Press. 350

Gormley, I. C., and Murphy, T. B. 2008. A mixture of experts model for rank data with applications in election studies. *Annals of Applied Statistics*, **2**(4), 1452–1477. 350

Gormley, I. C., and Murphy, T. B. 2010a. Clustering ranked preference data using sociodemographic covariates. Pages 543–569 of: Hess, S., and Daly, A. (eds.), *Choice Modelling: The State-of-the-Art and the State-of-Practice*. United Kingdom: Emerald. 315, 339, 350

Gormley, I. C., and Murphy, T. B. 2010b. A mixture of experts latent position cluster model for social network data. *Statistical Methodology*, **7**(3), 385–405. 350

Gormley, I. C., and Murphy, T. B. 2011. Mixture of experts models with social science applications. Pages 91–110 of: Mengersen, K., Robert, C., and Titterington, D. M. (eds.), *Mixture Estimation and Applications*. Wiley. 334, 339, 350

Gormley, I. C., and Murphy, T. B. 2018. *MEclustnet: Fits the Mixture of Experts Latent Position Cluster Model to Network Data*. R package version 1.2.1. 317

Govaert, G. 1977. Algorithme de classification d'un tableau de contingence. Pages 487–500 of: *First International Symposium on Data Analysis and Informatics*. Versailles: INRIA. 374

Govaert, G. 1983. *Classification croisée*. Thèse d'État, Université Paris 6, France. 172

Govaert, G., and Nadif, M. 2008. Block clustering with Bernoulli mixture models: Comparison of different approaches. *Computational Statistics and Data Analysis*, **52**, 3233–3245. 374, 377

Govaert, G., and Nadif, M. 2010. Latent block model for contingency table. *Communications in Statistics: Theory and Methods*, **39**(3), 416–425. 383

Govaert, G., and Nadif, M. 2014. *Co-clustering*. London: ISTE and Wiley. 374

Grandvalet, Y., and Bengio, Y. 2004. Semi-supervised learning by entropy minimization. Pages 529–536 of: *Proceedings of the 17th International Conference on Neural Information Processing Systems*. NIPS'04. Cambridge, MA, USA: MIT Press. 134

Greenacre, M., and Blasius, J. (eds.). 2006. *Multiple Correspondence Analysis and Related Methods.* Chapman & Hall/CRC. 178

Greene, E. L. 1909. *Landmarks of Botanical History: A Study of Certain Epochs in the Development of the Science of Botany. Part I. Prior to 1562 A.D.* Washington, D.C.: Smithsonian Institution. 2

Grün, B., and Leisch, F. 2007. Fitting finite mixtures of generalized linear regressions in R. *Computational Statistics & Data Analysis*, **51**(11), 5247–5252. 340

Grün, B., and Leisch, F. 2008. FlexMix Version 2: Finite mixtures with concomitant variables and varying and constant parameters. *Journal of Statistical Software*, **28**(4), 1–35. 339, 340

Guo, J., Levina, E., Michailidis, G., and Zhu, J. 2010. Pairwise variable selection for high-dimensional model-based clustering. *Biometrics*, **66**, 793–804. 208

Guyon, I., Matic, N., and Vapnik, V. 1996. Discovering informative patterns and data cleaning. *Advances in Knowledge Discovery and Data Mining*, 181–203. 161

Gyllenberg, M., Koski, T., Reilink, E., and Verlaan, M. 1994. Nonuniqueness in probabilistic numerical identification of bacteria. *Journal of Applied Probability*, **31**(2), 542–548. 167

Habbema, J. D. F., Hermans, J., and van den Broek, K. 1974. A stepwise discriminant analysis program using density estimation. Pages 101–110 of: G., Bruckman (ed.), *Compstat 1974: Proceedings in Computational Statistics.* Vienna: Physica-Verlag. 111

Hagenaars, J. A. 1988. Latent structure models with direct effects between indicators: Local dependence models. *Sociological Methods and Research*, **16**, 379–405. 198

Halbe, Z., Bortman, M., and Aladjem, M. 2013. Regularized mixture density estimation with an analytical setting of shrinkage intensities. *IEEE Transactions on Neural Networks and Learning Systems*, **24**, 460–470. 107

Hampel, F. R. 1971. A general qualitative definition of robustness. *Annals of Mathematical Statistics*, **42**, 1887–1896. 105

Handcock, M. S., Raftery, A. E., and Tantrum, J. M. 2007. Model-based clustering for social networks. *Journal of the Royal Statistical Society: Series A*, **170**(2), 1–22. 312, 314, 316, 350

Hanneke, S., Fu, W., and Xing, E. P. 2010. Discrete temporal models of social networks. *Electronic Journal of Statistics*, **4**, 585–605. 330

Hansen, L., Liisberg, C., and Salamon, P. 1997. The error-reject tradeoff. *Open Systems and Information Dynamics*, **4**, 159–184. 161

Harrison, P. J., and Stevens, C. F. 1971. Bayesian approach to short-term forecasting. *Operational Research Quarterly*, **22**, 341–362. 108

Hartigan, J. A. 1975. *Clustering Algorithms.* New York: John Wiley & Sons. 14

Hartigan, J. A., and Hartigan, P. M. 1985. The dip test of unimodality. *Annals of Statistics*, **13**, 70–84. 101

Hasnat, M. A., Velcin, J., Bonnevoy, S., and Jacques, J. 2017. Evolutionary clustering for categorical data using parametric links among multinomial mixture models. *Econometrics and Statistics*, **3**, 141–159. 198

Hastie, T., and Stuetzle, W. 1989. Principal curves. *Journal of the American Statistical Association*, **84**, 502–516. 229

Hastie, T., and Tibshirani, R. 1996. Discriminant analysis by Gaussian mixtures. *Journal of the Royal Statistical Society. Series B (Methodological)*, 155–176. 6, 126, 146, 152

Hastie, T., Buja, A., and Tibshirani, R. 1995. Penalized discriminant analysis. *The Annals of Statistics*, **23**, 73–102. 233, 236

Hastie, T., Tibshirani, R., and Friedman, J. 2009. *The Elements of Statistical Learning.* 2nd edn. New York: Springer. 111, 131, 145

Hathaway, R. J. 1985. A constrained formulation of maximum likelihood estimation for normal mixture distributions. *Annals of Statistics*, **13**, 795–800. 93, 106

Hathaway, R. J. 1986a. Another interpretation of the EM algorithm for mixture distributions. *Statistics and Probability Letters*, **4**(2), 53–56. 326

Hathaway, R. J. 1986b. A constrained EM algorithm for univariate normal mixtures. *Journal of Statistical Computation and Simulation*, **23**, 211–230. 93, 106

Haughton, D. 1988. On the choice of a model to fit data from an exponential family. *Annals of Statistics*, **16**, 342–355. 51

Hawkins, D., and McLachlan, G. J. 1997. High-breakdown linear discriminant analysis. *Journal of the American Statistical Association*, **92**(437), 136–143. 161

Heard, N. A., Holmes, C. C., and Stephens, D. A. 2006. A quantitative study of gene regulation involved in the immune response of anopheline mosquitoes: an application of Bayesian hierarchical clustering of curves. *Journal of the American Statistical Association*, **101**(473), 18–29. 353

Hellman, M. 1970. The nearest neighbour classification with a reject option. *IEEE Transactions on Systems Science and Cybernetics*, **6**(3), 179–185. 161

Hennig, C. 2004. Breakdown points for maximum likelihood-estimators of location-scale mixtures. *Annals of Statistics*, **32**, 1313–1340. 105, 106

Hennig, C. 2010. Methods for merging Gaussian mixture components. *Advances in Data Analysis and Classification*, **4**, 3–34. 99, 101, 103

Hennig, C. 2013. Discussion of "Model-based clustering with non-normal mixture distributions" by S. X. Lee and G. J. McLachlan. *Statistical Methods and Applications*, **22**, 455–458. 108

Hennig, C. 2015a. *fpc: Flexible Procedures for Clustering*. R package version 2.1-10. 12, 101, 340

Hennig, C. 2015b. What are the true clusters? *Pattern Recognition*, **64**, 53–62. 108

Hennig, C., and Coretto, P. 2008. The noise component in model-based cluster analysis. Pages 127–138 of: Preisach, C., Burkhardt, H., Schmidt-Thieme, L., and Decker, R. (eds.), *Data Analysis, Machine Learning and Applications*. Berlin: Springer. 106

Hennig, C., and Hausdorf, B. 2015. *prabclus: Functions for Clustering of Presence-Absence, Abundance and Multilocus Genetic Data*. R package version 2.2-6. 12, 83

Hennig, C., and Liao, T. F. 2013. How to find an appropriate clustering for mixed type variables with application to socio-economic stratification (with discussion). *Journal of the Royal Statistical Society. Series C (Applied Statistics)*, **62**, 309–369. 169, 188

Hennig, C., Meilă, M., Murtagh, F., and Rocci, R. (eds.). 2015. *Handbook of Cluster Analysis*. Chapman & Hall/CRC. 14

Henry, N. W. 1999. *Latent Structure Analysis at Fifty*. Paper presented at the 1999 Joint Statistical Meetings, Baltimore MD, August, 1999. `www.people.vcu.edu/ñhenry/LSA50.htm`. 72

Hoff, P. D., Raftery, A. E., and Handcock, M. S. 2002. Latent space approaches to social network analysis. *Journal of the American Statistical Association*, **97**(460), 1090–1098. 312, 313

Hofmann, T. 1999. Probabilistic latent semantic indexing. Pages 50–57 of: *Proceedings of the 22nd Annual International ACM SIGIR Conference on Research and Development in Information Retrieval*. ACM. 364

Holland, P. W., and Leinhardt, S. 1981. An exponential family of probability distributions for directed graphs. *Journal of the American Statistical Association*, **76**(373), 33–50. 329

Horaud, R., Forbes, F., Yguel, M., Dewaele, G., and Zhang, J. 2011. Rigid and articulated point registration with expectation conditional maximization. *IEEE Transactions on Pattern Analysis and Machine Intelligence*, **33**, 587–602. 105

Hosmer, D. W., Lemeshow, S., and Sturdivant, R. X. 2013. *Applied Logistic Regression*. 3rd edn. New York: Wiley. 109

Hotelling, H. 1931. The generalization of "Student's" ratio. *Annals of Mathematical Statistics*. 5

Hotelling, H. 1933. Analysis of a complex of statistical variables into principal components. *Journal of Educational Psychology*, **24**, 417–441. 228

Houdard, A., Bouveyron, C., and Delon, J. 2019. High-dimensional mixture models for unsupervised image denoising (HDMI). *SIAM Journal on Imaging Sciences, Society for Industrial and Applied Mathematics*, In press. 372

Howard, E., Meehan, M., and Parnell, A. 2018. Contrasting prediction methods for early warning systems at undergraduate level. *The Internet and Higher Education*, **37**, 66–75. 78

Howells, W. W. 1973. Cranial variation in man: A study by multivariate analysis of patterns of difference among recent human populations. *Papers of the Peabody Museum of Archaeology and Ethnology*, **67**, 1–259. 65

Howells, W. W. 1989. Skull shapes and the map: Craniometric analyses in the dispersion of modern homo. *Papers of the Peabody Museum of Archaeology and Ethnology*, **79**. 65

Howells, W. W. 1995. Who's who in skulls: Ethnic identification of crania from measurements. *Papers of the Peabody Museum of Archaeology and Ethnology*, **82**. 65

Howells, W. W. 1996. Howells' craniometric data on the internet. *American Journal of Physical Anthropology*, **101**, 441–442. 65

Howland, P., and Park, H. 2004. Generalizing discriminant analysis using the generalized singular decomposition. *IEEE Transactions on Pattern Analysis and Machine Learning*. 233

Huber, P. 1985. Projection pursuit. *Annals of Statistics*, **13**(2), 435–525. 226

Hubert, L., and Arabie, P. 1985. Comparing partitions. *Journal of Classification*, **2**, 193–218. 41, 172

Hurn, M., Justel, A., and Robert, C. P. 2003. Estimating mixtures of regressions. *Journal of Computational and Graphical Statistics*, **12**(1), 55–79. 331, 350

Ingrassia, S., and Rocci, R. 2007. Constrained monotone EM algorithms for finite mixture of multivariate Gaussians. *Computational and Statistical Data Analysis*, **51**, 5339–5351. 106

Ingrassia, S., Minotti, S. C., and Vittadini, G. 2012. Local statistical modeling via the cluster-weighted approach with elliptical distributions. *Journal of Classification*, **29**(3), 363–401. 350

Ingrassia, S., Punzo, A., Vittadini, G., and Minotti, S. C. 2015. The generalized linear mixed cluster-weighted model. *Journal of Classification*, **32**(1), 85–113. 350

Iscar, A. M., Garcia-Escudero, L. A., and Fritz, H. 2017. *tclust: Robust Trimmed Clustering*. R package version 1.3-1. 12

Jacobs, R. A., Jordan, M. I., Nowlan, S. J., and Hinton, G. E. 1991. Adaptive mixture of local experts. *Neural Computation*, **3**(1), 79–87. 333, 334

Jacques, J., and Biernacki, C. 2018. Model-based co-clustering for ordinal data. *Computational Statistics and Data Analysis*, **123**, 101–115. 383

Jacques, J., and Preda, C. 2013. Funclust: a curves clustering method using functional random variable density approximation. *Neurocomputing*, **112**, 164–171. 353

Jacques, J., and Preda, C. 2014a. Functional data clustering: A survey. *Advances in Data Analysis and Classification*, **8**(3), 231–255. 353

Jacques, J., and Preda, C. 2014b. Model-based clustering of multivariate functional data. *Computational Statistics and Data Analysis*, **71**, 92–106. 353, 359

James, G. M., and Sugar, C. A. 2003. Clustering for sparsely sampled functional data. *Journal of the American Statistical Association*, **98**(462), 397–408. 353, 354

Jeffreys, H. 1961. *Theory of Probability*. 3rd edn. Clarendon. 51

Jernite, Y., Latouche, P., Bouveyron, C., Rivera, P., Jegou, L., and Lamassé, S. 2014. The random subgraph model for the analysis of an ecclesiastical network in Merovingian Gaul. *Annals of Applied Statistics*, **8**(1), 377–405. 329

Jin, Z., Yang, J-Y., Hu, Z. S., and Lou, Z. 2001. Face recognition based on the uncorrelated optimal discriminant vectors. *Pattern Recognition*, **10**(34), 2041–2047. 233

Joachims, T. 1999. Transductive inference for text classification using support vector machines. Pages 200–209 of: *Proceedings of the Sixteenth International Conference on Machine Learning*. ICML '99. San Francisco, CA, USA: Morgan Kaufmann Publishers Inc. 134

John, G. H. 1995. Robust decision trees: Removing outliers from databases. Pages 174–179 of: *Proceedings of the First International Conference on Knowledge Discovery and Data Mining*. KDD'95. AAAI Press. 161

Jordan, M. I., and Jacobs, R. A. 1994. Hierarchical mixtures of experts and the EM algorithm. *Neural Computation*, **6**, 181–214. 336

Jöreskog, K. G. 1978. Structural analysis of covariance and correlation matrices. *Psychometrika*, **43**, 443–477. 229

Jörnsten, R., and Keleş, S. 2008. Mixture models with multiple levels, with application to the analysis of multifactor gene expression data. *Biostatistics*, **9**, 540–554. 108

Karlis, D. 2003. An EM algorithm for multivariate Poisson distribution and related models. *Journal of Applied Statistics*, **30**, 63–77. 192

Karlis, D., and Santourian, A. 2009. Model-based clustering with non-elliptically contoured distributions. *Statistics and Computing*, **19**(1), 73–83. 283

Kass, R. E., and Raftery, A. E. 1995. Bayes factors. *Journal of the American Statistical Association*, **90**, 773–795. 47, 51

Kass, R. E., and Wasserman, L. 1995. A reference Bayesian test for nested hypotheses and its relationship to the Schwarz criterion. *Journal of the American Statistical Association*, **90**, 928–934. 51

Keribin, C. 1998. Consistent estimate of the order of mixture models. *Comptes Rendues de l'Academie des Sciences, série I — Mathématiques*, **326**, 243–248. 53

Keribin, C., Brault, V., Celeux, G., and Govaert, G. 2015. Estimation and selection for the latent block model on categorical data. *Statistics and Computing*, **25**, 1201–1216. 378, 379, 383

Kim, D., and Seo, B. 2014. Assessment of the number of components in Gaussian mixture models in the presence of multiple local maximizers. *Journal of Multivariate Analysis*, **125**, 100–120. 107

Kim, S., Song, D. K. H., and DeSarbo, W. S. 2012. Model-based segmentation featuring simultaneous segment-level variable selection. *Journal of Marketing Research*, **49**, 725–736. 216

Kohonen, T. 1995. *Self-Organizing Maps*. New York: Springer-Verlag. 229

Kolaczyk, E. D. 2009. *Statistical Analysis of Network Data: Methods and Models*. New York: Springer. 294, 296

Krivitsky, P. N., and Handcock, M. S. 2008. Fitting latent cluster models for networks with latentnet. *Journal of Statistical Software*, **24**(5), 1–23. 317

Krivitsky, P. N., and Handcock, M. S. 2010. *latentnet: Latent position and cluster models for statistical networks*. R package version 2.4-4. 317, 320

Krivitsky, P. N., Handcock, M. S., Raftery, A. E., and Hoff, P. D. 2009. Representing degree distributions, clustering, and homophily in social networks with latent cluster random effects models. *Social Networks*, **31**(3), 204–213. 315

Krzanowski, W. 2003. *Principles of Multivariate Analysis*. Oxford: Oxford University Press. 233

Lance, G. N., and Williams, W. T. 1967. A general theory of classificatory sorting strategies. II. Clustering systems. *Computer Journal*, **10**, 271–277. 34

Langrognet, F., Lebret, R., Poli, C., and Iovleff, S. 2016. *Rmixmod: Supervised, Unsupervised, Semi-Supervised Classification with MIXture MODelling (Interface of MIXMOD Software)*. R package version 2.1-1. 12

Lasserre, J. A., Bishop, C. M., and Minka, T. P. 2006. Principled hybrids of generative and discriminative models. Pages 87–94 of: *IEEE Computer Society Conference on Computer Vision and Pattern Recognition (CVPR'06)*, vol. 1. IEEE Conference Publications. 111

Latouche, P., Birmelé, E., and Ambroise, C. 2010. Bayesian methods for graph clustering. Pages 229–239 of: Fink, A., Lausen, B., Seidel, W., and Ultsch, A. (eds.), *Advances in Data Analysis, Data Handling and Business Intelligence*. Studies in Classification, Data Analysis, and Knowledge Organization. Berlin, Heidelberg: Springer. 300

Latouche, P., Birmelé, E., and Ambroise, C. 2011. Overlapping stochastic block models with application to the French political blogosphere. *Annals of Applied Statistics*, **5**(1), 309–336. 329

Latouche, P., Birmelé, E., and Ambroise, C. 2012. Variational Bayesian inference and complexity control for stochastic block models. *Statistical Modelling*, **12**(1), 93–115. 302

Lavine, M., and West, M. 1992. A Bayesian method for classification and discrimination. *Canadian Journal of Statistics*, **20**, 451–461. 76, 107

Law, M. H., Figueiredo, M. A. T., and Jain, A. K. 2004. Simultaneous feature selection and clustering using mixture models. *IEEE Transactions on Pattern Analysis and Machine Intelligence*, **26**, 1154–1166. 200, 203, 204, 216

Lawrence, N., and Schölkopf, B. 2001. Estimating a kernel Fisher discriminant in the presence of label noise. Pages 306–313 of: *Proceedings of the Eighteenth International Conference on Machine Learning*. ICML '01. San Francisco, CA, USA: Morgan Kaufmann Publishers Inc. 146, 147

Lazarsfeld, P. F. 1950a. The logical and mathematical foundations of latent structure analysis. Chap. 10 of: Stouffer, S. A. (ed.), *Measurement and Prediction, Volume IV of The American Soldier: Studies in Social Psychology in World War II*. Princeton University Press. 3, 72, 73

Lazarsfeld, P. F. 1950b. The logical and mathematical foundations of latent structure analysis. Pages 362–412 of: Stouffer, S. A. (ed.), *Measurement and Prediction*. Princeton University Press. 165

Lazarsfeld, P. F. 1950c. Some latent structures. Chap. 11 of: Stouffer, S. A. (ed.), *Measurement and Prediction, Volume IV of The American Soldier: Studies in Social Psychology in World War II*. Princeton University Press. 3, 72, 73

Lazarsfeld, P. F., and Henry, N. W. 1968. *Latent Structure Analysis*. Boston: Houghton Mifflin. 197, 298

Lazebnik, S., Schmid, C., and Ponce, J. 2006. Beyond bags of features: Spatial pyramid matching for recognizing natural scene categories. Pages 2169–2178 of: *IEEE Computer Society Conference on Computer Vision and Pattern Recognition (CVPR'06)*, vol. 2. IEEE. 364

Lazega, E. 2001. *The Collegial Phenomenon: The Social Mechanisms of Cooperation Among Peers in a Corporate Law Partnership*. Oxford University Press. 298, 299

LeCun, Y., Bottou, L., Bengio, Y., and Haffner, P. 1998. Gradient-based learning applied to document recognition. *Proceedings of the IEEE*, **86**(11), 2278–2324. 5

Ledoit, O., and Wolf, M. 2003. Improved estimation of the covariance matrix of stock returns with an application to portfolio selection. *Journal of Empirical Finance*, **10**, 603–621. 107

Ledoit, O., and Wolf, M. 2004. A well-conditioned estimator for large-dimensional covariance matrices. *Journal of Multivariate Analysis*, **88**, 365–411. 107

Ledoit, O., and Wolf, M. 2012. Nonlinear shrinkage estimation of large-dimensional covariance matrices. *Annals of Statistics*, **40**, 1024–1060. 107

Lee, H., and Li, J. 2012. Variable selection for clustering by separability based on ridgelines. *Journal of Computational and Graphical Statistics*, **21**, 315–337. 216

Lee, S. X., and McLachlan, G. J. 2013a. EMMIXuskew: An R package for fitting mixtures of multivariate skew t distributions via the EM algorithm. *Journal of Statistical Software*, **55**(12), 1–22. 273, 275

Lee, S. X., and McLachlan, G. J. 2013b. *EMMIXuskew: Fitting Unrestricted Multivariate Skew t Mixture Models*. R package version 0.11-5. 275

Lee, S. X., and McLachlan, G. J. 2013c. Model-based clustering and classification with non-normal mixture distributions. *Statistical Methods and Applications. Journal of the Italian Statistical Society*, **22**(4), 427–454. 290

Lee, S. X., and McLachlan, G. J. 2013d. On mixtures of skew normal and skew t-distributions. *Advances in Data Analysis and Classification*, **7**(3), 241–266. 270, 290

Lee, S. X., and McLachlan, G. J. 2014. Finite mixtures of multivariate skew t-distributions: some recent and new results. *Statistics and Computing*, **24**(2), 181–202. 268, 272, 290

Lee, S. X., and McLachlan, G. J. 2016. Finite mixtures of canonical fundamental skew t-distributions: the unification of the restricted and unrestricted skew t-mixture models. *Statistics and Computing*, **26**, 573–589. 270, 291

Lee, S. X., and McLachlan, G. J. 2018. EMMIXcskew: an R package for the fitting of a mixture of canonical fundamental skew-t distributions. *Journal of Statistical Software*, **83**(3), 1–32. 291

Leisch, F. 2004. FlexMix: A general framework for finite mixture models and latent class regression in R. *Journal of Statistical Software*, **11**(8), 1–18. 12, 339, 340

Leroux, M. 1992. Consistent estimation of a mixing distribution. *Annals of Statistics*, **20**, 1350–1360. 53

Li, J. 2005. Clustering based on a multilayer mixture model. *Journal of Computational and Graphical Statistics*, **14**, 547–568. 108

Li, J., Ray, S., and Lindsay, B. G. 2007a. A nonparametric statistical approach to clustering via mode identification. *Journal of Machine Learning Research*, **8**, 1687–1723. 108

Li, J., Xia, Y., Shan, Z., and Liu, Y. 2015. Scalable constrained spectral clustering. *IEEE Transactions on Knowledge and Data Engineering*, **27**(2), 589–593. 160

Li, Q., Fraley, C., Bumgarner, R. E., Yeung, K. Y., and Raftery, A. E. 2005. Donuts, scratches and blanks: Robust model-based segmentation of microarray images. *Bioinformatics*, **21**, 2875–2882. 383

Li, Y., Wessels, L., de Ridder, D., and Reinders, M. 2007b. Classification in the presence of class noise using a probabilistic kernel Fisher method. *Pattern Recognition*, **40**(12), 3349–3357. 147

Lin, T.-C., and Lin, T.-I. 2010. Supervised learning of multivariate skew normal mixture models with missing information. *Computational Statistics*, **25**(2), 183–201. 290

Lin, T.-I. 2009. Maximum likelihood estimation for multivariate skew normal mixture models. *Journal of Multivariate Analysis*, **100**(2), 257–265. 268

Lin, T.-I. 2010. Robust mixture modeling using multivariate skew t distributions. *Statistics and Computing*, **20**(3), 343–356. 272

Lin, T.-I. 2014. Learning from incomplete data via parameterized t mixture models through eigenvalue decomposition. *Computational Statistics and Data Analysis*, **71**, 183–195. 289

Lin, T.-I., and Lin, T.-C. 2011. Robust statistical modelling using the multivariate skew t distribution with complete and incomplete data. *Statistical Modelling*, **11**(3), 253–277. 290

Lin, T.-I., Ho, H. J., and Chen, C. L. 2009. Analysis of multivariate skew normal models with incomplete data. *Journal of Multivariate Analysis*, **100**(10), 2337–2351. 268

Lin, T.-I., McNicholas, P. D., and Ho, H. J. 2014. Capturing patterns via parsimonious t mixture models. *Statistics and Probability Letters*, **88**, 80–87. 261

Lindsay, Bruce. 1995. *Mixture Models: Theory, Geometry and Applications*. Hayward, CA: Institute of Mathematical Statistics. 14

Linnaeus, C. 1735. *Systema Naturae*. 1st edn. Leiden, Netherlands: Theodorum Haak. 2

Linnaeus, C. 1753. *Species Plantarum*. 1st edn. Stockholm, Sweden: Laurentii Salvii. 2

Linnaeus, C. 1758. *Systema Naturae*. 10th edn. Stockholm, Sweden: Laurentii Salvii. 2

Linzer, D. A., and Lewis, J. B. 2011. poLCA: An R package for polytomous variable latent class analysis. *Journal of Statistical Software*, **42**(10), 1–29. 340

Liu, C. 1997. ML estimation of the multivariate t distribution and the EM algorithm. *Journal of Multivariate Analysis*, **63**, 296–312. 260

Liu, J. S. 1994. The collapsed Gibbs sampler in Bayesian computations with applications to a gene regulation problem. *Journal of the American Statistical Association*, **89**(427), 958–966. 305

Lo, K., and Gottardo, R. 2012. Flexible mixture modeling via the multivariate t distribution with the Box-Cox transformation: An alternative to the skew t distribution. *Statistics and Computing*, **22**(1), 33–52. 281

Lo, K., Brinkman, R. R., and Gottardo, R. 2008. Automated gating of flow cytometry data via robust model based clustering. *Cytometry A*, **73**, 321–332. 279

Lo, K., Hahne, F., Brinkman, R. R., and Gottardo, R. 2009. flowClust: a Bioconductor package for automated gating of flow cytometry data. *BMC Bioinformatics*, **10**, R145. 281

Lomet, A. 2012. *Sélection de modèle pour la classification croisée de données continues*. Ph.D. thesis, Compiègne. 383

Longford, N. T., and Bartošová, J. 2014. A confusion index for measuring separation and clustering. *Statistical Modelling*, **14**, 229–255. 108

Lorrain, F., and White, H. C. 1971. Structural equivalence of individuals in social networks. *Journal of Mathematical Sociology*, **1**(1), 49–80. 298

MacQueen, J. 1967. Some methods for classification and analysis of multivariate observations. Pages 281–297 of: LeCam, L. M., and Neyman, J. (eds.), *Proceedings of the Fifth Berkeley Symposium on Mathematical Statistics and Probability*, vol. 1. Berkeley, California: University of California Press. 75

Madeira, S. C., and Oliveira, A. L. 2004. Biclustering algorithms for biological data analysis: A survey. *IEEE/ACM Transactions on Computational Biology and Bioinformatics*, **1**, 24–45. 374

Mahalanobis, P. C. 1930. On tests and measures of group divergence. Part I. Theoretical formulae. *Journal and Proceedings of the Asiatic Society of Bengal*, **26**, 541–588. 5

Mangasarian, O. L., Street, W. N., and Wolberg, W. H. 1995. Breast cancer diagnosis and prognosis via linear programming. *Operations Research*, **43**, 570–577. 7

Manikopoulos, C., and Papavassiliou, S. 2002. Network intrusion and fault detection: a statistical anomaly approach. *IEEE Communications Magazine*, **40**(10), 76–82. 161

Marbac, M., Biernacki, C., and Vandewalle, V. 2015. Model-based clustering for conditionally correlated categorical data. *Journal of Classification*, **32**, 145–175. 198

Mariadassou, M., Robin, S., and Vacher, C. 2010. Uncovering latent structure in valued graphs: A variational approach. *Annals of Applied Statistics*, **4**(2), 715–742. 329

Markou, M., and Singh, S. 2003a. Novelty detection: A review - part 1: Statistical approaches. *Signal Processing*, **83**(12), 2481–2497. 161

Markou, M., and Singh, S. 2003b. Novelty detection: A review - part 2: Neural network based approaches. *Signal Processing*, **83**(12), 2499–2521. 161

Marriott, F. H. C. 1975. Separating mixtures of normal distributions. *Biometrics*, **31**, 767–769. 34

Masoudnia, S., and Ebrahimpour, R. 2014. Mixture of experts: A literature survey. *Artificial Intelligence Review*, **42**(2), 275–293. 350

Matias, C., and Miele, V. 2017. Statistical clustering of temporal networks through a dynamic stochastic block model. *Journal of the Royal Statistical Society, Series B*, **79**, 1119–1141. 329

Maugis, C., Celeux, G., and Martin-Magniette, M.-L. 2009a. Variable selection for clustering with Gaussian mixture models. *Biometrics*, **65**, 701–709. 200, 203, 212, 216

Maugis, C., Celeux, G., and Martin-Magniette, M.-L. 2009b. Variable selection in model-based clustering: A general variable role modeling. *Computational Statistics and Data Analysis*, **53**, 3872–3882. 201, 210

Maugis, C., Celeux, G., and Martin-Magniette, M.-L. 2011. Variable selection in model-based discriminant analysis. *Journal of Multivariate Analysis*, **102**, 1374–1387. 210

Mazza, A., Punzo, A., and Ingrassia, S. 2018. flexCWM: A flexible framework for cluster-weighted models. *Journal of Statistical Software*, **86**(2), 1–30. 350

McCullagh, P., and Nelder, J. A. 1983. *Generalized Linear Models*. London: Chapman and Hall. 334

McCutcheon, A. C. 1987. *Latent Class Analysis*. Beverly Hills: Sage Publications. 197

McDaid, A. F., and Hurley, N. J. 2010. Detecting highly overlapping communities with model-based overlapping seed expansion. Pages 112–119 of: Memon, N., and Alhajj, R. (eds.), *International Conference on Advances in Social Networks Analysis and Mining (ASONAM)*. IEEE Computer Society. 329

McDaid, A. F., Murphy, T. B., Friel, N., and Hurley, N. J. 2012. Model-based clustering in networks with Stochastic Community Finding. Pages 549–560 of: Colubi, A., Fokianos, K., Kontoghiorghes, E. J., and Gonzáles-Rodríguez, G. (eds.), *Proceedings of COMPSTAT 2012: 20th International Conference on Computational Statistics*. ISI-IASC. 300

McDaid, A. F., Murphy, T. B., Friel, N., and Hurley, N. J. 2013. Improved Bayesian inference for the stochastic block model with application to large networks. *Computational Statistics and Data Analysis*, **60**, 12–31. 300

McDaid, A. F., Hurley, N. J., and Murphy, T. B. 2014. Overlapping stochastic community finding. Pages 17–20 of: Wu, X., Ester, M., and Xu, G. (eds.), *Advances in Social Networks Analysis and Mining (ASONAM), 2014 IEEE/ACM International Conference on*. IEEE. 329

McLachlan, G. J. 1976. A criterion for selecting variables for the linear discriminant function. *Biometrics*, 529–534. 6

McLachlan, G. J. 1992. *Discriminant Analysis and Statistical Pattern Recognition.* John Wiley & Sons. 6, 115, 146

McLachlan, G. J., and Basford, K. E. 1988. *Mixture Models: Inference and Applications to Clustering.* New York: Marcel Dekker. 14, 76

McLachlan, G. J., and Ganesalingam, S. 1982. Updating a discriminant function on the basis of unclassified data. *Communications in Statistics-Simulation and Computation*, **11**(6), 753–767. 6

McLachlan, G. J., and Krishnan, T. 1997. *The EM Algorithm and Extensions.* Wiley. 23, 377

McLachlan, G. J., and Lee, S. X. 2016. Comment on "On nomenclature, and the relative merits of two formulations of skew distributions" by A. Azzalini, R. Browne, M. Genton, and P. McNicholas. *Statistics and Probability Letters*, **116**, 1–5. 267, 270

McLachlan, G. J., and Peel, D. 1998. Robust cluster analysis via mixtures of multivariate *t*-distributions. Pages 658–666 of: *Advances in pattern recognition (Sydney, 1998).* Lecture Notes in Comput. Sci., vol. 1451. Springer, Berlin. 260

McLachlan, G. J., and Peel, D. 2000. *Finite Mixture Models.* New York: Wiley. 14, 78, 107, 174, 183, 350

McLachlan, G. J., Peel, D., Basford, K. E., and Adams, P. 1999. The EMMIX software for the fitting of mixtures of normal and *t*-components. *Journal of Statistical Software*, **4**(2). 261

McLachlan, G. J., Peel, D., and Bean, R. 2003. Modelling high-dimensional data by mixtures of factor analyzers. *Computational Statistics and Data Analysis*, **41**(3–4), 379–388. 238, 240, 244, 246, 257

McLachlan, G. J., Bean, R. W., and Ben-Tovim Jones, L. 2007. Extension of the mixture of factor analyzers model to incorporate the multivariate *t*-distribution. *Computational Statistics and Data Analysis*, **51**(11), 5327–5338. 257, 261

McNicholas, P. D. 2016a. *Mixture Model-Based Clustering.* Boca Raton, Fl.: Chapman & Hall/CRC Press. 14, 78

McNicholas, P. D. 2016b. Model-based clustering. *Journal of Classification*, **33**, 331–373. 78

McNicholas, P. D., and Murphy, T. B. 2008. Parsimonious Gaussian mixture models. *Statistics and Computing*, **18**(3), 285–296. 244, 246

McNicholas, P. D., and Murphy, T. B. 2010a. Model-based clustering of longitudinal data. *Canadian Journal of Statistics*, **38**(1), 153–168. 77

McNicholas, P. D., and Murphy, T. B. 2010b. Model-based clustering of microarray expression data via latent Gaussian mixture models. *Bioinformatics*, **26**(21), 2705–2712. 244, 246, 257

McNicholas, P. D., ElSherbiny, A., McDaid, A. F., and Murphy, T. B. 2018. *pgmm: Parsimonious Gaussian Mixture Models.* R package version 1.2.2. 12

McParland, D., and Gormley, I. C. 2016. Model-based clustering for mixed data: clustMD. *Advances in Data Analysis and Classification*, **10**, 155–169. 186, 187

McParland, D., and Gormley, I. C. 2017. *clustMD: Model Based Clustering for Mixed Data.* R package version 1.2.1. 12

McParland, D., and Murphy, T. B. 2018. Mixture modelling of high-dimensional data. Pages 247–280 of: Celeux, G., Frühwirth-Schnatter, S., and Robert, C. P. (eds.), *Handbook of Mixture Analysis.* Chapman & Hall/CRC. 257

Meeds, E., and Roweis, S. 2007. *Nonparametric Bayesian Biclustering.* Tech. rept. UTML TR 2007-001. Department of Computer Science, University of Toronto. 374

Melnykov, V. 2016. ClickClust: An R package for model-based clustering of categorical sequences. *Journal of Statistical Software*, **74**(9). 198

Melnykov, V., Melnykov, I., and Michael, S. 2015. Semi-supervised model-based clustering with positive and negative constraints. *Advances in Data Analysis and Classification*, 1–23. 144

Meng, X.-L., and Van Dyk, D. 1997. The EM algorithm - an old folk song sung to a fast new tune. *Journal of the Royal Statistical Society. Series B (Methodological)*, **59**(3), 511–567. 246

Mengersen, K. L., Robert, C. P., and Titterington, D. M. (eds.). 2011. *Mixtures: Estimation and Applications*. Wiley. 14

Michael, S., and Melnykov, V. 2016. An effective strategy for initializing the EM algorithm in finite mixture models. *Advances in Data Analysis and Classification*, **10**, 563–583. 77

Miller, D., and Browning, J. 2003. A mixture model and em-based algorithm for class discovery, robust classification, and outlier rejection in mixed labeled/unlabeled data sets. *IEEE Transactions on Pattern Analysis and Machine Intelligence*, **11**(25), 1468–1483. 111, 155, 156, 157, 159

Mingers, J. 1989. An empirical comparison of pruning methods for decision tree induction. *Journal of Machine Learning*, **4**(2), 227–243. 161

Minka, T. P., Winn, J., Guiver, J., and Knowles, D. 2010. *Infer.NET*. Version 2.4. 306

Mkhadri, A., Celeux, G., and Nasrollah, A. 1997. Regularization in discriminant analysis: a survey. *Computational Statistics and Data Analysis*, **23**, 403–423. 117, 236

Montanari, A., and Viroli, C. 2010. Heteroscedastic factor mixture analysis. *Statistical Modeling*, **10**(4), 441–460. 243

Mosmann, T. R., Naim, I., Rebhahn, J., Datta, S., Cavenaugh, J. S., Weaver, J. M., and Sharma, G. 2014. SWIFT-scalable clustering for automated identification of rare cell populations in large, high-dimensional flow cytometry datasets, Part 2: Biological evaluation. *Cytometry, Part A*, **85**, 422–433. 108

Muise, R., and Smith, C. 1992. *Nonparametric Minefield Detection and Localization*. Technical Report CSS-TM-591-91. Coastal Systems Station, Panama City, Florida. 10

Mukherjee, S., Feigelson, E. D., Babu, G. J., Murtagh, F., Fraley, C., and Raftery, A. E. 1998. Three types of gamma ray bursts. *Astrophysical Journal*, **508**, 314–327. 77

Murphy, K., and Murphy, T. B. 2018a. *MoEClust: Gaussian Parsimonious Clustering Models with Covariates*. R package version 1.2.0. 340

Murphy, K., and Murphy, T. B. 2018b. Parsimonious model-based clustering with covariates. *arXiv preprint arXiv:1711.05632v2*. 340

Murphy, T. B., Raftery, A. E., and Dean, N. 2010. Variable selection and updating in model-based discriminant analysis for high-dimensional data with food authenticity applications. *Annals of Applied Statistics*, **4**, 396–421. 210

Murray, P. M., Browne, R. P., and McNicholas, P. D. 2017. A mixture of SDB skew-t factor analyzers. *Econometrics and Statistics*, **3**, 160–168. 290

Murtagh, F., and Raftery, A. E. 1984. Fitting straight lines to point patterns. *Pattern Recognition*, **17**, 479–483. 76

Murtagh, F., Raftery, A. E., and Starck, J. L. 2005. Bayesian inference for multiband image segmentation via model-based cluster trees. *Image and Vision Computing*, **23**, 587–596. 383

Nadif, M., and Govaert, G. 1998. Clustering for binary data and mixture models: Choice of the model. *Applied Stochastic Models and Data Analysis*, **13**, 269–278. 174

Nadolski, J., and Viele, K. 2004 (July). The role of latent variables in model selection accuracy. In: *International Federation of Classification Societies Meeting*. 174

Naim, I., Datta, S., Rebhahn, J., Cavenaugh, J. S., Mosmann, T. R., and Sharma, G. 2014. SWIFT-scalable clustering for automated identification of rare cell populations in large, high-dimensional flow cytometry datasets, Part 1: Algorithm design. *Cytometry, Part A*, **85**, 408–421. 108

Newman, M. E. J. 2016. Equivalence between modularity optimization and maximum likelihood methods for community detection. *Physical Review E*, **94**, 052315. 329

Nia, V. P., and Davison, A. C. 2012. High-dimensional Bayesian clustering with variable selection: The R package bclust. *Journal of Statistical Software*, **47**(5), 1–22. 215

Nigam, K., McCallum, A., Thrun, S., and Mitchell, T. 2000. Text classification from labeled and unlabeled documents using em. *Machine Learning*, **39**(2-3), 103–134. 364

Nobile, A., and Fearnside, A. T. 2007. Bayesian finite mixtures with an unknown number of components: The allocation sampler. *Statistics and Computing*, **17**, 147–162. 213

Nowicki, K., and Snijders, T. A. B. 2001. Estimation and prediction of stochastic blockstructures. *Journal of the American Statistical Association*, **96**(455), 1077–1087. 298, 300

Odin, T., and Addison, D. 2000. Novelty detection using neural network technology. Pages 731–743 of: *COMADEM 2000: 13th International Congress on Condition Monitoring and Diagnostic Engineering Management*. 161

Oh, M. S., and Raftery, A. E. 2001. Bayesian multidimensional scaling and choice of dimension. *Journal of the American Statistical Association*, **96**, 1031–1044. 77

Oh, M. S., and Raftery, A. E. 2007. Model-based clustering with dissimilarities: A Bayesian approach. *Journal of Computational and Graphical Statistics*, **16**, 559–585. 77

O'Hagan, A., and Ferrari, C. 2017. Model-based and nonparametric approaches to clustering for data compression in actuarial applications. *North American Actuarial Journal*, **21**(1), 107–146. 78

O'Hagan, A., and White, A. 2018. Improved model-based clustering performance using Bayesian initialization averaging. *Computational Statistics*, To appear. 77

O'Hagan, A., Murphy, T. B., and Gormley, I. C. 2012. Computational aspects of fitting mixture models via the expectation-maximization algorithm. *Computational Statistics and Data Analysis*, **56**(12), 3843–3864. 77

O'Hagan, A., Murphy, T. B., Gormley, I. C., McNicholas, P. D., and Karlis, D. 2016. Clustering with the multivariate normal inverse Gaussian distribution. *Computational Statistics and Data Analysis*, **93**, 18–30. 283

Pan, W., and Shen, X. 2007. Penalized model-based clustering with application to variable selection. *Journal of Machine Learning Research*, **8**, 1145–1164. 208

Papadimitriou, C., Tamaki, H., Raghavan, P., and Vempala, S. 1998. Latent semantic indexing: A probabilistic analysis. Pages 159–168 of: *Proceedings of the Seventeenth ACM SIGACT-SIGMOD-SIGART Symposium on Principles of Database Systems*. PODS '98. New York, NY, USA: ACM. 364

Papastamoulis, P., and Iliopoulos, G. 2010. An artificial allocations based solution to the label switching problem in Bayesian analysis of mixtures of distributions. *Journal of Computational and Graphical Statistics*, **19**, 313–331. 183

Pavlenko, T. 2003. On feature selection, curse of dimensionality and error probability in discriminant analysis. *Journal of Statistical Planning and Inference*, **115**, 565–584. 224, 225

Pavlenko, T., and Von Rosen, D. 2001. Effect of dimensionality on discrimination. *Statistics*, **35**(3), 191–213. 224, 225

Pearson, K. 1901. On lines and planes of closest fit to systems of points in space. *Philosophical Magazine*, **6**(2), 559–572. 228

Peel, D., and McLachlan, G. J. 2000. Robust mixture modelling using the *t* distribution. *Statistics and Computing*, **10**, 339–348. 106, 260, 261

Peng, F., Jacobs, R. A., and Tanner, M. A. 1996. Bayesian inference in Mixtures-of-Experts and Hierarchical Mixtures-of-Experts models with an application to speech recognition. *Journal of the American Statistical Association*, **91**(435), 953–960. 349

Pontikos, D. 2004. *Model-Based Clustering of World Craniometric Variation*. dienekes.50webs.com/arp/articles/anthropologica/clustering.html. September 2004, accessed January 27, 2016. 65

Pontikos, D. 2010. *World Craniometric Analysis with MCLUST Revisited*. dienekes.blogspot.com/2010/12/world-craniometric-analysis-with-mclust.html. December 5, 2010; accessed January 27, 2016. 65

Poon, L. K. M., Zhang, N. L., and Liu, A. H. 2013. Model-based clustering of high-dimensional data: Variable selection versus facet determination. *International Journal of Approximate Reasoning*, **54**, 196–215. 216

Prates, M. O., Cabral, C. R. B., and Lachos, V. H. 2013. mixsmsn: Fitting finite mixture of scale mixture of skew-normal distributions. *Journal of Statistical Software*, **54**(12), 1–20. 269

Punzo, A., and McNicholas, P. D. 2017. Robust clustering in regression analysis via the contaminated Gaussian cluster-weighted model. *Journal of Classification*, **34**, 249–293. 350

Pyne, S., Hua, X., Wang, K., Rossina, E., Lin, T.-I., Maiera, L. M., Baecher-Alland, C., McLachlan, G. J., Tamayoa, P., Haflera, D. A., De Jagera, P. L., and Mesirova, J. P. 2009. Automated high-dimensional flow cytometric data analysis. *Proceedings of the National Academy of Sciences USA*, **106**, 8519–8524. 271

Quandt, R. E., and Ramsey, J. B. 1978. Estimating mixtures of normal distributions and switching regressions. *Journal of the American Statistical Association*, **73**(364), 730–738. 340

Quinlan, J. R. 1996. Bagging, boosting, and C4.S. Pages 725–730 of: *Proceedings of the Thirteenth National Conference on Artificial Intelligence - Volume 1*. AAAI'96. AAAI Press. 161

R Development Core Team. 2010. *R: A Language and Environment for Statistical Computing*. R Foundation for Statistical Computing, Vienna, Austria. ISBN 3-900051-07-0. 339

Raftery, A. E. 1995. Bayesian model selection in social research (with discussion). *Sociological Methodology*, **25**, 111–193. 51, 173

Raftery, A. E. 1999. Bayes factors and BIC: Comment on 'A critique of the Bayesian Information Criterion for model selection'. *Sociological Methods and Research*, **27**, 411–427. 52

Raftery, A. E., and Dean, N. 2006. Variable selection for model-based clustering. *Journal of the American Statistical Association*, **101**, 168–178. 200, 203, 205, 210

Raftery, A. E., Niu, X., Hoff, P. D., and Yeung, K. Y. 2012. Fast inference for the latent space network model using a case-control approximate likelihood. *Journal of Computational and Graphical Statistics*, **21**, 901–919. 317

Ramsay, J. O., and Silverman, B. W. 2005. *Functional Data Analysis*. Second edn. Springer Series in Statistics. New York: Springer. 354, 359

Rand, W. M. 1971. Objective criteria for the evaluation of clustering methods. *Journal of the American Statistical Association*, **66**, 846–850. 40

Rao, C. R. 1948. The utilization of multiple measurements in problems of biological classification. *Journal of the Royal Statistical Society. Series B (Methodological)*, **10**(2), 159–203. 6

Rao, C. R. 1952. *Advanced Statistical Methods in Biometric Research*. Oxford, England: Wiley. 6

Rao, C. R. 1954. A general theory of discrimination when the information about alternative population distributions is based on samples. *Annals of Mathematical Statistics*, **25**(4), 651–670. 6

Rau, A., Maugis, C., Martin-Magniette, M.-L., and Celeux, G. 2015. Co-expression analysis of high-throughput transcriptome sequencing data with poisson mixture models. *Bioinformatics*, **31**, 1420–1427. 192, 194

Ray, S., and Lindsay, B. G. 2005. The topography of multivariate normal mixtures. *Annals of Statistics*, **33**, 2042–2065. 101

Ray, S., and Mallick, B. 2006. Functional clustering by Bayesian wavelet methods. *Journal of the Royal Statistical Society. Series B. Statistical Methodology*, **68**(2), 305–332. 353

Reaven, G. M., and Miller, R. G. 1979. An attempt to define the nature of chemical diabetes using a multidimensional analysis. *Diabetologia*, **16**, 17–24. 7

Redner, R. A., and Walker, H. F. 1984. Mixture densities, maximum likelihood and the EM algorithm. *SIAM Review*, **26**, 195–239. 32, 93, 106

Ripley, B. D. 1996. *Pattern Recognition and Neural Networks*. Cambridge University Press. 110, 140

Rivera-García, D., García-Escudero, L. A., Mayo-Iscar, A., and Ortega, J. 2018. Robust clustering for functional data based on trimming and constraints. *Advances in Data Analysis and Classification*, In press. 382

Roberts, S., and Tarassenko, L. 1994. A probabilistic resource allocating network for novelty detection. *Neural Computation*, **6**, 270–284. 161

Roberts, S., Husmeier, D., Rezek, I., and Penny, W. 1998. Bayesian approaches to Gaussian mixture modeling. *IEEE Transactions on Pattern Analysis and Machine Intelligence*, **20**, 1133–1142. 107

Roberts, S. J. 1999. Novelty detection using extreme value statistics. *IEE Proceedings - Vision, Image and Signal Processing*, **146**(3), 124–129. 161

Roeder, K., and Wasserman, L. 1997. Practical Bayesian density estimation using mixtures of normals. *Journal of the American Statistical Association*, **92**, 894–902. 53

Rosen, O., and Tanner, M. A. 1999. Mixtures of proportional hazards regression models. *Statistics in Medicine*, **18**, 1119–1131. 350

Rosenblatt, F. 1958. The perceptron: a probabilistic model for information storage and organization in the brain. *Psychological Review*, **65**(6), 386. 5

Rousseeuw, P. J., and Leroy, A. 1987. *Robust Regression and Outlier Detection*. New York: Wiley. 161

Ruan, L., Yuan, M., and Zou, H. 2011. Regularized parameter estimation in high-dimensional Gaussian mixture models. *Neural Computation*, **23**, 1605–1622. 107

Rubin, D. B., and Thayer, D. 1982. EM algorithms for ML factor analysis. *Psychometrika*, **47**(1), 69–76. 238

Runnals, A. 2007. A Kullback–Leibler approach to Gaussian mixture reduction. *IEEE Transactions on Aerospace and Electronic Systems*, **43**, 989–999. 108

Russell, N., Cribbin, L., and Murphy, T. B. 2014. *upclass: Updated Classification Methods using Unlabeled Data*. R package version 2.0. 136

Russell, N., Murphy, T. B., and Raftery, A. E. 2015. *Bayesian model averaging in model-based clustering and density estimation*. Technical Report 635. Department of Statistics, University of Washington. Also available at arXiv:1506.09035. 77

Sahu, S. K., Dey, D. K., and Branco, M. D. 2003. A new class of multivariate skew distributions with applications to Bayesian regression models. *The Canadian Journal of Statistics*, **31**(2), 129–150. 267, 269, 271

Sakakibara, Y. 1993. Noise-tolerant Occam algorithms and their applications to learning decision trees. *Journal of Machine Learning*, **11**(1), 37–62. 161

Salmond, D. J. 2009. Mixture reduction algorithms for point and extended object tracking in clutter. *IEEE Transactions on Aerospace and Electronic Systems*, **45**, 667–686. 108

Salter-Townshend, M. 2012. *VBLPCM: Variational Bayes Latent Position Cluster Model for Networks*. R package version 2.0. 317, 320

Salter-Townshend, M., and Murphy, T. B. 2009. Variational Bayesian inference for the latent position cluster model. In: *NIPS Workshop on Analyzing Networks and Learning with Graphs*. 317

Salter-Townshend, M., and Murphy, T. B. 2013. Variational Bayesian inference for the latent position cluster model for network data. *Computational Statistics and Data Analysis*, **57**(1), 661–671. 317

Salter-Townshend, M., and Murphy, T. B. 2015. Role analysis in networks using mixtures of exponential random graph models. *Journal of Computational and Graphical Statistics*, **24**, 520–538. 329

Salter-Townshend, M., White, A., Gollini, I., and Murphy, T. B. 2012. Review of statistical network analysis: Models, algorithms, and software. *Statistical Analysis and Data Mining*, **5**(4), 243–264. 294

Samé, A., Chamroukhi, F., Govaert, G., and Aknin, P. 2011. Model-based clustering and segmentation of times series with changes in regime. *Advances in Data Analysis and Classification*, **5**(4), 301–322. 353

Sampson, S. F. 1969. *Crisis in a Cloister*. Ph.D. thesis, Cornell University. 293, 294, 295, 302

Sanguinetti, G. 2008. Dimensionality reduction of clustered datasets. *IEEE Transactions on Pattern Analysis and Machine Intelligence*, **30**(3), 1–29. 252

Sarkar, P., and Moore, A. W. 2005a. Dynamic social network analysis using latent space models. Pages 1145–1152 of: *Proceedings of the 18th International Conference on Neural Information Processing Systems*. NIPS'05. Cambridge, MA, USA: MIT Press. 330

Sarkar, P., and Moore, A. W. 2005b. Dynamic social network analysis using latent space models. *SIGKDD Explorations*, **7**(2), 31–40. 330

Sarkar, P., Siddiqi, S. M., and Gordon, G. J. 2007. A latent space approach to dynamic embedding of co-occurrence data. Pages 420–427 of: *Proceedings of the Eleventh International Conference on Artificial Intelligence and Statistics, AISTATS 2007, San Juan, Puerto Rico, March 21-24, 2007*. 330

Schapire, R. 1990. The strength of weak learnability. *Machine Learning*, **5**, 197–227. 161

Schmutz, A., Bouveyron, C., Jacques, J., Martin, P., and Cheze, L. 2018. *Clustering multivariate functional data in group-specific functional subspaces*. Tech. rept. Preprint HAL 01652467. Université Côte d'Âzur. 353, 359

Schölkopf, B., Smola, A., and Müller, K. 1998. Non linear component analysis as a kernel eigenvalue problem. *Neural Computation*, **10**, 1299–1319. 229

Schölkopf, B., Williamson, R., Smola, A., Shawe-Taylor, J., and Platt, J. 1999. Support vector method for novelty detection. Pages 582–588 of: *Proceedings of the 12th International Conference on Neural Information Processing Systems*. NIPS'99. Cambridge, MA, USA: MIT Press. 161

Schwarz, G. 1978. Estimating the dimension of a model. *Annals of Statistics*, **6**, 461–464. 51, 133, 172

Scott, A. J., and Symons, M. J. 1971. Clustering methods based on likelihood ratio criteria. *Biometrics*, **27**, 387–397. 76

Scott, D. 1992. *Multivariate Density Estimation*. New York: Wiley & Sons. 222

Scott, D., and Thompson, J. R. 1983. Probability density estimation in higher dimensions. Pages 173–179 of: Gentle, J. E. (ed.), *Computer Science and Statistics: Proceedings of the Fifteenth Symposium on the Interface*. 225

Scrucca, L. 2010. Dimension reduction for model-based clustering. *Statistics and Computing*, **20**(4), 471–484. 257

Scrucca, L. 2016a. Genetic algorithms for subset selection in model-based clustering. Pages 55–70 of: Celebi, M. E., and Aydin, K. (eds.), *Unsupervised Learning Algorithms*. Springer International Publishing. 216

Scrucca, L. 2016b. Identifying connected components in Gaussian finite mixture models for clustering. *Pattern Recognition*, **93**, 5–17. 108

Scrucca, L., and Raftery, A. E. 2015. Improved initialisation of model-based clustering using a Gaussian hierarchical partition. *Advances in Data Analysis and Classification*, **9**, 447–460. 77

Scrucca, L., and Raftery, A. E. 2018. clustvarsel: a package implementing variable selection for Gaussian model-based clustering in R. *Journal of Statistical Software*, **84**, 1–28. 203

Scrucca, L., Fop, M., Murphy, T. B., and Raftery, A. E. 2016. mclust 5: clustering, classification and density estimation using Gaussian finite mixture models. *The R Journal*, **8**, 205–233. 12

Seo, B., and Kim, D. 2012. Root selection in normal mixture models. *Computational Statistics and Data Analysis*, **56**, 2454–2470. 107

Sewell, D. K., and Chen, Y. 2015. Latent space models for dynamic networks. *Journal of the American Statistical Association*, **110**, 1646–1657. 330

Shental, N., Bar-Hillel, A., Hertz, T., and Weinshall, D. 2003. Computing Gaussian mixture models with EM using equivalence constraints. Pages 465–472 of: *Proceedings of the 16th International Conference on Neural Information Processing Systems*. NIPS'03. Cambridge, MA, USA: MIT Press. 141, 144

Silvestre, C., Cardoso, M. G. M. S., and Figueiredo, M. A. T. 2015. Features selection for clustering categorical data with an embedded modelling approach. *Expert Systems*, **32**, 444–453. 216

Smídl, V., and Quinn, A. 2006. *The Variational Bayes Method in Signal Processing*. Springer. 337

Sneath, P. H. A. 1957. The application of computers to taxonomy. *Journal of General Microbiology*, **17**, 201–206. 2, 33

Snijders, T. A. B., and Nowicki, K. 1997. Estimation and prediction for stochastic blockmodels for graphs with latent block structure. *Journal of Classification*, **14**(1), 75–100. 298, 300

Sokal, R. R., and Michener, C. D. 1958. A statistical method for evaluating systematic relationships. *University of Kansas Scientific Bulletin*, **38**, 1409–1438. 2, 33

Sokal, R. R., and Sneath, P. H. A. 1963. *Principles of Numerical Taxonomy*. San Francisco: W. H. Freeman & Co. 2

Souza, F. A. A., and Araújo, R. 2014. Mixture of partial least squares experts and application in prediction settings with multiple operating modes. *Chemometrics and Intelligent Laboratory Systems*, **130**, 192–202. 350

Spearman, C. 1904. The proof and measurement of association between two things. *American Journal of Psychology*, **15**, 72–101. 229, 238

Stanford, D. C., and Raftery, A. E. 2000. Principal curve clustering with noise. *IEEE Transactions on Pattern Analysis and Machine Analysis*, **22**, 601–609. 53, 87, 105, 382

Steane, M. A., McNicholas, P. D., and Yada, R. Y. 2012. Model-based classification via mixtures of multivariate t-factor analyzers. *Communications in Statistics. Simulation and Computation*, **41**(4), 510–523. 261

Steele, R. J., and Raftery, A. E. 2010. Performance of Bayesian model selection criteria for Gaussian mixture models. Pages 113–130 of: Chen, M. H. (ed.), *Frontiers of Statistical Decision Making and Bayesian Analysis*. New York: Springer. 77

Steinley, D., and Brusco, M. J. 2008. Selection of variables in cluster analysis: An empirical comparison of eight procedures. *Psychometrika*, **73**, 125–144. 201

Stephens, M. 2000a. Bayesian analysis of mixture models with an unknown number of components—an alternative to reversible jump methods. *Annals of Statistics*, **28**(1), 40–74. 288

Stephens, M. 2000b. Dealing with label switching in mixture models. *Journal of the Royal Statistical Society. Series B (Statistical Methodology)*, **62**, 795–809. 107, 183

Stephenson, W. 1936. Introduction of inverted factor analysis with some applications to studies in orexia. *Journal of Educational Psychology*, **5**, 353–367. 2

Stone, M. 1974. Cross-validatory choice and assessment of statistical predictions. *Journal of the Royal Statistical Society. Series B (Methodological)*, **36**, 111–147. 132

Street, W. N., Wolberg, W. H., and Mangasarian, O. L. 1993. Nuclear feature extraction for breast tumor diagnosis. Pages 861–871 of: *Biomedical Image Processing and Biomedical Visualization*, vol. 1905. International Society for Optics and Photonics. 7

Sun, Y., Han, J., Gao, J., and Yu, Y. 2009. iTopicModel: Information network-integrated topic modeling. Pages 493–502 of: *Ninth IEEE International Conference on Data Mining. ICDM'09*. IEEE. 382

Tadesse, M. G., Sha, N., and Vannucci, M. 2005. Bayesian variable selection in clustering high-dimensional data. *Journal of the American Statistical Association*, **100**, 602–617. 200, 213

Tanner, Martin A., and Jacobs, R. A. 2001. Neural networks and related statistical latent variable models. Pages 10526–10534 of: Smelser, Neil J., and Baltes, Paul B. (eds.), *International Encyclopedia of the Social and Behavioral Sciences*. Elsevier. 350

Tantrum, J. M., Murua, A., and Stuetzle, W. 2003. Assessment and pruning of hierarchical model based clustering. Pages 197–205 of: *Proceedings of the Ninth ACM SIGKDD International Conference on Knowledge Discovery and Data Mining*. KDD '03. New York, NY, USA: ACM. 101

Tarassenko, L., Hayton, P., Cerneaz, N., and Brady, M. 1995. Novelty detection for the identification of masses in mammograms. Pages 442–447 of: *Fourth International Conference on Artificial Neural Networks*. 161

Tax, D., and Duin, R. 1999. Outlier detection using classifier instability. Pages 251–256 of: Amin, A., Dori, D., Pudil, P., and Freeman, H. (eds.), *Advances in Pattern Recognition*. Heidelberg: Springer. 161

Thompson, T. J., Smith, P. J., and Boyle, J. P. 1998. Finite mixture models with concomitant information: assessing diagnostic criteria for diabetes. *Journal of the Royal Statistical Society. Series C (Applied Statistics)*, **47**, 393–404. 349

Tibshirani, R., and Walther, G. 2005. Cluster validation by prediction strength. *Journal of Computational and Graphical Statistics*, **14**, 511–528. 101

Tiedeman, D. V. 1955. On the study of types. Pages 1–14 of: Sells, S. B. (ed.), *Symposium on Pattern Analysis*. Randolph Field, Tex.: USAF School of Aviation Medicine, Air University. 73

Tipping, M. E., and Bishop, C. M. 1997. *Probabilistic principal component analysis*. Tech. rept. NCRG-97-010. Neural Computing Research Group, Aston University. 229

Tipping, M. E., and Bishop, C. M. 1999. Mixtures of probabilistic principal component analysers. *Neural Computation*, **11**(2), 443–482. 246, 248, 249, 257

Titterington, D. M., Smith, A. F. M., and Makov, U. E. 1985. *Statistical Analysis of Finite Mixture Distributions*. Wiley. 14, 31, 92, 105, 106

Tortora, C., Franczak, B. C., Browne, R. P., and McNicholas, P. D. 2014. *Mixtures of Multiple Scaled Generalized Hyperbolic Distributions*. arXiv:1403.2332. 291

Tortora, C., Browne, R. P., Franczak, B. C., and McNicholas., P. D. 2015a. *MixGHD: Model Based Clustering, Classification and Discriminant Analysis Using the Mixture of Generalized Hyperbolic Distributions*. R package version 1.5. 285

Tortora, C., McNicholas, P. D., and Browne, R. P. 2015b. A mixture of generalized hyperbolic factor analyzers. *Advances in Data Analysis and Classification*, 1–18. 291

Tortora, C., Franczak, B. C., Browne, R. P., and McNicholas, P. D. 2018. A mixture of coalesced generalized hyperbolic distributions. *Journal of Classification*, To appear. 291

Toussile, W., and Gassiat, E. 2009. Variable selection in model-based clustering using multilocus genotype data. *Advances in Data Analysis and Classification*, **3**, 109–134. 212

Tryon, R. C. 1939. *Cluster Analysis: Correlation Profile and Orthometric (Factor) Analysis for the Isolation of Unities in Mind and Personality*. Edwards Brothers. 2

Turner, R. 2014. *mixreg: Functions to fit mixtures of regressions*. R package version 0.0-5. 340

Uebersax, J. S. 2010. *Latent Structure Analysis*. www.john-uebersax.com/stat/. 184, 197

Uebersax, J. S., and Grove, W. M. 1993. A latent trait finite mixture model for the analysis of rating agreement. *Biometrics*, **49**, 823–835. 166

van den Boogaart, K. G. 2009. *compositions: Compositional Data Analysis*. R package version 1.10-2. 306

Vandewalle, V. 2009. *Estimation et sélection en classification semi-supervisée*. Ph.D. thesis, Université de Lille 1. 171

Vandewalle, V., Biernacki, C., Celeux, G., and Govaert, G. 2013. A predictive deviance criterion for selecting a generative model in semi-supervised classification. *Computational Statistics and Data Analysis*, **64**, 220–236. 140, 141

Vannoorenbergue, P., and Denoeux, T. 2002. Handling uncertain labels in multiclass problems using belief decision trees. In: *Proceedings of IPMU'2002*. 161

Vapnik, V. 1998. *The Nature of Statistical Learning Theory*. New York: Springer. 111

Verleysen, M. 2003. Learning high-dimensional data. Pages 141–162 of: Ablameyko, S., Gori, M., Goras, L., and Piuri, V. (eds.), *Limitations and Future Trends in Neural Computations*. NATO Science Series, III: Computer and Systems Sciences, vol. 186. IOS Press. 222

Verleysen, M., and François, D. 2005. The curse of dimensionality in data mining and time series prediction. Pages 758–770 of: Cabestany, J., Prieto, A., and Sandoval, F. (eds.), *Computational Intelligence and Bioinspired Systems*. Berlin, Heidelberg: Springer. 222

Vermunt, J. K., and Magidson, J. 2005. *Technical Guide for Latent GOLD 4.0: Basic and Advanced*. www.statisticalinnovations.com. 185, 198

Volant, S., Bérard, C., Martin-Magniette, M.-L., and Robin, S. 2014. Hidden Markov models with mixtures as emission distributions. *Statistics and Computing*, **214**, 493–504. 108

Von Mises, R. 1945. On the classification of observation data into distinct groups. *Annals of Mathematical Statistics*, **16**(1), 68–73. 6, 110

Vrbik, I., and McNicholas, P. D. 2012. Analytic calculations for the EM algorithm for multivariate skew-t mixture models. *Statistics and Probability Letters*, **82**, 1169–1174. 290

Vrbik, I., and McNicholas, P. D. 2014. Parsimonious skew mixture models for model-based clustering and classification. *Computational Statistics and Data Analysis*, **71**, 196–210. 160, 290

Wald, A. 1939. Contributions to the theory of statistical estimation and testing hypotheses. *Annals of Mathematical Statistics*, **10**(4), 299–326. 6

Wald, A. 1944. On a statistical problem arising in the classification of an individual into one of two groups. *Annals of Mathematical Statistics*, **15**(2), 145–162. 6, 110

Wald, A. 1949. Statistical decision functions. *Annals of Mathematical Statistics*, 165–205. 6

Wallace, C. S., and Freeman, P. 1987. Estimation and inference via compact coding. *Journal of the Royal Statistical Society. Series B (Methodological)*, **49**(3), 241–252. 200

Wallace, M. L., Buysse, D. J., Germain, A., Hall, M. H., and Iyenbar, S. 2018. Variable selection for skewed model-based clustering: Application to the identification of novel sleep phenotypes. *Journal of the American Statistical Association*, **113**, 95–110. 216

Wang, K., Ng, A., and McLachlan., G. J. 2013. *EMMIXskew: The EM Algorithm and Skew Mixture Distribution*. R package version 1.0.1. 261, 268, 269, 272

Wang, N., and Raftery, A. E. 2002. Nearest neighbor variance estimation (NNVE): Robust covariance estimation via nearest neighbor cleaning (with discussion). *Journal of the American Statistical Association*, **97**, 994–1019. 106

Wang, S., and Zhu, J. 2008. Variable selection for model-based high-dimensional clustering and its application to microarray data. *Biometrics*, **64**, 440–448. 208

Wang, W.-L., and Lin, T.-I. 2013. An efficient ECM algorithm for maximum likelihood estimation in mixtures of *t*-factor analyzers. *Computational Statistics*, **28**(2), 751–769. 289

Wang, Y.-Q. 2013. E-PLE: An algorithm for image inpainting. *Image Processing On Line*, **3**, 271–285. 373, 374

Ward, J. H. 1963. Hierarchical groupings to optimize an objective function. *Journal of the American Statistical Association*, **58**, 234–244. 33, 75

Wasserman, S., and Faust, K. 1994. *Social Network Analysis: Methods and Applications*. Cambridge University Press. 294

Wasserman, S., Robins, G., and Steinley, D. 2007. Statistical models for networks: A brief review of some recent research. Pages 45–56 of: Airoldi, E. M., Blei, D. M., Fienberg, S. E., Goldenberg, A., Xing, E. P., and Zheng, A. X. (eds.), *Statistical Network Analysis: Models, Issues, and New Directions*. Lecture Notes in Computer Science, vol. 4503. Springer Berlin Heidelberg. 294

Wehrens, R., Buydens, L. M. C., Fraley, C., and Raftery, A. E. 2004. Model-based clustering for image segmentation and large datasets via sampling. *Journal of Classification*, **21**, 231–253. 35, 383

Welch, B. L. 1939. Note on discriminant functions. *Biometrika*, **31**(1/2), 218–220. 6

West, M., and Harrison, P. J. 1989. *Bayesian Forecasting and Dynamic Models*. New York: Springer-Verlag. 108

White, A., and Murphy, T. B. 2016a. Exponential family mixed membership models for soft clustering of multivariate data. *Advances in Data Analysis and Classification*, **10**(4), 521–540. 306

White, A., and Murphy, T. B. 2016b. Mixed membership of experts stochastic block model. *Network Science*, **4**(1), 48–80. 329, 350

White, A., Chan, J., Hayes, C., and Murphy, T. B. 2012. Mixed membership models for exploring user roles in online fora. Pages 599–602 of: Breslin, J., Ellison, N., Shanahan, J.G., and Tufekci, Z. (eds.), *Proceedings of the Sixth International AAAI Conference on Weblogs and Social Media (ICWSM 2012)*. AAAI Press. 306

White, A., Wyse, J., and Murphy, T. B. 2016. Bayesian variable selection for latent class analysis using a collapsed gibbs sampler. *Statistics and Computing*, **26**, 511–527. 211, 212, 213

Wilson, D. R., and Martinez, T. R. 1997. Instance pruning techniques. Pages 403–411 of: *Proceedings of the Fourteenth International Conference on Machine Learning*. ICML '97. San Francisco, CA, USA: Morgan Kaufmann Publishers Inc. 161

Witten, D. M., and Tibshirani, R. 2010. A framework for feature selection in clustering. *Journal of American Statistical Association*, **105**, 713–726. 77, 201

Wold, S., Sjöström, M., and Eriksson, L. 2001. PLS-regression: A basic tool of chemometrics. *Chemometrics and Intelligent Laboratory Systems*, **58**(2), 109–130. 235

Wolfe, J. H. 1963. *Object cluster analysis of social areas*. M.Phil. thesis, University of California, Berkeley. 3, 73

Wolfe, J. H. 1965. *A Computer Program for the Maximum-Likelihood Analysis of Types*. USNPRA Technical Bulletin 65-15. U.S. Naval Personnel Research Activity, San Diego. 3, 74, 75

Wolfe, J. H. 1967. *NORMIX: Computational Methods for Estimating the Parameters of Multivariate Normal Mixture Distributions of Types*. USNPRA Technical Bulletin 68-2. U.S. Naval Personnel Research Activity, San Diego. 3, 75

Wolfe, J. H. 1970. Pattern clustering by multivariate mixture analysis. *Multivariate Behavioral Research*, **5**, 329–350. 3, 75

Wolfe, J. H. 2018. *Personnal communication*. 73

Wu, C. F. J. 1983. On convergence properties of the EM algorithm. *Annals of Statistics*, **11**, 95–103. 23

Wyse, J., and Friel, N. 2012. Block clustering with collapsed latent block models. *Statistics and Computing*, **22**(2), 415–428. 374, 383

Wyse, J., Friel, N., and Latouche, P. 2017. Inferring structure in bipartite networks using the latent blockmodel and exact ICL. *Network Science*, **5**(1), 45–69. 383

Xie, B., Pan, W., and Shen, X. 2008. Penalized model-based clustering with cluster-specific diagonal covariance matrices and grouped variables. *Electronic Journal of Statistics*, **2**, 168–212. 208

Xing, E. P., Fu, W., and Song, L. 2010. A state-space mixed membership blockmodel for dynamic network tomography. *Annals of Applied Statistics*, **4**(2), 535–566. 329, 330

Xu, K. S., and Hero, A. O. 2014. Dynamic stochastic blockmodels for time-evolving social networks. *IEEE Journal of Selected Topics in Signal Processing*, **8**(4), 552–562. 329

Yamamoto, M., and Hwang, H. 2017. Dimension-reduced clustering of functional data via subspace separation. *Journal of Classification*, **34**(2), 294–326. 382

Yang, T., Chi, Y., Zhu, S., Gong, Y., and Jin, R. 2011. Detecting communities and their evolutions in dynamic social networks—a Bayesian approach. *Machine Learning*, **82**(2), 157–189. 329

Yeung, D.-Y., and Chow, C. 2002. Parzen window network intrusion detectors. Pages 385–388 of: *Object recognition supported by user interaction for service robots*. 161

Yeung, K. Y., Fraley, C., Murua, A., Raftery, A. E., and Ruzzo, W. L. 2001. Model-based clustering and data transformations for gene expression data. *Bioinformatics*, **17**, 977–987. 78

Yi, J., Zhang, L., Yang, T., Liu, W., and Wang, J. 2015. An efficient semi-supervised clustering algorithm with sequential constraints. Pages 1405–1414 of: *Proceedings of the 21st ACM SIGKDD International Conference on Knowledge Discovery and Data Mining*. ACM. 160

Yoshida, R., Higuchi, T., and Imoto, S. 2004. A mixed factor model for dimension reduction and extraction of a group structure in gene expression data. Pages 161–172 of: *Proceedings of the 2004 IEEE Computational Systems Bioinformatics Conference*, vol. 8. 241

Yoshida, R., Higuchi, T., Imoto, S., and Miyano, S. 2006. Array cluster: An analytic tool for clustering, data visualization and model finder on gene expression profiles. *Bioinformatics*, **22**, 1538–1539. 241

Young, W. C., Raftery, A. E., and Yeung, K. Y. 2017. Model-based clustering with data correction for removing artifacts in gene expression data. *Annals of Applied Statistics*, **11**, 1998–2026. 78

Yu, G., Sapiro, G., and Mallat, S. 2012. Solving inverse problems with piecewise linear estimators: From Gaussian mixture models to structured sparsity. *IEEE Transactions on Image Processing*, **21**(5), 2481–2499. 372, 373

Yuksel, S. E. 2012. Twenty years of mixture of experts. *Neural Networks and Learning*, **23**(8), 1177–1193. 350

Zachary, W. 1977. An information flow model for conflict and fission in small groups. *Journal of Anthropological Research*, **33**(4), 452–473. 11, 302

Zanghi, H., Ambroise, C., and Miele, V. 2008. Fast online graph clustering via Erdös-Rényi mixture. *Pattern Recognition*, **41**, 3592–3599. 300

Zanghi, H., Volant, S., and Ambroise, C. 2010. Clustering based on random graph model embedding vertex features. *Pattern Recognition Letters*, **31**, 830–836. 329

Zeng, H., and Cheung, Y. M. 2014. Learning a mixture model for clustering with the completed likelihood minimum message length criterion. *Pattern Recognition*, **47**, 2011–2030. 108

Zeng, X., and Martinez, T. 2003. A noise filtering method using neural networks. Pages 26–31 of: *IEEE International Workshop on Soft Computing Techniques in Instrumentation, Measurement and Related Applications*. 161

Zhang, N. L. 2004. Hierarchical latent class models for cluster analysis. *Journal of Machine Learning Research*, **5**, 697–723. 198

Zhang, Z., Chan, K. L., Wu, Y., and Chen, C. 2004. Learning a multivariate Gaussian mixture model with the reversible jump MCMC algorithm. *Statistics and Computing*, **14**, 343–355. 107

Zhao, J., Jin, L., and Shi, L. 2015. Mixture model selection via hierarchical BIC. *Computational Statistics and Data Analysis*, **88**, 139–153. 107

Zhou, H., Pan, W., and Shen, X. 2009. Penalized model-based clustering with unconstrained covariance matrices. *Electronic Journal of Statistics*, **3**, 1473–1496. 202, 208, 209

Zhu, X., Wu, X., and Chen, Q. 2003. Eliminating class noise in large datasets. Pages 920–927 of: *Proceedings of the Twentieth International Conference on Machine Learning*. ICML'03. AAAI Press. 161

Zou, H., Hastie, T., and Tibshirani, R. 2007. On the "degrees of freedom" of the lasso. *Annals of Statistics*, **35**(5), 2173–2192. 209

Zreik, R., Latouche, P., and Bouveyron, C. 2017. The dynamic random subgraph model for the clustering of evolving networks. *Computational Statistics*, **32**, 501–533. 330

Zubin, J. 1938. A technique for measuring likemindedness. *Journal of Abnormal Psychology*, **33**, 508–516. 2

Author Index

Subject Index